Reading the *Principia*
The Debate on Newton's Mathematical Methods for Natural Philosophy from
1687 to 1736

Isaac Newton's *Philosophiae Naturalis Principia Mathematica* (known as the *Principia* for short) is considered one of the masterpieces in the history of science. However, it was written in a mathematical language that would be considered unusual by a modern reader.

The mathematical methods employed by Newton in the *Principia* stimulated much debate among his contemporaries, especially Leibniz, Huygens, Johann Bernoulli and Euler, who debated their merits and drawbacks. Among the questions they asked were the following. How should natural philosophy be mathematized? Is it legitimate to use uninterpreted symbols? Is it possible to depart from the established Archimedean or Galilean/Huygenian tradition of geometrizing nature? What is the value of elegance and conciseness? What is the relation between Newton's geometrical methods and the calculus? This book explains how Newton addressed these issues. The author takes into consideration the values that directed the research of Newton and his contemporaries. An advantage of this approach is that the reader gains new insight into the *Principia*, so that this book can be used as an historically motivated introduction to Newton's masterpiece.

This book will be of interest to researchers and advanced students in departments of history of science, philosophy of science, physics, mathematics and astronomy, and especially anyone with a particular interest in Newton's *Principia*.

NICCOLÒ GUICCIARDINI holds a degree in physics and a degree in philosophy (both from the University of Milan). He teaches history of science at the University of Bologna. He is the author of *The Development of Newtonian Calculus in Britain, 1700–1800* and of an introduction to Newton entitled *Newton: un filosofo della natura e il sistema del mondo* (Le Scienze, 1998) which sold 30 000 copies in Italy and has been translated into German. He has published research papers devoted to Newton's *Principia* in *Annals of Science, Centaurus, Historia Mathematica* and *Studies in History and Philosophy of Science*. He has published essays aimed at a wide audience in the *Companion Encyclopaedia of the History and Philosophy of the Mathematical Sciences* (Routledge), the *New Dictionary of National Biography* (Oxford University Press), the Cambridge *Companion to Newton* and a volume on the history of calculus entitled *Geschichte der Analysis* (Spektrum Verlag). He is interested in the technical and cultural aspects of the history of the exact sciences in the seventeenth and eighteenth century.

Reading the *Principia*
The Debate on Newton's Mathematical Methods
for Natural Philosophy from 1687 to 1736

NICCOLÒ GUICCIARDINI

PUBLISHED BY THE PRESS SYNDICATE OF THE UNIVERSITY OF CAMBRIDGE
The Pitt Building, Trumpington Street, Cambridge, United Kingdom

CAMBRIDGE UNIVERSITY PRESS
The Edinburgh Building, Cambridge CB2 2RU, UK
40 West 20th Street, New York NY 10011–4211, USA
477 Williamstown Road, Port Melbourne, VIC 3207, Australia
Ruiz de Alarcón 13, 28014 Madrid, Spain
Dock House, The Waterfront, Cape Town 8001, South Africa

http://www.cambridge.org

First published 1999
First paperback edition 2003

Typeface Times 11/14pt *System* LATEX 2$_\varepsilon$ [DBD]

A catalogue record for this book is available from the British Library

Library of Congress Cataloguing in Publication Data

Guicciardini, Niccolò
Reading the *Principia*: the debate on Newton's mathematical methods for natural philosophy from 1687 to 1736
/ Niccolò Guicciardini.
p.   cm.
Includes bibliographical references and index.
ISBN 0 521 64066 0 hardback
1.   Newton, Isaac, Sir, 1642-1727. Principia.   2.   Mechanics – Early works to 1800.
3.   Celestial mechanics – Early works to 1800.   I. Title. QA803.G85 1999   98-51726 CIP

ISBN 0 521 64066 0 hardback
ISBN 0 521 54403 3 paperback

Contents

1

Purpose of this book

The *Principia* was to remain a classic fossilized, on the wrong side of the frontier between past and future in the application of mathematics to physics.†

1.1 The problem
1.1.1 Principia *as a plural*

Newton's *magnum opus* bears a title which is both imposing and perplexing. Undoubtedly, the great achievement referred to in the title consists in the application of 'mathematical principles' to the physical world, or better to 'natural philosophy'. However, when we ask ourselves which, and how many, are Newton's mathematical principles, the answer does not come so easily. We know much more about the natural philosophy. Many scholars have taught us about the laws of motion, absolute time and space, the law of universal gravitation, the cosmology of void and matter. The names of I. Bernard Cohen, Richard Westfall, Rupert Hall and John Herivel come immediately to the mind of any historian of science. But, from the point of view considered in my research, Tom Whiteside comes first.

In his many papers, but especially in the critical apparatus of the sixth volume of 'his' *Mathematical Papers*, Whiteside has given a profound and detailed analysis of Newton's mathematical natural philosophy. As a study of Newton's mathematical achievements in the *Principia* Whiteside's studies will endure and this book has not been written to replace them. Actually, I began the research which led to this book by reading and following Whiteside's studies. Reading the preparatory manuscripts for the *Principia*, aided by Whiteside's commentaries, I realized how complex, varied and stratified are Newton's mathematical methods for natural philosophy. Newton did not possess a universal mathematical tool which he applied to natural philosophy. He was rather in possession of a whole variety of mathematical instruments. These methods, sometimes in conflict between

† Hall (1958): 301.

1

themselves, originate from different periods of Newton's mathematical life and play different roles in the *Principia*. I thus learnt that '*Principia*' deserves to be treated as a plural. Newton's masterpiece, under its classic appearance given by division into Lemmas, Propositions, Corollaries and Scholia, hides a plurality of mathematical methods. This diversity, however, is somewhat hidden by Whiteside himself, who, for the benefit of the modern reader, translates Newton's demonstrations into modern mathematical language. Paradoxically, the effect of Whiteside's penetrating mathematical insight is that of suppressing diversity and plurality. From this point of view, he did not help me in finding an answer to my search for the mathematical methods of Newtonian natural philosophy.

The plurality and complexity of the methods employed in the *Principia* is reflected by the diversity of judgments that have been given during the centuries. A passage from the preface (which is now believed to come from the pen of Fontenelle) to l'Hospital's *Analyse* published in 1696 has remained justly famous since Newton himself quoted it in the 1710s:

Furthermore, it is a justice due to the learned M. Newton, and that M. Leibniz himself accorded to him: That he has also found something similar to the differential calculus, as it appears in his excellent book entitled *Philosophiae Naturalis Principia Mathematica*, published in 1687, which is almost entirely about this calculus.†

This quotation shows us how Newton's mathematical natural philosophy was perceived by some late-seventeenth-century natural philosophers. It was perceived as based on modern techniques, geometrical limits or infinitesimals, and therefore ready to be translated into the language of the fluxional or differential/integral algorithms. A completely different evaluation was given in 1837 by William Whewell who wrote in *History of the inductive sciences*:

The ponderous instrument of synthesis, so effective in [Newton's] hands, has never since been grasped by one who could use it for such purposes; and we gaze at it with admiring curiosity, as on some gigantic implement of war, which stands idle among the memorials of ancient days, and makes us wonder what manner of man he was who could wield as a weapon what we can hardly lift as a burden.‡

A few decades later Maximilien Marie seems to reply to Fontenelle, writing:

In the *Principia* one finds excellent infinitesimal geometry, but I could not find any infinitesimal analysis: I add that to those who wish to see the calculus of fluxions in the *Principia*, one could then also show the differential calculus in Huygens's *Horologium*.§

---

† 'C'est encore une justice dûë au sçavant M. Newton, & que M. Leibniz lui a renduë lui-même: Qu'il avoit aussi trouvé quelque chose de semblable au calcul différentiel, comme il paroît par l'excellent Livre intitulé *Philosophiae Naturalis Principia Mathematica*, qu'il nous donna en 1687, lequel est presque tout de ce calcul.' L'Hospital (1696): xiv. For Newton's quotation of these lines see Cohen (1971): 294.

‡ Whewell (1837): 167. Quoted in Whiteside (1970): 132n.

§ 'On trouve dans les *Principes de Philosophie naturelle* d'excellente Géométrie infinitésimale, mais je n'y ai pas découvert d'Analyse infinitésimale: j'ajoute qu'à celui qui voudrait voir le calcul des fluxions dans le *Livre des Principes*, on pourrait aussi bien montrer le calcul différentiel dans l'*Horologium*.' Marie (1883–88), **6**: 13.

While Fontenelle stresses the modernity of Newton's mathematical methods in the *Principia*, underlining their equivalence with the new calculus, Whewell and Marie are impressed by the distance which separates Newtonian mathematical natural philosophy from modern analytical mechanics. Marie would completely disagree with Fontenelle and define Newton's method as one which is closer to the 'infinitesimal geometry' of Huygens, rather than to Leibniz's differential and integral calculus.

A text cannot be read out of context. The homogenization effect of Whiteside's notes and the diversity of judgments of Fontenelle, Whewell and Marie are caused by exposing the *Principia* to different audiences, who approach it with different background knowledge and for different purposes. It is thus that demarcation lines between method and method, tensions and equivalences, have now faded away: we cannot see them any longer. A possible way to re-establish the hidden structure of a text is to follow the reactions of its contemporary readers.

1.1.2 *The debate on the mathematical methods of the* Principia, *1687–1736*

At the beginning of the eighteenth century the *Principia* was read, and approved or criticized, by a number of *savants* all over Europe. The debate concerned a whole spectrum of issues: from theology to philosophy, from astronomy and cosmology to the principles of dynamics. The mathematical methods employed by Newton drew considerable attention, particularly in the first two decades of the eighteenth century. A motivation for this interest is easily found in the priority dispute on the invention of the calculus. The Newtonians and Leibnizians became divided during this notorious squabble. After a series of attacks and counterattacks, which began in the late 1690s, in 1713 a Committee of the Royal Society, secretly guided by Newton, formally accused Leibniz of plagiarism. The German diplomat was accused of having plagiarized Newton's fluxional algorithm, publishing in 1684 and 1686 his differential and integral calculus. The debate on the mathematics of the *Principia* is part of the priority debate. Actually Newton tried to use the *Principia* as a proof of his knowledge of calculus prior to 1684, the year in which he sent a first version to the Royal Society.

Newton himself was, however, well aware of the distance that separated the calculus of fluxions (the 'analytical method of fluxions', as he named it) from the geometrical methods employed in the *Principia*. In fact, he was not able to make much use of the *Principia* as proof of his knowledge of the algorithm of fluxions. He could only refer to a few propositions. The bulk of the work, he had to admit, was 'demonstrated synthetically'. He sought to maintain that he had discovered 'most of the propositions' with the help of fluxional calculus, the 'new analysis', but that it was now difficult to see the analysis utilized in the process of discovery.

He further suggested an analogy between his mathematical procedures and those of the 'Ancients'. In 1714, speaking of himself in the third person, Newton wrote:

By the help of this new Analysis Mr Newton found out most of the Propositions in his *Principia Philosophiae*. But because the Ancients for making things certain admitted nothing into Geometry before it was demonstrated synthetically, he demonstrated the Propositions synthetically that the systeme of the heavens might be founded upon good Geometry. And this makes it now difficult for unskillful men to see the Analysis by wch those Propositions were found out.†

However, external and internal evidence seems to be against Newton's statement. The preparatory manuscripts of the *Principia* reveal little use of calculus and seem to indicate that Newton wrote it in the form in which it was published.‡ An internal analysis of the structure of the demonstrations in the *Principia* furthermore reveals that Newton's geometrical natural philosophy is, in many significant cases at least, to a certain extent independent of calculus techniques. This independence should be neither overstated nor overlooked. Defining Newton's mathematical methodology employed in the *Principia* is a complex task, not only because of the plurality of Newton's geometrical techniques, but also because of their contiguity with the analytical method of fluxions. In some cases, the geometrical demonstrations of the *Principia* lend themselves to an almost immediate translation into calculus concepts; in other cases, this translation is complicated, unnatural, or even problematic. Needless to say, notwithstanding Newton's rhetorical declaration of continuity between his methods and the methods of the 'Ancients', his geometrical natural philosophy is a wholly seventeenth-century affair.

It is generally maintained that Newton's statements, such as the one quoted above, were not honest. In his recent biography of Newton, Rupert Hall has defined Newton's manoeuvres aimed at stating a use of calculus in the art of discovery of the *Principia* as the 'fable of fluxions'.§ Newton would have been desperately trying to use this fable as proof of his knowledge of calculus before the publication of Leibniz's *Nova methodus* in 1684. It is certainly true that the above quotation has a role in Newton's rhetorical manoeuvres against Leibniz. It is an ambiguous statement and should not be taken as a reliable historical record. However, as I will show in this book, there is some truth in it. At the end of the nineteenth century a manuscript dating from the years of composition of the *Principia* was found: in these pages in Newton's hand the result on the solid of least resistance, expressed in geometric terms in the second book of the *Principia*, is achieved via a fluxional equation.¶ But this manuscript has been considered the exception which confirms

---

† *Mathematical Papers*, **8**: 598–9. This quotation comes from Newton's review of the *Commercium epistolicum* that appeared anonymously in the *Philosophical Transactions* in early 1715.
‡ Whiteside (1970).
§ Hall (1992): 212–13.
¶ *Mathematical Papers*, **6**: 456–80.

the rule. I will add evidence that Newton had his own good reasons when he stated that the 'new analysis' was one of the ingredients which allowed the writing of the *magnum opus*.

Today, we take it for granted that calculus is a more suitable tool than geometry, particularly in the case of applications to dynamics. But at the beginning of the eighteenth century, the choice of mathematical methods to be applied to the science of motion and force was problematic. Firstly, a plurality of geometrical methods had to be compared not with a calculus, but with a plurality of calculi. Calculus came in at least two forms: it could be based either on infinitesimal concepts or on limits. Furthermore, there were several competing notations broadly speaking falling in two groups, the fluxional and the differential/integral notation. Secondly, the way in which dynamical concepts should be represented was not (and is not) obvious. Thirdly, the calculus (at least up to Euler) was never thought of as completely independent from geometrical representation. Calculus was far from being understood as an abstract uninterpreted formalism: there were geometric 'equivalents' of the algorithm (Archimedes' exhaustion methods, the geometry of fluent and fluxional quantities, the geometry of the infinitely smalls) which were deployed in different ways and for different puroposes. Early-eighteenth-century mathematicians debated such issues.

The question of the equivalence of the *Principia*'s mathematical methods with the calculus was faced in a systematic way in the first two decades of the eighteenth century. Mathematicians belonging most notably to the Basel and to the Paris schools initiated a programme of translation of the *Principia* into the language of the Leibnizian calculus. Their efforts were followed and contrasted all over Europe. If the *Principia*'s mathematical procedures were simply calculus in disguised form, as Fontenelle seems to maintain, translation into the language of Leibniz's differential/integral or Newton's fluxional algorithms would be a *routine* exercise. This, however, was not the case. The Continental mathematicians who, at the beginning of the eighteenth century, set themselves the task of applying the calculus to Newton's *Principia* (most notably Pierre Varignon, Jacob Hermann and Johann Bernoulli) had to surmount difficult problems. This task was part of a wider programme of application of calculus to dynamics, a programme to which the Leibnizians gave priority and publicity.

In this book I will prove that the programme of translation of the *Principia* into calculus language was not an exclusively Continental affair. Newton and a restricted group among his disciples (David Gregory, De Moivre, Cotes, Keill, Fatio de Duillier) were able to apply the analytical method of fluxions to some problems concerning force, motion and acceleration. By reading the correspondence and notes exchanged within the close circle of Newton's more mathematically competent associates we discover that they applied fluxional equations to central forces and motion in a resisting medium. However, unlike the Leibnizians, the Newtonians gave very little publicity to these researches. When studying the Newtonian school one has to draw a distinction between what was public and what was meant to remain private. It is by following the private exchanges amongst the Newtonians that we can begin to discover the validation criteria that they accepted, criteria that determined a publication policy which differed dramatically from that of the Leibnizians.

The debate on the mathematical methods of the *Principia* was concerned with a number of specific issues, such as projectile motion in resisting media, the inverse problem of central forces, and the representation of trajectories. This book is devoted to that debate. I hope that by following it, we will be able to appreciate distinctions and tensions between methods – distinctions and tensions that are difficult to discern with the modern eyes of the mathematician accustomed to present-day analytical mechanics.

1.2 Plan of the work

The book is divided into three parts. In Part 1 I introduce the reader to Newton's *Principia*. In Part 2 I consider the reactions to the *Principia* of three giant readers: Newton himself, Huygens and Leibniz. In Part 3 I devote attention to the two schools that divided Europe in the period considered: the British Newtonian and the Continental Leibnizian. It should be made clear that by 'school' I mean here those very small groups of competent mathematicians that gathered around the two great heroes, defended them and shared with them certain values and methods. The time span goes from the year of publication of the *Principia* to the year of publication of Euler's *Mechanica*. As I will argue in Chapter 9, the advent of Euler's calculus and mechanics marks a watershed. After Euler the *Principia*'s mathematical methods belong definitely to what is past and obsolete.

In Chapter 2 the reader will find a concise presentation of Newton's method of series and fluxions. My aim is to give an idea of the mathematical method, the 'new analysis', that Newton had already developed in its full extent before writing the *Principia*. The experts on Newton's mathematics can skip this chapter. They should, however, take into consideration §2.3 where I contrast what Newton called the 'analytical' and the 'synthetic' methods of fluxions. In fact, in the 1670s Newton distanced himself from his early analytical method, an algorithm based on infinitesimals, in order to develop a geometrical method based on limits. It is the latter which is employed in most of the *Principia*'s demonstrations. Newton's refusal of the analytical fluxional method (which in Leibnizian terms we would call the differential and integral calculus) is indeed one of the most spectacular processes in the history of mathematics, comparable to Einstein's refusal of quantum mechanics. Einstein, after having contributed to the birth of quantum theory, distanced himself from it because of epistemological concerns. Similarly, Newton after having participated in one of the most fruitful discoveries in the history of mathematics, the calculus, refused to rely on it.

Chapter 3, which completes the introductory Part 1, is devoted to the mathematical methods employed by Newton in the *Principia*. It is not an easy chapter and many might find it prolix. Again the experts, i.e. people who have mastered Newton's *magnum opus*, or who have studied commentaries by Chandrasekhar, Cohen, Densmore, De Gandt or Brackenridge, may read it only cursorily. I have very little to add in terms of depth of analysis to these recent commentaries or to Whiteside's notes to Volume 6 of Newton's *Mathematical Papers*. I am heavily indebted to all these technical introductions to the *Principia*.†

My book is not to be taken as an introduction to, or a critical analysis of, the *Principia*, but rather as a work devoted to the reception of Newton's *magnum opus*. Chapter 3 is, however, necessary in the economy of my book since I must present some of Newton's mathematical techniques, which are referred to in later chapters, to my readers. In Chapter 3 I have thus tried to convey to the reader an awareness of the plurality of the mathematical methods deployed by Newton. I do not aim at completeness. Rather, I have been selective, picking those demonstrations that seem to me more exemplary of Newton's methods. The analysis of Newton's methods will deepen in Parts 2 and 3, when I leave the word to Newton's contemporaries: it is *their* analysis, not mine, that constitutes the subject matter of my book. In fact, what I have done in Chapter 3 is simply to paraphrase and explain some of the basic parts of the first two books, while I devote just a few comments to the third book. Generally I use the third edition of the *Principia*, and I refer to the previous editions when relevant variants occur. In the last stage of preparation of my book I was able to use the recent English translation by I. Bernard Cohen and Anne Whitman. In my commentary I adhere almost entirely to Newton's notation (e.g. the symbol $\propto$ has been used for phrases such as 'is proportional to'). The demonstrations have not, however, been reproduced in entirety. I have tried to reproduce only the structure of Newton's proofs. There are, in fact, in Newton's imposing and difficult masterpiece, many lines which are devoted to rather dull and uninteresting passages. Of course, these lines are

---

† Chandrasekhar (1995) is an analysis of *almost* all the *Principia*. It is mathematically masterful, but rather a-historical. I was able to use, in the final stage of preparation of my book, the very erudite *Guide* which accompanies Cohen and Whitman's new translation of the *Principia* (cited as Newton (1999)). Densmore (1995) is a complete, thorough and historically accurate analysis of the basic Propositions of Books 1 and 3 related to gravitation theory. De Gandt (1995) considers in every detail the *De motu* and Propositions 39–41, Book 1. Brackenridge in (1988), (1990) and (1995) studies the first three sections of Book 1 and has a great deal to say on the variants between the three editions of the *Principia*. On Whiteside's notes see above. Herivel (1965) charts Newton's studies in dynamics preliminary to the *Principia*. A classic introduction to the *Principia* which is still useful is Brougham & Routh (1855). See also Jourdain (1920).

necessary to complete the proof, but in following them there is a risk of losing sight of the ingenuity and elegance of Newton's demonstrations. These seem to many, especially today when we are not trained in geometry, very difficult and intricate. But when you grant, say, some properties which depend on long trains of proportions between the sides of similar triangles and go to the core of the demonstration, there does emerge a demonstrative structure which is often striking in its simplicity.

Any vague discourse about the 'geometrical method' of the *Principia* should vanish after an analysis of Newton's demonstrations. It will appear that 'geometrical method' is a misuse for two reasons. Firstly, in several important instances Newton abandons geometry and deploys quadratures and infinite series. Second, it is a singular: we should speak about the geometrical *methods*! We have to distinguish several *geometries* in the *Principia*. Some geometrical demonstrations are in fact very close to, or implicitly refer to, the calculus, others are real alternatives to calculus techniques. Section 3.16 might be read with interest (or disapproval) also by experts in the Newtoniana.

In Part 2 I consider the three giant 'readers' of the *Principia*. Chapter 4 concerns Newton's evaluation of his own published masterpiece. After having published the *Principia*, Newton took into consideration radical restructurings. These projects are interesting, since they reveal what he thought of his *Principia*. He was divided between the idea of revealing the analytical methods necessary to complete some of the demonstrations (typically advanced quadrature techniques), and the idea of stressing the classical appearance of his work. These worries became even more acute during the priority dispute with Leibniz. Newton found himself in an anomalous position. On the one hand, he wanted to use the *Principia* as proof of his knowledge of calculus. On the other hand, he wished to underline the superiority of his methods when compared with the Leibnizian algorithm. Thus, in this chapter we will follow Newton's attempts to defend the mathematical methods of the *Principia* against the criticisms of the Leibnizians. Most notably he had to justify the almost total absence of calculus. In manuscripts and letters related to the *Principia* Newton gives us some clues for understanding the values that directed his mathematical research during his mature life. We will see evidence of his concern for rendering his mathematics compatible with Ancient geometry. He went so far as to state that his mathematical work was mainly a development of the lost analysis of the Ancients' geometers. It is well known that Newton believed in the *prisca sapientia* of the Ancients: gravitation, the Copernican system and atoms were known to the priests of Israel, Egypt and Mesopotamia. These priests were also in possession of a hidden analysis which they did not reveal outside a circle of initiates. Newton thought of himself as a rediscoverer of this lost mathematical wisdom (may we call it a *prisca geometria*?). Of course, his mathematical methods are a wholly seventeenth-century affair, but the conviction of a continuity with ancient exemplars played a role in Newton's mature life and was transmitted to some of his disciples. The classical appearance of the *Principia* thus had a meaning which goes beyond a search for elegance and rigour. But in Chapter 4, we will see that Newton also had other concerns. We will show his interest in the mathematical competence of his readers and we will find him complaining about the fact that this competence had changed in the decades following 1687. According to Newton, most of the *Principia* was written in a style understandable for 'philosophers steeped in geometry', who by the early eighteenth century had disappeared. The early-eighteenth-century readers of the *Principia*, he regretted, were rather versed in 'algebra' and 'modern analysis'. We will also show that Newton, in the 1690s, took into consideration the possibilities of applying the analytical method of fluxions to central forces. In some manuscripts and letters he gave evidence of his ability in handling central forces via fluxional (i.e. differential) equations. This should refute a widespread belief that Newton was not able to write differential equations of motion![†] It is beyond doubt that he developed some propositions of the *Principia* in terms of fluxional equations. However, he was reluctant to publish his natural philosophy in the language of the analytical method of fluxions. He rather shared this language in private communications with his close accolytes.

In Chapter 5 I discuss Huygens' reaction to the *Principia*. The great Hollander died in 1695: he just had the time to read and comment on Newton's achievement. He was immediately aware of Newton's stature as a mathematician. We have some evidence, however, that he was dissatisfied with Newton's use of proportion theory. In an interesting commentary to Proposition 6, Book 1, he criticized the use there of this classic ingredient of Ancient geometry. It is highly probable that Newton wrote the *Principia* having the *Horologium oscillatorium* in mind as a model. Despite his admiration for Huygens, Newton was departing from the high standards of rigour of the Dutch natural philosopher.

In Chapter 6 I complete the trio with Leibniz. Much has been written by Aiton, Bertoloni Meli and Bos on Leibniz's calculus and mathematization of dynamics and I rely heavily on their historical work.[‡] In this chapter the reader will find some information on Leibniz's first attempts to apply the calculus to planetary orbits and resisted motion: i.e. to some of the main problems faced by Newton in the *Principia*. In §6.4 I discuss Leibniz's approach to some foundational aspects of his mathematical methods. It has been shown by Aiton and Bertoloni Meli that important differences exist between Leibniz and Newton regarding the use of infinitesimals and the representation of trajectories. In my opinion, in

† See, for example, Truesdell (1960): 9, and Costabel (1967): 125–6.
‡ Aiton (1989a), Bertoloni Meli (1993a), Bos (1974).

the comparison between Leibniz and Newton it is useful to employ a category first introduced by Feynman and rediscovered in historical context by Sigurdsson. According to Sigurdsson, Newton's and Leibniz's calculi can be defined as 'not equivalent in practice'.† I understand this as meaning that, despite the fact that the two calculi are equivalent syntactically and semantically, it would be wrong to think that they were equivalent, since they were *used* in different ways. The two algorithms were translatable one into the other. Furthermore, at a foundational level (as Knobloch's recent studies have proved), Leibniz and Newton agreed on many basic issues.‡ However, they differed in mathematical practice: they oriented two equivalent mathematical tools towards different directions. In order to appreciate this pragmatic difference, we have to take into consideration the values that orient research along different lines. In contrast to Newton, Leibniz stressed as positive values the novelty of his method, its mechanical algorithmic character, and the possibility of freeing mathematical reasoning from the burden of geometrical interpretation. Leibniz's pragmatic approach, so different from that of Newton studied in Chapter 4, leads to a critical attitude towards the mathematical methods of the *Principia*.

In Part 3 I move on to chart the mathematically minded readers of the *Principia* up to the publication of Euler's *Mechanica* in 1736. As a matter of fact there are already detailed studies on the reception of the *Principia*: let me just mention here the monumental *Introduction* by Cohen.§ The mathematical side of this story has, however, been somewhat neglected. We will follow the formation of two schools – of two small groups of mathematicians who were in close contact with their masters – the Newtonian British (Chapter 7) and the Leibnizian Continental (Chapter 8), which differed in mathematical practice. We will see that some of the values that directed Newton and Leibniz along different lines were accepted by their followers. The contrasting approaches of the two schools to the mathematical methods of the *Principia* reflect these differences in formation, skills and expectations. This confrontation became part of the priority dispute on the invention of the calculus. We can refer to Hall's book on the topic for definitive information on this famous squabble.¶ However, once again, the mathematical content of this polemic has been given little attention. The tendency has been to describe the quarrel between Newtonians and Leibnizians only as a sociological event. This is in part legitimate. However, despite the fact that the waters were muddied, the combatants debated issues whose content is worth considering. The cluster of problems addressed which concern the *Principia*'s mathematical methods is particularly interesting.

† Sigurdsson (1992).
‡ Knobloch (1989).
§ Cohen (1971).
¶ Hall (1980).

How should natural philosophy be mathematized? Is it legitimate to use in this context uninterpreted symbols? Can we depart from the established Archimedean or Galilean/Huygenian tradition of geometrizing Nature? What is the value of elegance and conciseness? What is the relation between Newton's geometrical methods of the *Principia* and calculus? What is the advantage of using the latter? These questions were often faced in addressing specific issues, such as the inverse problem of central forces. The fact that the answers to these questions given by the Leibnizians and the Newtonians differed determined a divergence in their publication strategies. It determined a different public image of the Newtonian and Leibnizian mathematizations of Nature. Thus if we restrict our attention to the public texts, the two schools appear quite separate: a huge divide seems to exist between those who followed the style of the *Principia*, and those who published in the *Acta Eruditorum* or in the *Mémoires* of the Paris Academy. But when we consider the private side of the story, when we consider the correspondence and notes which circulated amongst the Newtonians, we realize how wide was the overlapping of competences on algorithms, proof methods, and even conceptions on foundational matters.

Part 3 is concluded by a short characterization of Euler's *Mechanica* and Le Seur and Jacquier's edition of the *Principia* – an edition which was enriched by an extensive commentary. These two works mark the end of the Newtonian methods for natural philosophy: the Eulerian mathematical style imposed itself as the preferred way to face the problems that Newton had mathematized some fifty years before.

A thesis that I state in the Conclusion (§9.3.3) is that the Newtonians could not rely exclusively on the algorithmic techniques of the 'new analysis', not only because of their metamathematical concerns, but also because of the cosmology that they accepted. The Newtonian and the Leibnizian algorithms were not developed enough to allow one to mathematize fine details of gravitation theory. The mathematization of these details (e.g. tides, or planetary shapes) was essential for the Newtonian cosmology. Newtonians knew how to apply the analytical method of fluxions to some dynamical systems, but in order to tackle more difficult mathematical constructs, to use Cohen's terminology, they were compelled to deploy qualitative geometrical techniques.† Geometrical language appeared to them more powerful: it was a language which responded to the needs of their cosmological interests. It was only much later, at the middle of the eighteenth century, that calculus developed into a mathematical theory which could be applied to fine details of planetary perturbations, the attraction of ellipsoids, the tides, etc.

---

† Cohen (1980). Recent research by Chandrasekhar and Nauenberg has taught me that I had been underestimating Newton's capacity in handling the three-body problem in calculus terms. See Chandrasekhar (1995): 219–68, Nauenberg (1995).

In the concluding chapter I also summarize and discuss some of the results achieved in this book and I give a glance ahead after 1736. I devote some brief remarks to the development of the mathematization of natural philosophy in the eighteenth century. I maintain that it would be wrong to think about the eighteenth century as a period of extension and application of mathematical theories discovered in the previous century: the Newtonian/Leibnizian calculus and Newtonian mechanics. As a matter of fact, the Age of Reason was an extremely creative period. The calculus underwent deep changes, made possible mainly by Euler. From 1730 to 1750 the concept of function emerged, and multivariate functions, partial differential equations, and the calculus of variations were first defined and studied. Mechanics too, as Truesdell has convincingly maintained, underwent a profound transformation into a science, not envisaged by Newton, based on minimum principles and applied to extended (rigid, flexible and fluid) bodies.† The birth of what can be called the 'Eulerian calculus and mechanics', a process which was due mainly, but not exclusively, to the Continentals, marks a deep transformation of the mathematical sciences. After this change, the *Principia* and the debate surrounding it belong to the past.

1.3 Conventions: abbreviations, quotations and mathematical notation

For the great majority of the references I have followed the author–date citation system. Only a few works which are often cited are referred to by short titles. I have preferred 'Newton *Correspondence*, **4**: 205' to 'Newton (1959–1977), **4**: 205'. In the final *List of References* the reader will find under an author's name an alphabetical list of short titles (when applicable) and then a chronological list of publication dates.

Quotations in the main text are translated into English. If not otherwise stated all translations are mine. The original text can be found in footnotes. Whiteside's translations from Latin in Newton's *Mathematical Papers* occur frequently. They are so reliable that I have thought it necessary to reproduce the Latin in footnote only when the original affords the reader with some relevant conceptual insight. The same holds true for Cohen and Whitman's new translation of the *Principia*.

I hesitated over the policy to follow with the quotations from the *Principia*. The subject of my book requires a distinction between the three editions. The debate over the mathematical methods of the *Principia* was concerned with all the editions and, in some cases, the variations introduced by Newton are extremely relevant. However, there is no complete English translation of the first and second edition.‡

---

† See e.g. Truesdell (1960).
‡ An English translation of the first three sections of the first edition is provided by Mary Ann Rossi in Brackenridge (1995).

To be purist, for each quotation from the *Principia*, I should have reproduced all the variants taken from Koyré and Cohen's edition.† I should also have provided an English translation of the first two editions. But what about the third? Should I have followed the Motte–Cajori translation?‡ And how should I have integrated the Motte–Cajori with my own translations? This was the poor condition in which I found myself until Bernard Cohen came to my assistance. With a courtesy which is paralleled only by his scholarship, he sent me proofs of his and Anne Whitman's forthcoming new translation of the *Principia*. It seemed wise to me to adopt this new translation, not only because it will long replace the Motte–Cajori in the English-speaking world, but also because it includes a translation of the most relevant variants between the three editions. All English quotations from the *Principia* are thus based on this forthcoming translation.§ The reader will therefore be aware of the most relevant variants. I have added in footnotes reference to the corresponding pages of the *Variorum* edition and to the Motte–Cajori translation.

It should be noted that Cohen and Whitman in their translation of the *Principia*, for the benefit of the modern reader, alter some of Newton's notations and technical terms. So AB^2 stands for Newton's ABq or $ABquad.$, and AB^3 stands for Newton's $ABcub.$, etc.: the above expressions mean 'the area of the square (the volume of the cube) whose side is AB'. Furthermore, Newton's usage of expressions – unusual for the modern reader – which belong to the theory of proportions are translated quite freely. For instance, 'duplicata ratione' is rendered as 'squared ratio', rather than a more literal 'duplicate ratio'. I have not accepted these departures from a literal rendering of Newton's notation and text, since in my book the exact form of the various mathematical styles must be rendered as closely as possible. Further notational conventions that I have adopted in my commentary of the *Principia* are described in §3.1.¶

In modern texts a clear notational distinction between scalar and vector quantities is made. In Newton's times this distinction was performed in words by specifying that a certain magnitude is oriented towards a point: with the exception of a few footnotes I will follow the old practice.

A last warning concerns mass. Some readers might be surprised to find no occurrence of mass in so many formulas related to topics such as central force motion, resisted motion, and so on. The fact is that in Newton's times mathematicians thought in terms of proportions, even when they wrote formulas which look very much like equations. Constant factors were thus not always

---

† Cited as Newton *Variorum*.
‡ Cited as Newton *Principles*.
§ This translation is to appear as Newton (1999).
¶ Of course Cohen and Whitman are better aware than I am of the 'perils' of their linguistic choices, which are perfectly justified by the fact that the audience of their translation will be much wider than the audience of my book.

made explicit. Dimensional analysis of many formulas appearing in this book
might leave many readers perplexed. Most notably force is often equated with
acceleration.

1.4 Acknowledgments

Writing this book would have been impossible without the help of the many
scholars that I have met at conferences and seminars in Italy and abroad. In
particular I would like to thank my colleagues Silvia Bergia, Anna Guagnini,
Giuliano Pancaldi and Pietro Redondi, and the organizers of several meetings
at Oberwolfach (Germany) and Luminy (France), and of the First Seven Pines
Symposium (Lewis, Wisconsin). I owe a great debt to all the people who patiently
gave me advice. I should also mention the help and insights that I gained
from printed sources: I refer in particular to the works by Eric Aiton, Richard
Arthur, Domenico Bertoloni Meli, Michel Blay, Henk Bos, Bruce Brackenridge,
Subrahmanyan Chandrasekhar, Bernard Cohen, Dana Densmore, François De
Gandt, Herman Erlichson, Eberhard Knobloch, Michael Nauenberg, Enrico Pasini,
Bruce Pourciau, Richard Westfall and Tom Whiteside cited in the footnotes and in
the list of references.

Several scholars have contributed with great generosity to improving my text.
Richard T. W. Arthur, during a stay in Bologna as visiting professor, read the whole
manuscript giving me important advice on Leibniz and on matters related to the
organization and presentation of my book. Umberto Bottazzini has shown interest
in my research project since its inception and reacted with many useful suggestions.
Michael Nauenberg has given great help in correcting several misjudgments of
mine related to some advanced aspects of Newton's mathematical procedures: the
correspondence that I entertained with him has radically changed several parts
of this book. Discussions with Maurizio Mamiani, and an exchange of letters
with him related to my introduction to Newton, helped me to approach Newton
from a much broader point of view.† My student Paolo Palmieri read critically
several parts of the manuscript. Correspondence with Bernard Cohen has been of
invaluable help in understanding the *Principia* and in shaping my book: some of
his comments arrived on my desk in the form of three mini-tapes, which I will
treasure for the rest of my life. I have already mentioned his contribution with the
translation of the *Principia*. For this alone my debt to Cohen would have been very
great. I acknowledge also the help that I received from Herman Erlichson.

A particular mention has to be reserved for Tom Whiteside. I had the privilege
of discussing Newtonian matters with him in Cambridge many years ago. I still
keep the notes that I took after our encounters. The last time I saw him was

---

† *Newton: un filosofo della natura e il sistema del mondo.* Guicciardini (1998c).

in 1988, on my way back to Italy. Tom loaded my suitcase, my protestations notwithstanding, with three heavy presents: the *editio variorum* of the *Principia*, volume 8 of 'his' *Mathematical Papers* and Hofmann's *Leibniz in Paris*. I took it as a commitment: this book is an attempt to study Newton the mathematician – revealed in the *Mathematical Papers* – at work in the *Principia*, and the criticisms that Leibniz and his adherents directed against him. Thus there is a precise sense in which it can be stated that my book has been inspired by Whiteside: I conceive it as a footnote to the edition of Newton's *Mathematical Papers*, perhaps the greatest achievement in this century in the field of history of science.

The following permissions are gratefully acknowledged: from Taylor & Francis to reproduce in §8.6.2 parts of my 'Johann Bernoulli, John Keill and the inverse problem of central forces', *Annals of Science* (1995) **52**: 537–75; from Academic Press to reuse in §8.5 my 'An Episode in the History of Dynamics: Jakob Hermann's Proof (1716–1717) of Proposition 1, Book 1, of Newton's *Principia*', *Historia Mathematica* (1996) **23**: 167–81; and from the Royal Society of London to quote in §7.3 from MS210. Material which will appear in 'Newtons Methode und Leibniz' Kalkül', forthcoming in *Geschichte der Analysis*, N. Jahnke (ed.), has been used in §2.2, §6.2 and §6.4.1. Some of the theses defended in my book have been published in my 'Did Newton use his calculus in the *Principia*?', *Centaurus* (1998) **40**: 303–44, parts of which I have reused with kind permission of Munkgaards International Publishers Ltd.

I am grateful for their help to the librarians of the Departments of Astronomy, Mathematics and Philosophy of the University of Bologna, of the Library of the Royal Society of London, of the University Library of Cambridge (in particular to Stephen Lees), and of the Library of Christ Church, Oxford, and to Martin Mattmüller and Fritz Nagel of the Bernoulli-Edition in Basel.

Part one
Newton's methods

2

Newton's methods of series and fluxions

2.1 Purpose of this chapter

Section 2.2 is concerned with a mathematical method which Newton developed from the year 1664. He labelled it the 'analytical method of fluxions'. It can be defined as the Newtonian equivalent, or counterpart, of the Leibnizian differential and integral calculus, and briefly – but somewhat improperly – called the 'fluxional calculus'. From 1664 to the 1690s Newton elaborated several versions of it. Furthermore, Newton distinguished between an analytical and a synthetic method of fluxions (§2.3). In this chapter I will attempt a periodization of these versions, paying attention to concepts, rather than to results. What follows cannot be taken as an exposition or introduction to Newton's mathematical discoveries. This would take us too much space, and indeed, after Whiteside's works, would also be useless. What I will do rather is to sketch the nature of the mathematical method for drawing tangents and 'squaring curves' that Newton had developed before the composition of the *Principia*.

2.2 The analytical method

2.2.1 Mathematical background

Newton's interest in mathematics began around 1664 when he read François Viète's works (1646), the second Latin edition of René Descartes' *Geometria* (1659–61), William Oughtred's *Clavis mathematicae* (1631), and John Wallis's *Arithmetica infinitorum* (1656). It was in reading this selected group of works that Newton learned about the most exciting discoveries in analytical geometry, algebra, tangents, maxima and minima, quadratures, infinite series and infinite products. From Viète and Oughtred he learned how arithmetical operations could be dealt with in a general way through symbols which stand for constant or for variable quantities. In Descartes' *Geometria* he found a correspondence between algebraic equations and curves. There he learned about the Cartesian algebraic method for drawing a tangent to a curve. He was also able to become acquainted with the recent results on maxima and minima of the Dutch school exposed in the commentaries to the second Latin edition by Van Schooten, Hudde and Sluse. Wallis's work was a treasure on quadratures via summations of infinite series. Newton adhered enthusiastically to these new methods, which he usually grouped under the label 'new analysis'.†

2.2.2 Power series

After a few months of self-instruction, Newton was able, during winter 1664–65, to establish his first mathematical discovery: the binomial theorem. He obtained this result by generalizing Wallis's 'inductive' method for squaring the unit circle. Starting from a table of the binomial coefficients for positive integer powers, he interpolated for fractional powers and extrapolated for negative powers through complex and rather shaky guesswork. For instance, he obtained:

$$(1 - x^2)^{1/2} = 1 - \frac{1}{2}x^2 - \frac{1}{8}x^4 - \frac{1}{16}x^6 - \frac{5}{128}x^8 \cdots, \qquad (2.1)$$

and

$$(1 + x)^{-1} = 1 - x + x^2 - x^3 + x^4 - \cdots. \qquad (2.2)$$

Newton verified that the power series obtained by applying the binomial theorem agreed with those obtained by algebraical procedures. For instance, he applied to $(1 - x^2)^{1/2}$ standard techniques of root extraction and to $(1 + x)^{-1}$ standard techniques of 'long division' and observed that the power series obtained were in agreement with the general law of formation of the binomial coefficients that he had previously established.

Newton applied the binomial theorem to quadrature problems. He knew, by generalization of results by Wallis, that the area $\mathcal{A}$ under $y = x^n$ and over the interval $[0, x]$ is:

$$\mathcal{A} = \frac{x^{n+1}}{n + 1}. \qquad (2.3)$$

Applying (2.3) termwise to power series, Newton could 'square' a variety of

† On the influence of the analytical British school on Newton see Pycior (1997).

curves. In seventeenth-century mathematical jargon 'squaring' a curvilinear figure meant finding a square whose area is equal to the curvilinear area. For instance, expanding via the binomial theorem $(1 - x^2)^{1/2}$ and applying (2.3) termwise he could square the unit circle, while the same procedure applied to $(1 + x)^{-1}$ allowed him to find the area under the hyperbola. This last result was deployed in order to calculate logarithms.†

The use of the binomial series allowed Newton to penetrate into a new mathematical world. Previous 'integration' techniques were based on taking limits of sums of circumscribed or inscribed rectilinear areas. These procedures were very laborious and required a great deal of intuition on how to find the best partition of the subtended curvilinear area. Newton, instead, by termwise application of formula (2.3) to binomial expansions, could 'square', in a completely analytical and general way, many of the most difficult curves.

We note three aspects of Newton's work on the binomial series. First of all he introduced, following Wallis's suggestion, negative and fractional exponents. Without this innovative notation ($x^{a/b}$) no interpolation or extrapolation of the binomial theorem from positive integers to the rationals would have been possible. Second, Newton obtained a method for expanding into power series a large class of 'curves'. For him curves are thus expressed not only by finite algebraical expressions (as for Descartes), but also by infinite series (preferably power series). In 1665 mathematicians had just begun to appreciate the usefulness of infinite series as representations of 'difficult' curves. Transcendental curves, such as the logarithmic curve, could therefore receive an 'analytical' representation to which the rules of algebra (i.e. 'analysis') could be applied. Newton also obtained power series representations of the exponential curve and of the trigonometric curves. Before the advent of infinite series such transcendental 'functions' (to use a later terminology) did not have an analytical representation, but were generally defined in geometric terms. Infinite series, were indeed understood by Newton and by his contemporaries as 'infinite equations', to which the rules of algebra could be applied. The realm of analysis could thus be extended to all the known curves. Newton wrote:

From all this it is to be seen how much the limits of analysis are enlarged by such infinite equations: in fact by their help analysis reaches, I might almost say, to all problems.‡

And he further noted:

And whatever common analysis performs by equations made up of a finite number of terms

---

† In modern notation one can express (2.3) as $\int_0^x x^n \mathrm{d}x = x^{n+1}/(n + 1)$.

‡ Newton *Correspondence*, **2**: 39.

(whenever it may be possible), this method may always perform by infinite equations: in consequence, I have never hesitated to bestow on it also the name of analysis.†

Third and last point: it should be noted that Newton, as all his contemporaries, had a rather intuitive concept of convergence – a concept that nowadays would be considered as unrigorous. For instance, he thought it sufficient to say that the binomial series can be applied when x is 'small'.

2.2.3 The fundamental theorem

Termwise application of (2.3) allowed Newton to find the area subtended by many curves. Another approach to quadrature problems was possible thanks to the so-called 'fundamental theorem'. Around 1665 Newton realized that tangent and area problems are the inverse of one another. We will see in §2.2.6.2 one of the demonstrations that he gave of the fundamental theorem.‡ The solution of the former and easier problem could be deployed to solve the latter and more difficult one. In modern terms: the fundamental theorem allowed Newton to reduce quadrature problems to the search for 'primitive functions'. He actually built 'catalogues of curves', what in Leibnizian terms we would call 'tables of integrals'. He made use of techniques of substitution of variables in order to reduce a curve with a given equation to a curve whose 'primitive' was known from the table.

2.2.4 The three rules of Newton's analysis

Newton considered power series a very important mathematical tool. It is significant that his first systematic mathematical tract bears the title *De analysi per aequationes numero terminorum infinitas* (On analysis by equations unlimited in the number of their terms).§ Newton began this short summary of his discoveries with the enunciation of three rules, which can be rendered as follows:

Rule 1: if $y = ax^{m/n}$, then the area under y is $an/(n + m)x^{1+(m/n)}$,

Rule 2: if y is given by the sum of more terms (also an infinite number of terms), then the area under y is given by the sum of the areas of all the terms,

Rule 3: in order to calculate the area under a curve $f(x, y) = 0$, one must expand y as a sum of terms of the form $ax^{m/n}$ and apply Rule 1 and Rule 2.¶

---

† 'Et quidquid Vulgaris Analysis per aequationes ex finito terminorum numero constantes (quando id sit possibile) perficit, haec per aequationes infinitas semper perficiat: Ut nil dubitaverim nomen Analysis etiam huic tribuere'. *Mathematical Papers*, **2**: 240–1. Translation from Latin by Whiteside.

‡ *Mathematical Papers*, **1**: 302–1 and Westfall (1980): 127–8 for a different demonstration which remained unpublished until recently.

§ *Mathematical Papers*, **2**: 206–47.

¶ *Mathematical Papers*, **2**: 206ff.

Rule 1 was stated by Wallis, who extended formula (2.3) to n rational $\neq -1$. As we will see, Newton provided a proof of Rule 1 based on the fundamental theorem. The binomial series proved to be an important tool in implementing Rule 3.

2.2.5 Fluents, fluxions and moments

2.2.5.1 Definitions

Whereas the *De analysi* was devoted mainly to series expansions and the use of series in quadratures, in the so-called *De methodis serierum et fluxionum*, an untitled tract written in 1670–71, Newton concerned himself with the use of an algorithm that he had developed in 1665–6.†

The central idea is the introduction of quantities which are 'infinitely' or 'indefinitely' small in comparison with finite quantities. Such infinitesimal quantities had been already widely used in the seventeenth century. For these infinitely little quantities a principle of cancellation holds. If α is infinitely little and A finite then:

$$A + \alpha = A. \tag{2.4}$$

In some places Newton spoke about these infinitely small quantities as if they were actual fixed constituents of finite quantities. In other places he introduced the concept of 'moment': an infinitesimal increment acquired by a finite quantity which varies in time.

The idea of deploying a kinematical conception of quantity dates back to some manuscripts written by Newton in 1666. According to this concept, the objects to which Newton's algorithm is applied are quantities which 'flow' in time. For instance the motion of a point generates a line, the motion of a line generates a surface. The quantities generated by 'flow' are called 'fluents'. Their instantaneous speeds are called 'fluxions'. The 'moments' of the fluent quantities are 'the infinitely small additions by which those quantities increase during each infinitely small interval of time'.‡ Therefore, consider a point which flows with variable speed along a straight line. The distance covered at time t is the fluent, the instantaneous speed is the fluxion, the 'infinitely' (or 'indefinitely') small increment acquired after an 'infinitely' (or 'indefinitely') small period of time is the moment. Newton further observed that the moments of the fluent quantities 'are as their speeds of flow' (i.e., as the fluxions).§ His reasoning is based on the idea that during an 'infinitely small period of time' the fluxion remains constant, therefore the moment is proportional to the fluxion. Newton warns the reader not

† *Mathematical Papers*, **3**: 32–353.
‡ 'Additamenta infinite parva quibus illae quantitates per singula temporis infinite parva intervalla augentur'. *Mathematical Papers*, **3**: 80–1. Translation by Whiteside.
§ 'sunt ut fluendi celeritates'. *Mathematical Papers*, **3**: 78–9.

to identify the 'time' of the fluxional method, with real time. Any fluent quantity whose fluxion is assumed constant plays the role of fluxional 'time'.

2.2.5.2 Notations

In this context Newton developed the following notation. He employed a, b, c, d for constants, v, x, y, z for the fluents and l, m, n, r for the respective fluxions, so that, e.g., m is the fluxion of x. The 'indefinitely' (or 'infinitely') small interval of time was denoted by o. So the moment of y is no. It was only in the 1690s that Newton introduced the now standard notation: the fluxion of x is thus denoted by $\dot{x}$, and the moment of x by $\dot{x}o$. The fluxions themselves can be considered as fluent quantities, therefore one can seek for their fluxions. In the 1690s Newton denoted the 'second' fluxion of x with $\ddot{x}$. Newton did not consistently use a notation for the area under a curve. Generally he put words such as 'the area of' (and in one instance a capital Q†) before the analytical expression of the curve. In some cases he used ' $\boxed{a/x^2}$ ' for 'the area under the curve of equation $y = a/x^2$' (in Leibnizian terms one would have $\int (a/x^2)\mathrm{d}x$). In the 1690s Newton also employed $\acute{x}$ to denote a fluent quantity whose fluxion is x.

2.2.5.3 Algorithm

The basic algorithm for calculating fluxions is given by Newton with an example. He considers the equation:

$$x^3 - ax^2 + axy - y^3 = 0. \tag{2.5}$$

He substitutes $x + \dot{x}o$ for x and $y + \dot{y}o$ for y, and develops the powers. Deleting $x^3 - ax^2 + axy - y^3$ as equal to zero and after division by o he obtains an equation from which he cancels the terms which have o as a factor. In fact these terms 'will be equivalent to nothing in respect to the others', since 'o is supposed to be infinitely small'.‡ At last Newton arrives at:

$$3\dot{x}x^2 - 2a\dot{x}x + a\dot{x}y + a\dot{y}x - 3\dot{y}y^2 = 0. \tag{2.6}$$

This result is achieved by employing the rule of cancellation of infinitesimals (2.4). Notice that in the above example the rules for the fluxions of xy and of x^n are simultaneously stated.

Even though Newton presents his 'direct' algorithm applied to particular cases, his procedure can be generalized. Given a curve expressed by a function in parametric form, $f(x(t), y(t)) = 0$, the relation between the fluxions $\dot{x}$ and $\dot{y}$

---

† *Mathematical Papers*, **7**: 17.

‡ 'respectu caeterorum nihil valebunt', 'supponitur esse infinite parvum'. *Mathematical Papers*, **3**: 80–81. Translation by Whiteside.

is obtained by calculating:

$$f(x + \dot{x}o, y + \dot{y}o) = (\partial f/\partial x)\dot{x}o + (\partial f/\partial y)\dot{y}o + o^2(\cdots) = 0. \qquad (2.7)$$

After division by o, the remaining terms in o are cancelled. Such a modern reconstruction clearly says more than Newton could express. I have used concepts and notations not available to Newton for a function $f(x(t), y(t))$ and for partial derivatives. However, with due caution, it can be used to highlight the following points.

- Newton assumes that, during the infinitesimal interval of time o, the motion is uniform, such that when x flows to $x + \dot{x}o$, y flows to $y + \dot{y}o$. Therefore, $f(x, y) = f(x + \dot{x}o, y + \dot{y}o)$;
- he applies the principle of cancellation of infinitesimals, so in the last step the terms in o are dropped.

Newton's justification for his algorithmic procedure is not much more rigorous than in previous seventeenth-century works, such as those by Pierre Fermat or by Hudde. As we will see below he was soon to face seriously foundational questions.

2.2.6 Problem reduction

2.2.6.1 Two problems

In the *De methodis* Newton gives the solutions of a series of problems. The main problems are to find maxima and minima, tangents, curvatures, areas, surfaces, volumes, arc lengths and centres of gravity. Thanks to the representation of quantities as generated by continuous flow, all these problems can be reduced to the following two *Problems*:

1. Given the length of the space continuously (that is, at every time), to find the speed of motion at any time proposed.

2. Given the speed of motion continuously, to find the length of the space described at any time proposed.†

This is one of the greatest generalizations in the history of mathematics. The great majority of problems faced by Newton's contemporaries were reduced to two large classes.

The problems of finding tangents, extremal points and curvatures can be reduced to the first *Problem*. In fact, imagine a plane curve $f(x, y) = 0$ to be generated by the continuous flow of a point $P(t)$. If (x, y) are the Cartesian coordinates of the curve, $\dot{y}/\dot{x}$ will be equal to $\tan \gamma$ where γ is the angle formed by the tangent in $P(t)$

---

† '1. Spatij longitudine continuo (sive ad omne tempus) data, celeritatem motus ad tempus propositum invenire. 2. Celeritate motus continuo data longitudinem descripti spatij ad tempus propositum invenire.' *Mathematical Papers*, **3**: 70–1.

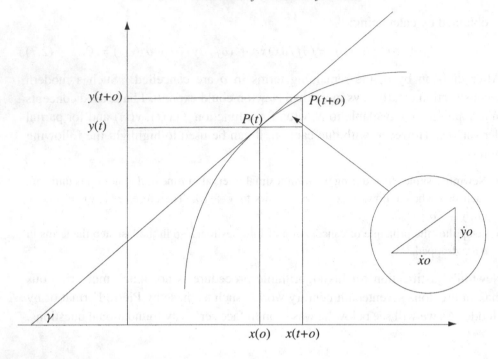

Fig. 2.1. A curve generated by the continuous flow of a point

with the x-axis (see fig. 2.1). Newton expressed the above concept in connected prose. In fact the point will move during the 'indefinitely small period of time' with uniform rectilinear motion from $P(t)$ to $P(t + o)$. The infinitesimal triangle indicated in fig. 2.1 has catheti equal to $\dot{y}o$ and $\dot{x}o$, therefore

$$\dot{y}o/\dot{x}o = \dot{y}/\dot{x} \qquad (2.8)$$

is a measure of the inclination of the tangent at P. The ratio $\dot{y}/\dot{x}$ can be obtained via the algorithm exemplified by formulas (2.5) and (2.6). An extremal point will have:

$$\dot{y}/\dot{x} = 0. \qquad (2.9)$$

Newton proved that the radius of curvature ρ of a plane curve is given by:

$$\rho = \frac{(1 + (\dot{y}/\dot{x})^2)^{3/2}}{(\ddot{y}/\dot{x}^2)}. \qquad (2.10)$$

The fact that the finding of areas can be reduced to the second *Problem* is stated by the fundamental theorem. Let z be the area generated by continuous uniform flow ($\dot{x} = 1$) of ordinate y (see fig. 2.2). The speed of motion is given continuously,

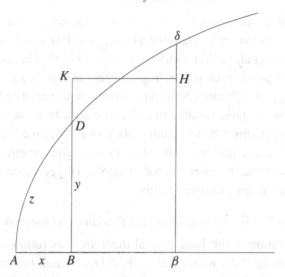

Fig. 2.2. The fundamental theorem. After *Mathematical Papers*, **2**: 242

i.e., $\dot{z}$ is given. The fundamental theorem states that:

$$y = \dot{z}. \tag{2.11}$$

In order to find the area, a method is required for obtaining z from $y = \dot{z}$, which is *Problem 2*.†

The reduction of arclength problems to *Problem 2* depends on application of Pythagoras's theorem to the moment of arc length s (see fig. 2.1):

$$\dot{s}o = \sqrt{(\dot{x}o)^2 + (\dot{y}o)^2}. \tag{2.12}$$

Therefore

$$s = Q\sqrt{\dot{x}^2 + \dot{y}^2}, \tag{2.13}$$

where Q is one of the Newtonian equivalents of the integral sign.

It should be stressed how the conception of quantities as generated by continuous flow allowed Newton to conceive the problem of determining the area under a curve as a special example of *Problem 2*. Newton writes:

By means of the preceding catalogues ['integral tables' in Leibnizian terms] not merely the areas of curves but also other quantities of any kind generated at an analogous rate of flow may be derived from their fluxions [...] if a fluxion of whatever kind be expressed by an ordinate line [...] the quantity generated by that fluxion will be expressed by the area described by that ordinate.‡

† In Leibnizian notation: if $z = \int y\,dx$, then $dz/dx = d/dx \left(\int y\,dx\right) = y$.
‡ *Mathematical Papers*, **3**: 284–5. Translation by Whiteside.

Problem 2 is, of course, much more difficult. Given a 'fluxional equation' $f(x, y, \dot{x}, \dot{y}) = 0$, Newton seeks a relation $g(x, y, c) = 0$ (c constant) such that the application of the direct algorithm yields $f(x, y, \dot{x}, \dot{y}) = 0$. That is, in Leibnizian terms, he poses the problem of integrating differential equations. Newton had a very general strategy, which allowed him to solve a great variety of such 'inverse problems': either he changed variable in order to reduce to a known catalogue of fluents (in Leibnizian terms a 'table of integrals'), or he deployed series expansion techniques (termwise integration). His strategy is a great improvement on the geometrical quadrature techniques of, for example, Huygens, or the techniques of direct summation of, for example, Wallis.

2.2.6.2 A demonstration of the fundamental theorem

Newton's demonstration of the fundamental theorem is as follows. He considers the curve $AD\delta$ (see fig. 2.2), where $AB = x$, $BD = y$ and the area $ABD = z$. He defines $B\beta = o$ and $BK = v$, such that 'the rectangle $B\beta HK$ ($= ov$) is equal to the space $B\beta\delta D$'. Furthermore, Newton assumes that $B\beta$ is 'infinitely small'. With these definitions one has that $A\beta = x + o$ and the area $A\delta\beta$ is equal to $z + ov$. At this point Newton writes: 'from any arbitrarily assumed relationship between x and z I seek y.' This problem is solved for $an/(n + m)x^{1+(m/n)} = z$. For instance, let $2/3x^{3/2} = z$ or $4/9x^3 = z^2$. Substituting $x + o$ for x and $z + ov$ for z, after some manipulations (including a division by o of each term) one gets:

$$\frac{4}{9}\left(3x^2 + 3xo + o^2\right) = 2zv + v^2 o. \tag{2.14}$$

Newton continues his demonstration:

If we now suppose $B\beta$ to be infinitely small, that is, o to be zero, v and y will be equal and terms multiplied by o will vanish and there will consequently remain $(4/9)3x^2 = 2zv$ or $2/3x^2(= zy) = 2/3x^{3/2}y$, that is, $x^{1/2}(= x^2/x^{3/2}) = y$.†

Then he adds a statement which expresses the inverse relationship of what, in Leibnizian terms, one would call differentiation and integration:

Conversely therefore if $x^{1/2} = y$, then will $2/3x^{3/2} = z$.‡

Although Newton's proof of the fundamental theorem refers to particular instances, his reasoning is meant to be completely general.

2.2.7 Some general results on quadratures

Newton's results on quadratures of the *De methodis* allow one to integrate several classes of functions. In the 1670s Newton generalized and extended these

† *Mathematical Papers*, **2**: 242–5. Translation by Whiteside.
‡ *Mathematical Papers*, **2**: 245. Translation by Whiteside.

results. His effort was to generalize with a single 'prime theorem' the quadratures expressed by the long 'catalogues' of the *De methodis*. For instance in the so-called *epistola posterior* written to Leibniz in 1676 he gave, without proof, a method for 'squaring' the curve

$$y = x^\theta (e + fx^\eta)^\lambda, \tag{2.15}$$

where e and f are constants.[†]

Newton achieved this very general result. Let

$$
\begin{aligned}
R &= e + fx^\eta + gx^{2\eta} + hx^{3\eta} + \cdots \\
S &= a + bx^\eta + cx^{2\eta} + dx^{3\eta} + \cdots.
\end{aligned}
$$

Furthermore, we set $r = \theta/\eta$, $s = r + \lambda$, $t = s + \lambda$, $v = t + \lambda$, …. Then the area under the curve

$$x^{\theta-1} R^{\lambda-1} S \tag{2.16}$$

is:

$$
\begin{aligned}
x^\theta R^\lambda \Bigg(\frac{a/\eta}{re} &+ \frac{b/\eta - sfA}{(r+1)e} x^\eta + \\
&\frac{c/\eta - (s+1)fB - tgA}{(r+2)e} x^{2\eta} + \\
\frac{d/\eta - (s+2)fC - (t+1)gB - vhA}{(r+3)e} x^{3\eta} &+ \cdots \Bigg),
\end{aligned}
$$

where each $A, B, C \ldots$ is the coefficient of the preceding power of x.

It should be noted that Newton was able to square curves of the form

$$y = c_1 x^{\eta-1} / \sqrt{c_2 + c_3 x^\eta + c_4 x^{2\eta}}. \tag{2.17}$$

These results should give an idea of the extent of Newton's techniques in the inverse method of fluxions.[‡]

2.3 The synthetic method

2.3.1 The Ancients versus the Moderns

As we have seen, in his early writings Newton employed methods characteristic of the seventeenth-century 'new analysis': i.e. he used series and an algorithm based on infinitesimal quantities. In some instances, infinitesimals entered as fixed constituents of finite quantities. They were more often defined as 'moments', infinitesimal increments of a 'flowing' variable quantity. The kinematical approach

---

[†] *Correspondence*, **2**: 115. This quadrature method was published in Wallis (1693–99), **2**: 390–6.
[‡] In §2.2.7 I have followed Edwards (1979): 226–30. For further details see Whiteside's notes in *Mathematical Papers*, **3**: 237–65.

to the calculus was therefore prevalent, even though not exclusively followed, in Newton's work from the very beginning. For him reference to our intuition of continuous 'flow' provided a means to define the referents of the calculus: fluents, fluxions and moments.

Up to the composition of the *De methodis* (1670–1), Newton described himself with pride as a promoter of the seventeenth-century 'new analysis'. For instance, the *De methodis* opens with the following statement:

Observing that the majority of geometers, with an almost complete neglect of the ancients' synthetical method, now for the most part apply themselves to the cultivation of analysis and with its aid have overcome so many formidable difficulties that they seem to have exhausted virtually everything apart from the squaring of curves and certain topics of like nature not yet fully elucidated: I found not amiss, for the satisfaction of learners, to draw up the following short tract in which I might at once widen the boundaries of the field of analysis and advance the doctrine of series.†

In the 1670s Newton was led to distance himself from his early mathematical researches: he abandoned the calculus of fluxions in favour of a geometry of fluxions where infinitesimal quantities were not employed. He labelled this new method the 'synthetic method of fluxions' as opposed to his earlier 'analytical method of fluxions'. In the late 1710s he wrote:

Quantities increasing in a continuous flow we call fluents, the speeds of flowing we call fluxions and the momentary increments we call moments, and the method whereby we treat quantities of this sort we call the method of fluxions and moments: this method is either synthetic or analytical.‡

Newton's self-instruction in mathematics was based on the most up-to-date works in modern analytics. As a young mathematician he trod in the footsteps of Descartes, Wallis and Barrow.§ From the mid-1670s Newton began to criticize modern mathematicians. For instance, commenting on Descartes' solution of Pappus's problem in *Géométrie*, he stated with vehemence:

Indeed their [the Ancients'] method is more elegant by far than the Cartesian one. For he [Descartes] achieved the results by an algebraic calculus which, when transposed into words (following the practice of the Ancients in their writings), would prove to be so tedious and entangled as to provoke nausea, nor might it be understood. But they accomplished it by certain simple propositions, judging that nothing written in a different

---

† *Mathematical Papers*, **3**: 33. Translation by Whiteside.

‡ 'Quantitates continuo fluxu crescentes vocamus fluentes & velocitates crescendi vocamus fluxiones, & incrementa momentanea vocamus momenta, et methodum qua tractamus ejusmodi quantitates vocamus methodus fluxionum et momentorum: estque haec methodus vel synthetica vel analytica'. *Mathematical Papers*, **8**: 454–5. Translation by Whiteside.

§ On Barrow's influence on Newton see Arthur (1995) and Feingold (1993).

style was worthy to be read, and in consequence concealing the analysis by which they found their constructions.†

Later in the 1690s he wrote:

And if the authority of the new Geometers is against us, nonetheless the authority of the Ancients is greater.‡

In the 1670s Newton studied in depth the seventh book of Pappus's *Collectiones*. In particular he commented on the lost books on *Tangencies* and on the *Inclinations* by Apollonius and compared algebraical and geometrical solutions. Later on, from 1691 to 1695, he devoted attention to the introduction of the seventh book of the *Collectiones*. He concentrated especially on the lost books on *Porisms* by Euclid and the *Loci Plani* by Apollonius. His results have a place in the history of seventeenth-century projective geometry.§ These new interests in the works of Apollonius led Newton to re-evaluate the use of geometry. In the 1670s he devoted particular attention to Pappus's 'four lines locus'. His geometric solution of this problem – clearly framed in opposition to Descartes' own solution, a case study in *Géométrie* – attained 'not by a calculus [as Descartes had done] but a geometrical composition, such as the Ancients required' was to appear in print in Section 5, Book 1, of the *Principia*.¶

Newton in the 1670s characterized the 'geometry of the Ancients' as simple, elegant, concise, fitting for the problem posed, always interpretable in terms of existing objects. In particular, geometrical demonstrations have, according to Newton, a safe referential content. On the other hand, Newton stressed the mechanical character of the algebraical methods of the 'Moderns', their utility only as heuristic tools and not as demonstrative techniques, the lack of referential clarity of the concepts employed, their redundance. Newton's admiration for the Ancients' geometrical methods, and his critical approach towards the algebraical methods of the 'Moderns', have their roots in the 1670s. Newton never abandoned this position: if possible, he strengthened it as the years passed. Henry Pemberton, a privileged witness – as editor of the third (1726) edition of the *Principia* – of Newton's last years wrote in *View of Sir Isaac Newton's Philosophy* :

I have often heard him censure the handling of geometrical subjects by algebraic calculations; [. . .] he frequently praised Slusius, Barrow and Huygens for not being influenced by the false taste, which then began to prevail. He used to commend the laudable attempt of

---

† *Mathematical Papers*, **4**: 277. Translation by Westfall in Westfall (1980): 379. See also a memorandum by David Gregory dated 22 October 1704: 'Mr. Newton says that Apollonius de Ratione sectione was in order to the solution of the Problem that Euclid proposed, & which Des Cartes pretended to have solved.' Quoted in Hiscock (1937): 20.

‡ 'Et si authoritas novorum Geometrarum contra nos facit, tamen major est authoritas Veterum'. *Mathematical Papers*, **7**: 185n.

§ There is evidence that Newton read Fermat's reconstruction (which appeared posthumously in 1679) of the *Loci Plani* and of five propositions of the *Porisms*.

¶ *Variorum*: 150. Cf. *Principles*: 81. On Newton's solution of Pappus's problem see Di Sieno & Galuzzi (1989).

Hugo de Omerique to restore the ancient analysis, and very much esteemed Apollonius's book *De sectione rationis* for giving us a clearer notion of that analysis that we had before [...] Sir Isaac Newton has several times particularly recommended to me Huygens's style and manner. He thought him the most elegant of any mathematical writer of modern times, and the most just imitator of the ancients. Of their taste and form of demonstration Sir Isaac always professed himself a great admirer: I have heard him even censure himself for not following them yet more closely than he did; and speak with regret of his mistake at the beginning of his mathematical studies, in applying himself to the works of Des Cartes and other algebraic writers before he had considered the elements of Euclide with that attention, which so excellent a writer deserves.†

It would certainly be excessive to say that Newton abandoned completely the 'new analysis' that he had developed in his *anni mirabiles* by mastering and extending fine details of the Latin edition of Descartes' *Géométrie* and Wallis's *Arithmetica infinitorum*. Newton's relationship with the Cartesian tradition is ambiguous. Descartes was, in many ways, his starting point (not only in mathematics!). As Newton matured he developed a tense relationship with Cartesianism: he rejected it, but, at the same time, remained more Cartesian than he would have wanted to admit. However, we can safely say that during the 1670s Newton contrasted geometrical methods with algebraical ones, with the purpose of showing the superiority of the former to the latter.

The reasons that induced this champion of series, infinitesimals and algebra to distance himself from his early researches are complex. They have to do with foundational worries about the nature of infinitesimal quantities. They are related to his desire to find in geometry unifying principles of techniques which were growing wildly in his early writings. They have to do also with his aversion towards Descartes, indeed towards anything Cartesian, and with his admiration for Huygens.

In 1673 Huygens sent a complimentary copy of his *Horologium oscillatorium* to Newton. Newton must have been strongly impressed by it because of the geometrical rigour followed by the great Dutch natural philosopher. In the *Horologium* proofs were given in accordance with the *ad absurdum* proofs of Archimedes' so-called method of exhaustion. Furthermore, the geometrical language of proportion theory was applied to the study of the motion of pendula. Some of Huygens' methodology survived in Newton's works on the 'synthetic method of fluxions' of the 1670s, and, most notably, in some sections of the *Principia*. However, as we will see in Chapter 5, relevant differences between the styles followed in the *Horologium* and in the *Principia* exist.

It should be noticed furthermore that, from the mid-1670s, Newton began looking at ancient texts not only for mathematical interests. As Dobbs and Figala have shown, from the early 1670s, he was led to look for a restoration of an

---

† Pemberton (1728): *Preface*.

ancient knowledge in alchemy.† Newton began, in the 1670s, to conceive himself as a man who belonged to a remnant of interpreters who could restore, through the deciphering of Biblical texts, the original natural philosophy and religion of mankind, revealed by God to Noah.‡ He believed that this knowledge had passed, through the sages of Israel, to Egypt. He also maintained that it was corrupted there by priest and rulers. From time to time, enlightened interpreters were able to restore part of this lost wisdom. Mathematics has long enjoyed a close relationship with mysticism. For Newton 'mathematics was God's language'.§ It is striking that in the same years Newton began attributing to Jews, Egyptians and Pythagoreans a lost knowledge concerning alchemy, God *and* mathematics. It is plausible that in Newton's mind the restoration of the lost books of the ancient geometers of Alexandria was linked to his attempt to re-establish a *prisca sapientia*. We will see in Chapter 4 that, after the publication of the *Principia*, Newton's beliefs concerning the Ancients became even more pervasive.

Newton's approach to classic geometry is by no means unique in the history of mathematics. Many Renaissance mathematicians shared the view that the Ancients had possessed a 'hidden' analysis, a secret method of discovery, that would have allowed them to achieve their extraordinary results. These mathematicians devoted themselves to the 'divination' of the Ancients' lost analysis. During Newton's lifetime this Renaissance tradition, which has its roots in the sixteenth-century editions of the Greek classics, was declining. Leibniz in his *iter italicum* of 1689–90 could still find defenders of the geometrical methods of the Ancients, such as Vincenzo Viviani, who fiercely resisted the introduction of the new calculus in Italy.¶

However, Newton's attitude towards the Ancients is peculiar of the English *milieu* in which he lived. A critical attitude towards the new mechanistic natural philosophy was shared by many English philosophers and theologians, such as Henry More, who might have had some influence on Newton. In the 1670s Newton began to state that Cartesian mechanicism was a mistaken view: Nature could not be reduced to matter and motion. Some spiritual agent was active, being responsible for chemical, optical and possibly also gravitational phenomena. Newton shared with many of his English contemporaries the idea that the new mechanistic philosophers were defending an image of Nature which was partial and potentially dangerous from a theological point of view. He also accepted the idea, deeply rooted in Renaissance thought, and defended by alchemists such as Elias Ashmole and interpreters of prophecies such as Joseph Mede (whose works

---

† Dobbs (1975), Figala (1977).
‡ Mandelbrote (1993).
§ Mandelbrote (1993): 301. See also Mamiani's introduction in Newton (1995).
¶ Robinet (1988).

Newton studied in depth), that it was necessary to look in the ancient texts for the remnants of a lost knowledge superior to that accepted by the 'Moderns'. It is in this context that Newton began his researches in ancient alchemical texts and in biblical criticism. He viewed the history of mankind as a regress, a process of corruption, rather than a progress. In his alchemical, theological and chronological works, which he began composing in secrecy in the 1670s, the 'Moderns' are always depicted as inferior to the 'Ancients'. Newton's rejection of Cartesian algebra, his distancing himself from the analytical method of fluxions, and his interests in the geometrical works of Apollonius and Pappus are in resonance with other facets of his intellectual endeavour. If it is true, as many Newtonian scholars have emphasized, that Newton brought into the fields of alchemy and chronology some of the methodological rigour of the mathematician, it is equally reasonable to state that some attitudes typical of the alchemist and theologian (e.g. the myth of the Ancients, the reluctance in publishing the methods of discovery) might have had some impact on the mathematician.† We will return to these themes in Chapter 4.

2.3.2 Geometry and Nature: first and ultimate ratios

Just after completing the *De methodis* Newton composed an *addendum* where he developed a 'more natural approach' to the fluxional method – an approach 'based on the genesis of surfaces by their motion of flow' and 'which will come to be still more perspicuous and resplendent if certain foundations are, as is customary with the synthetic method, first laid'.‡ Whereas in his early writings Newton represented the fluents with algebraical symbols, in this new approach he referred directly to geometrical figures. These figures, however, are not static, as in classic geometry: they must be conceived to be 'in motion'. In the *addendum* Newton did not employ algebraic expressions, but referred to geometric figures which flow in time and to their fluxions, or rates of flow. He was interested in determining the fluxions of geometrically defined quantities, typically four 'fluent' lines such that $A/B = C/D$. Most notably he determined the fluxion of a product AB as $\mathrm{fl}(A)B + A\mathrm{fl}(B)$.§

In order to determine these fluxions he utilized the idea, elevated to the rank of axiom, that 'contemporaneous moments are as their fluxions' (since motion during a moment of time is uniform). He had already employed this 'axiom' in

---

† On Newton's religious thought Manuel (1963) and (1974) are still an indispensable starting point. Westfall (1980): 281–401 provides a magisterial synopsis of Newton's work in alchemy, theology and mathematics carried on in the 1670s. Further information can be derived from the more recent studies by Force & Popkin (1990), Gascoigne (1991) and Mandelbrote (1993).

‡ *Mathematical Papers*, **3**: 282–3, 328–31. Translation by Whiteside.

§ fl(*A*) stands for 'fluxion of *A*'.

his symbolical reasonings. Furthermore, because of the 'infinite smallness of the moments', terms multiplied by them are rejected compared with finite quantities. The *addendum* is thus a geometric version of Newton's fluxional algorithm. It is furthermore based on axioms 'as is customary with the synthetic method'.

In a successive work entitled *Geometria curvilinea*, dated tentatively by Whiteside around 1680, Newton introduced the idea of 'first and ultimate ratios'.† The *Geometria curvilinea* again is concerned with geometric flowing quantities. Whereas in the *addendum* infinitesimal quantities generated by motion are referred to as 'moments', here Newton talks about 'nascent or vanishing parts'. In Axiom 6 he states that:

Fluxions of quantities are in the first ratio of their nascent parts or, what is exactly the same, in the ultimate ratio of those parts as they vanish by defluxion.‡

This Axiom seems to imply 'variable infinitesimals': quantities, which, in the process of coming into existence from nothing, or vanishing into nothing, pass through a state in which they are neither finite nor nothing. However, in the Preface of the *Geometria curvilinea* Newton appears to distance himself from the use of infinitesimals. He writes:

Those who have taken the measure of curvilinear figures have usually viewed them as made up of infinitely many infinitely small parts; I, in fact, shall consider them as generated by growing, arguing that they are greater, equal or less according as they grow more swiftly, equally swiftly or more slowly from their beginning. And this swiftness of growth I shall call the fluxion of a quantity. So when a line is described by the movement of a point, the speed of the point – that is, the swiftness of the line's generation – will be its fluxion. I should have believed that this is the natural source for measuring quantities generated by continuous flow according to a precise law, both on account of the clarity and brevity of the reasoning involved and because of the simplicity of the conclusions and the illustrations required.§

The demonstrations in the *Geometria organica* are not based on moments, they rather depend on the determination of limits of ratios of 'vanishing parts'.

In this manuscript the determination of the fluxions of trigonometric quantities is given particular attention. Newton determines the limits of ratios of vanishing geometrical quantities related to a circle. For instance he states that:

In a given circle the fluxion of an arc is to the fluxion of its sine as the radius to its cosine; to the fluxion of its tangent as its cosine is to its secant; and to the fluxion of its secant as its cosine to its tangent.¶

A 'synthetic' demonstration can be provided as follows. Consider a circle (see fig. 2.3) with radius OA. Imagine that the radius flows from position OB to

† 'rationes primae et ultimae' is sometime translated as 'prime and ultimate ratios'.
‡ *Mathematical Papers*, **4**: 427.
§ *Mathematical Papers*, **4**: 423. Translation by Whiteside.
¶ *Mathematical Papers*, **4**: 441. Translation by Whiteside.

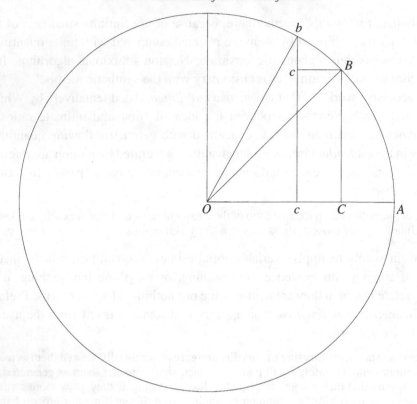

Fig. 2.3. Fluxions of trigonometric quantities

position Ob. It is easy to show that, as b 'flows back' to B: (i) the triangles bcB and OCB approach similarity. In fact one can observe that, as b 'flows back' until b and B 'come together': (ii) the 'ultimate ratio' of the arc bB to the chord bB is equal to 1; (iii) the ultimate ratio of angles bBc/OBC is equal to 1.† These intuitive and 'visual' considerations on the limiting form of triangle bcB allow one to obtain the fluxions of the trigonometric quantities. In fact we have that the ratio of the fluxion of the arc to the fluxion of the sine is equal (because of Axiom 6) to the ultimate ratio of bB to bc, which is equal (because of (i)) to the ratio of the radius to the cosine (OB/OC).‡

It should be observed that the *Geometria curvilinea* is opened by a long declaration about the lack of rigour and elegance of the methods followed by those 'men of recent times' who have abandoned the geometrical methods of the

---

† As far as (ii) cf. Lemma 7, Section 1, Book 1 of the *Principia* discussed below in §3.3. I leave the proof of (iii) to the reader.
‡ In modern usage CB is $OB\sin(AOB)$.

Ancients.† It is also clear that Newton conceived his synthetic method of first and ultimate ratios as an improvement on the methods of the Ancients, whose *ad absurdum*, indirect, procedures were often 'tedious'.‡

In the *De quadratura curvarum*, a treatise whose composition began in 1691–92 (a version was published in 1704), Newton presented a symbolical version of the synthetic method of first and ultimate ratios of the *Geometria curvilinea* and the *Principia*. He made it clear that the symbolical demonstrations of *De quadratura curvarum* are safely grounded in the geometry of the synthetic method of fluxions. In the Preface it is stated that the calculus is referred only to finite flowing quantities. A continuity with ancient tradition is also underlined:

Mathematical quantities I here consider not as consisting of least possible parts, but as described by a continuous motion. [...] These geneses take place in the reality of physical nature and are daily witnessed in the motion of bodies. And in much this manner the ancients, by 'drawing' mobile straight lines into the length of stationary ones, taught the genesis of rectangles.§

Notice that in these opening lines Newton introduces, beside the reference to the Ancients, another theme which was to become a *leitmotiv* during the priority controversy with Leibniz: the ontological content of the method of fluxions. Fluent and fluxions are really exhibited in *rerum natura*. In the 1710s, when opposing Leibniz, Newton contrasted the safe referential content of his method with the lack of meaning of the differential calculus. In Leibniz's calculus, according to Newton, 'indivisibles' occur, but the use of such quantities is not only a departure from ancient tradition, it also leads to the use of symbols devoid of referential content. In the anonymous account to the *Commercium epistolicum* Newton wrote:

We have no ideas of infinitely little quantities & therefore Mr Newton introduced fluxions into his method that it might proceed by finite quantities as much as possible. It is more natural & geometrical because founded on *primae quantitatum nascentium rationes* wch have a being in Geometry, whilst *indivisibles* upon which the Differential method is founded have no being either in Geometry or in nature. [...] Nature generates quantities by continual flux or increase, & the ancient Geometers admitted such a generation of areas & solids [...]. But the summing up of indivisibles to compose an area or solid was never yet admitted into Geometry.¶

Nature and geometry are the two key concepts: they allow Newton to defend his method because of its continuity with ancient tradition as well as its ontological content.

† *Mathematical Papers*, **4**: 420–5.
‡ See *Variorum*: 86, *Principles*: 38.
§ 'Quantitates Mathematicas non ut ex partibus quam minimis constantes sed ut motu continuo descriptas hic considero [...] Hae Geneses in rerum natura locum vere habent & in motu corporum quotidie cernuntur. Et ad hunc modum Veteres ducendo rectas mobiles in longitudinem rectaurum immobilium genesin docuerunt rectangulorum.' *Mathematical Papers*, **8**: 122–3. Translation by Whiteside.
¶ *Mathematical Papers*, **8**: 597–8.

An example of the symbolical version of the method of first and ultimate ratios can be chosen from the *De quadratura*. In order to find the fluxion of $y = x^n$ Newton proceeded as follows:

Let the quantity x flow uniformly and the fluxion of the quantity x^n needs to be found. In the time that the quantity x comes in its flux to be $x + o$, the quantity x^n will come to be $(x + o)^n$, that is [when expanded] by the method of infinite series

$$x^n + nox^{n-1} + \frac{1}{2}(n^2 - n)o^2 x^{n-2} + \cdots; \tag{2.18}$$

and so the augments o and $nox^{n-1} + \frac{1}{2}(n^2 - n)o^2 x^{n-2} + \cdots$ are one to the other as 1 and $nx^{n-1} + \frac{1}{2}(n^2 - n)ox^{n-2} + \cdots$. Now let those augments come to vanish and their last ratio will be 1 to nx^{n-1}; consequently the fluxion of the quantity x is to the fluxion of the quantity x^n as 1 to nx^{n-1}.[†]

Notice that the increment o is finite and that the calculation aims at determining the limit of the ratio $((x + o)^n - x^n)/o$, as o tends to zero. In the lines which immediately follow this demonstration, Newton states that such limit procedures are in harmony with the geometry of the Ancients:

In finite quantities, however, to institute analysis in this way and to investigate the first or last ratios of nascent or vanishing finites is in harmony with the geometry of the ancients, and I wanted to show that in the method of fluxions there should be no need to introduce infinitely small figures into geometry.[‡]

After the methodological turn of the 1670s Newton did not completely abandon the 'analytic art'. He was still busy in algebra, series and quadratures. In 1683 he deposited – in compliance with his duties as Lucasian Professor – his lectures on arithmetic and algebra (these were to be published in 1707 as the *Arithmetica universalis*). In the 1690s he was to improve his quadrature methods and write several versions of what became the *De quadratura*. Other topics such as the enumeration of cubic curves and series were still important for him for the rest of his life. However, from the evidence that will be presented in the next chapters, it emerges that analytics occupied a lower place in Newton's hierarchy of values.

In his maturity Newton was rethinking his early analytical methods from a new point of view, a point of view which had matured in the geometrical researches of the 1670s and in the *Principia*. In the *De quadratura* he presented a new version of the analytical method of fluxions. But he avoided infinitesimals and moments. Rather, he had recourse to limit procedures. He also made it clear that the symbols and concepts employed could be exhibited in geometric form: 'For fluxions are finite quantities and real, and consequently ought to have their own symbols; and each time it can conveniently so be done, it is preferable to express them by finite

---

[†] *Mathematical Papers*, **8**: 126–9. Translation by Whiteside.
[‡] *Mathematical Papers*, **8**: 129. Translation by Whiteside.

lines visible to the eye rather than by infinitely small ones'.† The analytical method of *De quadratura* was understood by Newton as always translatable in terms of the finite geometric fluent quantities and limits of *Geometria curvilinea*.

The circumstances of publication of *Arithmetica universalis* are revealing of Newton's convictions on the subordination of analytics to geometry. He agreed to publish his lectures on algebra, which appeared in 1707. However, this highly original work appeared anonymously. Newton made it clear that he was compelled to publish it in order to obtain the support of his Cambridge colleagues in the election to the 1705 Parliament. In the opening 'To the Reader' it was stated that the author had 'condescended to handle' the subject.‡ The *Arithmetica universalis* also ended with often quoted statements in favour of pure geometry and against the 'Moderns' who have lost the 'Elegancy of Geometry'.§

2.4 Conclusion: Newton's classicism

Even though this chapter has been extremely superficial, I hope that it has succeeded in conveying to the reader some idea about the extent of Newton's method of series and fluxions. Most of the problems about tangents and quadratures left open in 1665 found a solution in Newton's mathematical work during the late 1660s and early 1670s. The historian will never cease to be surprised by two facts: the variety of notations, interpretations and methods that Newton experimented with until the early 1690s, and the circumstance that this method was not published in its entirety until 1704, when the *De quadratura* appeared as an appendix to the *Opticks*. These two facts could be linked to one another. Newton did not publish a method which always seemed to him in need of reformulation.

It might also be argued that Newton's publication strategy was not so unusual at the middle of the seventeenth century. Journals such as the *Philosophical Transactions* were just beginning to be issued and it was not clear in what form a mathematical paper could be published there. Mathematicians generally published their results in letters addressed to their colleagues. They also circulated manuscripts or published books. It was also common to keep the methods of discovery somewhat hidden. Newton conformed to these practices. He allowed some of his mathematical discoveries to be divulged through letters, manuscripts and inclusion in published books.¶ However, in Newton's strategy there is some-

---

† *Mathematical Papers*, **8**: 113–15. Translation by Whiteside.
‡ *Arithmetick*: 4.
§ *Arithmetick*: 119–20. See Pycior (1997): 200–8.
¶ For example, Newton rendered his method of fluxions public, at least in part, through his correspondence with Collins and Oldenburg. Also, Wallis's works contained some of Newton's mathematical discoveries. Namely, excerpts from the 1676 epistolae to Leibniz in Wallis (1685): 318–20, 330–3, 338–47; prime theorem on quadratures (cf. formula (2.15)) as part of a draft of *De quadratura* in Wallis (1693–99), **2**: 390–6; full text of the 1676 epistolae to Leibniz in Wallis (1693–99), **3**: 622–9, 634–45.

thing irreducible to seventeenth-century publication practices. He was peculiarly secretive and reluctant to publish his method of fluxions, especially its analytical version. Newton's contemporaries judged his reluctance to publish strange and, often, frustrating.

A major divide, in the complex story of developments and amendments that the method of series and fluxions underwent at Newton's hands, is that between the analytical and the synthetic versions. As we will see in the next chapter, both versions had a role in the composition of the *Principia*, but the use of the former was almost completely hidden. The analytical version is nothing less than the infinitesimal calculus, one of the greatest mathematical discoveries in the history of science. The extraordinary fact is that, from the 1670s, Newton distanced himself from his early analytical style. For a complex series of reasons he was led to take a critical attitude towards the mathematics of the 'Moderns'. He thus lost interest in a theory, the analytical method of fluxions, which, had it been published in the 1670s, would have established him as the top mathematician in Europe.

3

The mathematical methods of the *Principia*

3.1 Purpose of this chapter

In this chapter I will try to give the reader some information about the mathematical methods employed by Newton in the *Principia*. We will have to consider with some patience a variety of lemmas, propositions, corollaries and scholia. I have found it convenient to postpone a more detailed analysis of some propositions to later chapters.†

I have tried to render the structure of Newton's demonstrations without wasting too much space on trivial details. Therefore, I have skipped the lines of demonstration in which appeal is made to simple geometrical properties (e.g. similarity of triangles). We would adopt the same strategy in presenting Laplace's or Poincaré's demonstrations. When dealing with proofs given in a more familiar symbolic language we do not expect that every substitution of variable, or that every elementary integration, should be made explicit and commented on. The *Principia* has a reputation for being a very difficult, if not tedious, work. This is mainly due to the fact that nowadays we are not used to geometrical techniques. However, when the trivialities are skipped, the structure of most demonstrations emerges as remarkably simple. I hope that the reader will thus be able to follow the essential steps.

In the quotations from the *Principia* I am using Cohen and Whitman's forthcoming translation.‡ I deemed it necessary not to follow their policy of altering some of Newton's notations and technical terms (see §1.3). I have departed from this excellent translation in the following details:

(1) Newton's usage of expressions proper of proportion theory has been maintained. Most notably, I have rendered 'duplicata ratione' as 'duplicate ratio', rather than 'squared ratio'.

(2) Newton in some cases uses the old notation for raising powers (which he most

---

† i.e. Book 1: Prop. 1 (§8.5), Prop. 6 (§5.5), Cor. 1 to Props. 11–13, Props. 39–41 (§8.6.2.1). Book 2: Prop. 10 (§8.6.3.1 and §8.6.3.3).
‡ Newton (1999).

probably learned in Viète's and Oughtred's works), e.g. '*ABquad*' meaning 'the area of the square whose side is AB'. In other cases he uses the notation AB^2. In my book all the quotations follow closely Newton's changing style for raising powers.

(3) Cohen and Whitman translate 'in infinitum' as 'indefinitely'. I prefer to follow the Motte–Cajori translation and leave the Latin phrase 'in infinitum' unaltered.

In my commentary the following notation conventions have been adopted:
(1) I have rendered in symbolic form some statements about proportionality which Newton presents in connected prose. So, instead of statements such as 'A is as B directly', or 'A is as B inversely', 'A is to B as C is to D', I have written '$A \propto B$', '$A \propto 1/B$', '$A/B = C/D$'.†
(2) I have also taken the liberty of using symbols such as F for 'force', or v for 'velocity'. So where Newton writes 'the force is as the square of the velocity', in my commentary I write '$F \propto v^2$'. In modern texts a clear notational distinction between scalar and vector quantities is made. In Newton's times this distinction was performed in words by specifying that a certain magnitude is oriented towards a point: with the exception of a few footnotes I will follow the old practice. I am aware that the use of these symbolic abbreviations and modernizations in the commentary is open to criticism. I had to decide between being purist and considering the reader's convenience: I chose the latter alternative. I hope that the anachronistic effect caused by the conventions adopted in the commentary is somewhat mitigated by the quotations where Newton's style is scrupulously reproduced.

Of course Newton's mathematical demonstrations in the *Principia* are based on the first three 'axioms or laws of motion'. I assume that the reader is aware of these laws and of the literature devoted to them.‡ A last warning concerns mass. Some readers might be surprised to find no occurrence of mass in so many formulas related to topics such as central force motion, resisted motion, and so on. The fact is that in Newton's times mathematicians thought in terms of proportions, even when they wrote formulas which look very much like equations. Constant factors were thus not made explicit. Dimensional analysis of many formulas appearing in this book might leave many readers perplexed. Most notably force is often equated with acceleration.

---

† This seems to me in line with Newton's intentions since in the second edition of the *Principia* he added the following explanation: 'If A is said to be as B directly and C directly and D inversely, the meaning is that A is increased or decreased in the same ratio as $B \times C \times 1/D$, that is, that A and BC/D are to each other in a given ratio'. This quotation comes from a Scholium to Lemma 10, Section 1, Book 1, which is lacking in the first edition. *Variorum*: 83. Cf. *Principles*: 35.

‡ See, for instance, Westfall (1971).

3.2 Book 1: an overview

Book 1 consists of fourteen sections. It deals with the 'motion of bodies' in a space free of resistance. The mathematical theory of planetary motions is thus laid down in this first Book of *Principia*. This theory will be applied to astronomical phenomena in Book 3, devoted to the 'System of the World'.

In Section 1, Book 1, Newton presents the 'method of first and ultimate ratios', already developed in *Geometria curvilinea*, as the foundation for the geometrical limit procedures which are deployed in the *Principia*.

In Sections 2 and 3 (Propositions 1–17) Newton lays down the foundation of his treatment of the direct problem of central forces. By *central force* Newton means a force whose magnitude varies with distance from a given point (the *centre of force*) and which is directed towards or away from the centre of force. In these sections Newton considers a 'body' (we would say a 'point mass') accelerated by such a force. By *direct problem of central forces* late-seventeenth-century natural philosophers meant the problem of determining the law of force, i.e. the variation of force magnitude as a function of distance from its centre, when the trajectory (a plane curve) and the centre of force (a point lying on the plane of the trajectory) are given.

In Sections 2, 3, 6, 7, 8 and 9 Newton considers a mathematical model constituted by a body situated at P (for 'planet') accelerated by a central force with centre of force S (for 'Sun'). The body is supposed to have a mass and to be pointlike. Newton knows that this mathematical model can be applied only approximately to the planetary system. In practice when one considers a system composed of two bodies 1 and 2, sufficiently far from other disturbing bodies and sufficiently far one from the other, and 1 has a much greater mass then 2, then one can consider 1 as an immovable centre of force and 2 as a point mass. This simplified model occurs also in Section 9, devoted to the motion of the line of apsides. It is only in Section 11 that Newton will consider the motion of two, or more than two, bodies which mutually attract each other. It is only in Sections 12 and 13 that he will pay attention to the shape of the bodies, and to the effects of it on the motions caused by their mutual attractions.†

In Propositions 1 and 2 Newton deals with the dynamical implications of Kepler's area law. The area law, published by Kepler in 1609, states that the *radius vector* (the line joining the Sun with a planet) lies on a fixed plane and sweeps equal areas in equal times. Newton proves that a force is central if and only if the area law holds.‡

In Propositions 1 and 2 Newton singles out two quantities which are *conserved*, i.e. they do not change in time, in central force motion. First: the *plane of orbital*

---

† On Newton's use of mathematical models see Cohen (1980).

‡ We may mention here that for us the area law is equivalent to conservation of angular momentum. Newton did not have the concept of angular momentum.

motion is conserved. If one considers a reference system in which S is at rest, the radius vector SP moves always on the same plane. Second: the velocity with which the area swept by SP increases (the *areal velocity*) is constant. The conservation of physical quantities allows one to make predictions. In Section 6 Newton will utilize the conservation of the plane of orbital motion and areal velocity to predict the future position of a body which orbits in an inverse square force field.

The realization of the dynamical implications of Kepler's area law is one of Newton's greatest scientific ideas. Like all great discoveries, it is remarkably simple, and, as we will see, full of consequences. One can be noted here: as pointed out very elegantly by François De Gandt, *the area law allows Newton to represent geometrically time by area.*† The area swept by SP is a fluent quantity which flows uniformly. Therefore time can be represented geometrically by it. The uniform flow of this geometrical quantity is not stipulated by a convention, it depends on dynamics. This is different from what Newton did in his method of fluxions. In his mathematical writings Newton considered algebraical or geometrical quantities (*fluents*) linked by specified relations: $f(x, y, z, \ldots) = 0$. As we have seen in §2.2.5, he stipulated that one of the fluents has a constant *fluxion* (say $\dot{x} = 1$), while the others can vary with nonuniform velocity. The chosen fluent can be taken, according to Newton, to represent time 'formaliter'. In his *Principia* Newton is not that free to stipulate the laws of variation of the geometrical quantities involved in the problem. In order to reduce dynamics to geometry he has to find a geometrical representation of time: i.e. he has to deduce from the laws of motion that a dynamical quantity flows uniformly. This is the quantity which can represent time. One should distinguish between a 'fluxional' and a 'physical' time. The former is any fluent which is supposed arbitrarily to flow uniformly. The latter is real absolute time, which in Nature flows uniformly and continuously. As we will see below (see §8.6.3.1 and §8.6.3.5) Newton did not always choose to identify fluxional with physical time: there are cases in which a fluent (x) which represents real time *does not* have a uniform fluxional flow ($\dot{x} \neq constant$).

In Proposition 6 Newton reaches a result which will be of fundamental importance: namely, he gives a geometrical representation of central force. Proposition 6 allows Newton to reduce to geometry the dynamical problems on central force motion. This is what Newton does in Propositions 7–13 for some particular direct problems of central force. Most notably, in Propositions 11–13 Newton shows that if the trajectory is a conic section and the force centre is at one focus, then the central force varies inversely with the square of distance.

In Sections 2 and 3 Newton deals also with the *inverse problem of central forces*. In this problem, when the initial position and initial velocity of a body accelerated

---

† De Gandt (1995): 22ff.

by a known central force are given, it is required to determine the trajectory. In Corollary 1 to Propositions 11–13 Newton states that if the force is inverse square, then the trajectory is a conic section. Proposition 17 is also related to the inverse problem. The inverse problem is tackled in a more general way in Propositions 39–41 of Sections 7 and 8.

It should be noted that Sections 4 and 5 remain quite separate from the rest of the *Principia*. They do not play a major role in Newton's mathematization of natural philosophy. They are concerned with problems in pure geometry which had occupied Newton in the 1670s: most notably Pappus's four-line locus. As we know from §2.3.1, in *Géométrie* Descartes had given a solution of this problem in order to show the superiority of his analytical methods over those of the ancient geometers. In fact, according to Descartes, the Ancients could not solve this problem. Newton in Sections 4 and 5 aims at showing that the four-line locus can be determined by pure geometry. These two sections are thus an anti-Cartesian manifesto.

3.3 First and ultimate ratios: Section 1, Book 1

The method of first and ultimate ratios, first developed in *Geometria curvilinea*, is the mathematical foundation of most of the demonstrations in the *Principia*. The method is presented in the eleven Lemmas of Section 1, Book 1.†

In this section we read a clear statement about the use of infinitesimals:

whenever in what follows I consider quantities as consisting of particles or whenever I use curved line-elements in place of straight lines, I wish it always to be understood that I have in mind not indivisibles but evanescent divisibles, and not sums and ratios of definite parts but the limits of such sums and ratios, and that the force of such proofs always rests on the method of the preceding lemmas.‡

Newton also points out that the method of first and ultimate ratios rests on the following Lemma 1:

Quantities, and also ratios of quantities, which in "any finite time" constantly tend to equality, and which before the end of that time approach so close to one another that their difference is less than any given quantity, become ultimately equal.§

Newton's *ad absurdum* proof runs as follows:

If you deny this, *b*let them become ultimately unequal, and*b* let their ultimate difference be *D*. Then they cannot approach so close to equality that their difference is less than the given difference *D*, contrary to the hypothesis.¶

---

† The reader may consult De Gandt (1986) and Pourciau (1998) for a clear exposition of Newton's theory of limits.
‡ *Variorum*: 87. Cf. *Principles*: 38.
§ In the first edition *aa* reads 'a given time'. *Variorum*: 73. Cf. *Principles*: 29.
¶ The first edition lacks *bb*. *Variorum*: 73. Cf. *Principles*: 29.

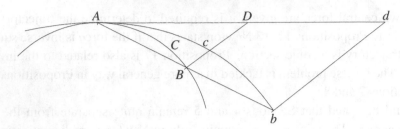

Fig. 3.1. Limiting ratio of chord, tangent and arc. After *Variorum*: 79. Some lines have been omitted

This principle might be regarded as an anticipation of Cauchy's theory of limits, but this would certainly be a mistake, since Newton's theory of limits is referred to a geometrical rather than a numerical model. The objects to which Newton applies his 'synthetic method of fluxions' or 'method of first and ultimate ratios' are geometrical quantities generated by continuous flow.

A typical mathematical problem which occurs in the *Principia* is the study of the limit to which the ratio of two geometrical fluents tends when they simultaneously vanish (Newton uses the expression the 'limit of the ratio of two vanishing quantities'). For instance, in Lemma 7 Newton shows that given a curve (see fig. 3.1):

the ultimate ratio of the arc [*ACB*], the chord [*AB*], and the tangent [*AD*] to one another is a ratio of equality.†

In order to give the reader an idea of the method of first and ultimate ratios let us read in some detail the demonstration of Lemma 7. This demonstration has the following structure. Let us consider two geometrical quantities X and Y which vanish simultaneously when A and B 'come together'. When the two quantities are finite the ratio can in principle be determined by standard geometrical techniques. The problem is to determine the limit of the ratio when 'B approaches point A'. Newton constructs two other quantities, x and y, which remain always finite, and such that $X/Y = x/y$. As B tends to A, the ratio X/Y tends to 0/0, but the ratio x/y tends to a finite value which is to be taken as the 'first or ultimate ratio of the vanishing quantities X and Y'. Here is Newton's proof of Lemma 7:

For while point B approaches point A, let AB and AD be understood always to be produced to the distant points b and d; and let bd be drawn parallel to secant BD. And let arc Acb be always similar to arc ACB. Then as points A and B come together, the angle dAb will vanish, by the preceding Lemma, and thus the straight lines Ab and Ad (which are always finite) and the intermediate arc Acb will coincide and therefore will be equal. Hence, the straight lines AB and AD and the intermediate arc ACB (which are always

† *Variorum*: 78. Cf. *Principles*: 32.

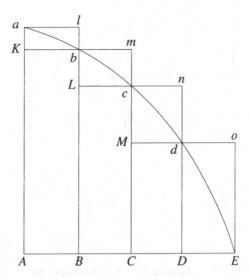

Fig. 3.2. Approximating curvilinear areas. After *Variorum*: 74. Some lines have been omitted

proportional to the lines *Ab* and *Ad* and the arc *Acb* respectively) will also vanish and will have to one another an ultimate ratio of equality.†

In Lemma 2 Newton shows that a curvilinear area *AabcdE* (see fig. 3.2) can be approached as the limit of inscribed *AKbLcMdD* or circumscribed *AalbmcndoE* rectilinear areas. Each rectilinear area is composed of a finite number of rectangles with equal bases *AB*, *BC*, *CD*, etc. The proof is magisterial in its simplicity. Its structure is still retained in present day calculus textbooks in the definition of the Riemann integral. It consists in showing that the difference between the areas of the circumscribed and the inscribed figures tends to zero, as the number of rectangles is 'increased *in infinitum*'. In fact this difference is equal to the area of rectangle *ABla* which, 'because its width *AB* is diminished *in infinitum*, becomes less than any given rectangle'.‡

Notice how in Lemma 2 and Lemma 7 Newton provides proofs of two assumptions that were made in the seventeenth-century 'new analysis'. The 'new analysts' (Newton himself in his early writings!) had assumed that a curve can be conceived as a polygonal of infinitely many infinitesimal sides, and that a curvilinear area can be conceived as composed of infinitely many infinitesimal strips. According to Newton, the method of first and ultimate ratios provides a foundation for such infinitesimal procedures. In the *Geometria curvilinea* and in the *Principia* curves

† *Variorum*: 78. Cf. *Principles*: 32.

‡ Notice that Cohen and Whitman translate 'augeatur in infinitum' and 'in infinitum minuitur' with 'increased indefinitely' and 'diminished indefinitely'. *Variorum*: 74. Cf. *Principles*: 29–30.

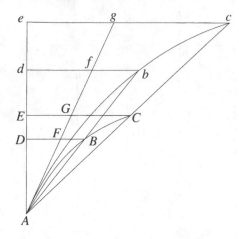

Fig. 3.3. Lemma 9. After *Variorum*: 80

are smooth, and curvilinear areas are not resolved into infinitesimal elements. In the synthetic method of fluxions one always works with finite quantities and limits of ratios of finite quantities.

Since Newton has banished infinitesimals and moments from the *Principia* in favour of limits, he has to justify the limits themselves. In modern terms, he has to provide existence and uniqueness proofs, and in order to do so he makes use once again of geometrical and kinematical intuition. It is worth quoting from Section 1 at some length on this particular point:

It may be objected that there is no such thing as an ultimate proportion of vanishing quantities, inasmuch as before vanishing the proportion is not ultimate, and after vanishing it does not exist at all. But by the same argument it could equally be contended that there is no ultimate velocity of a body reaching a certain place *a*at which the motion ceases*a*; for before the body arrives at this place, the velocity is not the ultimate velocity, and when it arrives there, there is no velocity at all. But the answer is easy; to understand the ultimate velocity as that with which a body is moving, neither before it arrives at its ultimate place and the motion ceases, nor after it has arrived there, but at the very instant when it arrives, that is, the very velocity with which the body arrives at its ultimate place and with which the motion ceases. And similarly the ultimate ratio of vanishing quantities is to be understood not as the ratio of quantities before they vanish or after they have vanished, but the ratio with which they vanish.†

We turn now to Lemma 9 (see fig. 3.3) which states:

If the straight line AE and the curve ABC, both given in position, intersect each other at a given angle A, and if BD and CE are drawn as ordinates to the straight line AE at another given angle and meet the curve in B and C, and if then points B and C *a*simultaneously*a*

---

† First and second editions lack *aa*. *Variorum*: 87. *Principles*: 38–9.

approach point A, I say that the areas of the triangles ABD and ACE will ultimately be to each other [b]in the duplicate ratio of the sides[b].†

As in Lemma 7, the essential step in the proof of Lemma 9 consists in local linearization. The curve ABC can be identified in the neighbourhood of A with its tangent AG. Therefore, if one takes the point B close to A, the curvilinear area ABD subtended by the curve grows very nearly as the square of the 'side' AD.

Lemma 9 has very important consequences for Newton's science of motion. These consequences are spelled out in Lemma 10:

The spaces which a body describes when urged by any [a]finite[a] force, [b]whether that force is determinate and immutable or is continually increased or continually decreased,[b] are at the very beginning of the motion [c]in the duplicate ratio of the times[c].‡

From Galileo's writings one knew that, when the force is 'determined and immutable', the space travelled from rest is proportional to the square of time. Newton states that this result is applicable to variable forces 'at the very beginning of the motion'. Newton's proof is as follows:

Let the times be represented by lines AD and AE, and the generated velocities by ordinates DB and EC; then the spaces described by these velocities will be as the areas ABD and ACE described by those ordinates, that is, at the very beginning of the motion (by Lemma 9) in the [a]duplicate ratio[a] of the times AD and AE.§

The first or ultimate ratio of displacement from rest, or generally displacement from inertial motion,¶ over the square of the time is constant: it is proportional to instantaneous acceleration. In fact, Newton will state in a Corollary to Lemma 10 added to the second edition (1713) that 'the forces are as the spaces described at the very beginning of the motion directly and as the squares of the times inversely'.‖

As we will see in §3.4.2, in the *Principia* Newton deals with variable forces. Thus he needs a geometric representation of force. Lemma 10 says that locally the velocity can be considered as varying linearly with time. Under this local approximation one can state that the displacement from inertial motion is proportional to the force times the square of time.††

† First edition lacks *aa*. Cohen and Whitman translate *bb*, 'in duplicata ratione laterum', as 'as the squares of the sides'. *Variorum*: 80. Cf. *Principles*: 34.

‡ In first edition *aa* reads 'regular', *bb* is lacking. Cohen and Whitman translate *cc*, 'in duplicata ratione temporum', as 'in the squared ratio of the times'. *Variorum*: 80. Cf. *Principles*: 34.

§ i.e. $ABD/ACE = AD^2/AE^2$. Notice that in Cohen and Whitman's translation *aa*, 'duplicata ratione', is rendered as 'squared ratio'. *Variorum*: 81. Cf. *Principles*: 34–5.

¶ In Newton's words 'ipso motus initio' meaning 'at the very beginning of the motion'.

‖ *Variorum*: 82. Cf. *Principles*: 35.

†† Nowadays we would write: $s - s_0 - v_0 \mathrm{d}t \propto F \mathrm{d}t^2$.

3.4 The geometry of central forces: Sections 2 and 3, Book 1

3.4.1 Synthetic method of first and ultimate ratios applied to the law of areas: Props. 1–2

Proposition 1 reads as follows:

The areas which bodies made to move in orbits describe by radii drawn to an unmoving centre of forces lie in unmoving planes and are proportional to the times.†

We will consider in some detail the demonstration of this proposition in §8.5. Here I will just sketch the structure of Newton's proof.

A body is fired at *A* with given initial velocity in the direction *A B* (see fig. 3.4). The centripetal force acting on the body must be first imagined as consisting of a series of impulses which act after equal finite intervals of time ('Let the time be divided into equal parts'). The trajectory will then be a polygonal *ABCDEF*. The body moves, during the first interval of time, from *A* to *B* with uniform rectilinear inertial motion. If the impulse did not act at *B* the body would continue its rectilinear uniform motion: it would reach *c* at the end of the second interval of time, so that *AB = Bc*. But because of the first impulse the body will be instantaneously deflected: it will reach *C* at the end of the second interval of time. Applying the first two laws of motion it is possible to show that triangles *SAB* and *SBC* have the same area and lie on the same plane. Similarly, all the triangular areas *SCD*, *SDE*, *SEF*, etc. spanned by the radius vector in equal times are equal and planar. In order to prove Proposition 1 Newton takes a limit. When the time interval tends to zero, the impulsive force approaches a continuous centripetal force and the trajectory approaches a smooth plane curve. The result (equal planar areas spanned in equal times by the radius vector) obtained for the polygonal trajectory generated by the impulsive force is extrapolated to the limiting smooth trajectory generated by the continuous force.

Following a similar procedure Newton proves Proposition 2 which states the inverse of Proposition 1. In Propositions 1 and 2 Newton has shown that a force is central if and only if the area law holds: the plane of orbital motion is constant and the radius vector sweeps equal areas in equal times. Notice that Newton in his proofs of these Propositions makes recourse to limit arguments, according to the method of first and ultimate ratios. Are these arguments well grounded? This question is not at all easy, since in Propositions 1 and 2 we have to consider not only geometrical quantities, e.g. areas and curves, but also forces. Can a continuous force be approximated as a limit of a discontinuous impulsive force, as the time intervals shrink to zero? As we will see in the next subsection §3.4.2, in Proposition 6 Newton instead approximates the trajectory via a continuous infinitesimal

---

† *Variorum*: 88. Cf. *Principles*: 40.

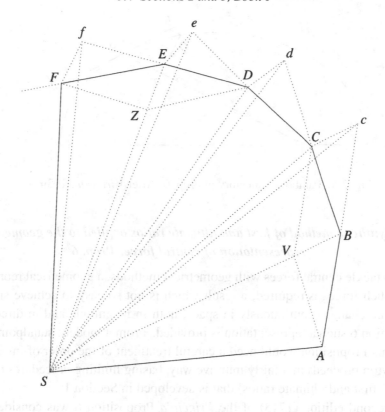

Fig. 3.4. Polygonal trajectory. After *Variorum*: 90

parabolic path. Which representation is preferable? This question was faced by
Leibniz and Pierre Varignon in the early eighteenth century (see §6.3.2).† In 1716
Hermann provided an alternative proof of Proposition 1 based 'on a completely
different foundation', i.e. on the integral calculus (see §8.5). Hermann's proof
is a good example of how different could be a demonstration given in terms of
calculus from one given in terms of the synthetic method of fluxions. In Newton's
demonstration of Propositions 1 and 2 no higher mathematics is required: only
the first two laws of motion, some elementary geometry and geometric limits are
utilized. We have here an example of how, in some instances, the synthetic method
of first and ultimate ratios can be seen as independent from the higher techniques
of the analytical method of fluxions.

† Aiton in (1989a) and Whiteside in *Mathematical Papers*, **6**: 37n maintain that a translation of Newton's
geometrical demonstration of Proposition 1 into calculus language leads to a proof of Kepler's law not for
finite, but only for infinitesimal arcs. I am aware that some Newtonian scholars are now casting doubts on the
correctness of the Aiton–Whiteside criticism. See Erlichson (1992b) and Nauenberg (1998c): note 4.

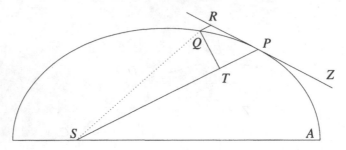

Fig. 3.5. Parabolic trajectory in Prop. 6. After *Variorum*: 103n

3.4.2 *Synthetic method of first and ultimate ratios applied to the geometric representation of central force: Prop. 6*

In order to tackle central forces with geometrical methods, a geometrical representation of such forces is required, a result which is not so easy to achieve since a central force changes continuously in space, both in magnitude and in direction. In Proposition 6 such a representation is provided. From a modern standpoint, the proof of this Proposition would need a careful treatment of the order of infinitesimals. Newton proceeds in a fairly intuitive way, basing limiting procedures on the method of 'first and ultimate ratios' that is developed in Section 1.

In the second edition (1713) of the *Principia* Proposition 6 was considerably modified, as Newton added a new geometrical representation of force (see §4.5).†
Here I will sketch the structure of Proposition 6 as it was published in the first edition. The reader will find a more detailed discussion of Proposition 6 (1687 version) in §5.5.1.

A body accelerated by a centripetal force directed towards S (the centre of force) describes a trajectory as shown schematically in fig. 3.5. PQ is the arc traversed in a finite interval of time. The point Q is fluid in its position on the orbit, and one has to consider the limiting situation 'when points Q and P come together'. The line ZPR is the tangent to the orbit at P, while QR (the *versed sine*) tends to be parallel to SP (the *radius vector*) as Q approaches P. QT is normal to SP. As we know from Lemma 10 (see §3.3), 'at the very beginning of the motion' the force can be considered as constant. In the case considered in fig. 3.5, this implies that, *as Q approaches P, the displacement QR is proportional to force times the square of time*. In fact, in the limiting situation, QR can be considered as a small Galilean fall caused by a constant force.

Newton can now obtain the required geometrical representation of force. Since

---

† The most detailed study on Proposition 6 and in general on the first three Sections of the *Principia* is Brackenridge (1995). On the alternative measure of central force see pp. 35–7. See also Needham (1993).

Kepler's area law holds (the force is central, cf. Proposition 1 in §3.4.1), the area SPQ (a triangle since the limit of the ratio between the vanishing chord PQ and arc PQ is 1; cf. Lemma 7 in 3.3) is proportional to time. The area of triangle SPQ is $\frac{1}{2}(SP \cdot QT)$. Therefore, the geometrical measure of force is:

$$F \propto \frac{QR}{(SP \cdot QT)^2},\tag{3.1}$$

where the above ratio has to be evaluated in the limiting situation 'when points P and Q come together'.† Notice that, according to the convention spelled out in §3.1, I have used the symbol F for Newton's 'central force' and $\propto$ for 'is proportional to'.‡

Proposition 6 is a good example of application of the synthetic method of first and ultimate ratios. The limit to which tends the ratio $QR/(SP \cdot QT)^2$ is to be evaluated by purely geometric means. As Mahoney puts it:

Newton concentrated on the orbit or trajectory of motion. Keeping before his and the reader's eyes the path of the body's motion, he located within the geometric configuration structural elements proportionally representative of the dynamical, nongeometrical parameters determining that motion. The task was made all the more difficult, of course, by the fact that in most cases those parameters are continuously variable.§

3.4.3 Proposition 6 applied to the direct problem: Props. 7–13

Formula (3.1) is employed by Newton in the solution of several instances of the direct problem of central forces. These problems have the following structure: given a plane curve O and a point S in the plane of that curve, determine the central force directed towards, or away from, the force centre S which accelerates a 'body' (nowadays we would say a 'point mass') making it describe the orbit O.

Newton's standard technique for solving the direct problem consists in determining the limit of the ratio of QR/QT^2 as Q tends to P, and in applying formula (3.1).¶

In Proposition 7 the orbit is a circle and S is placed on its circumference. The solution is $F \propto 1/SP^5$, the force varies inversely as the fifth power of distance. In

---

† 'si modo solidi illius ea semper sumatur quantitas quae ultimo fit ubi coeunt puncta P & Q'. *Variorum*: 103. Cf. *Principles*: 49.

‡ A completely 'modern' translation is obtained by making explicit the proportionality constants and using the limit notation:

$$F = \lim_{Q \to P} \frac{2h^2}{m} \frac{QR}{(SP \cdot QT)^2},$$

where h is angular momentum, and m the mass.

§ Mahoney (1993): 187.

¶ Notice that in formula (3.1) SP remains constant.

fact Newton shows that in this case the ultimate ratio as Q tends to P is

$$QR/QT^2 \propto 1/SP^3. \tag{3.2}$$

When the orbit is an equiangular spiral (in polar coordinates $\ln r = a\theta$) and S is placed at the centre (Proposition 9), as Q tends to P:

$$QR/QT^2 \propto 1/SP, \tag{3.3}$$

and the force varies inversely with the cube of distance.

When the orbit is an ellipse and the centre of force is at the centre of the ellipse (Proposition 10) the force varies directly with distance. Let us see how Newton proves Proposition 10:

Let a body revolve in an ellipse; it is required to find the law of centripetal force tending toward the centre of the ellipse.†

Consider fig. 3.6. Here, points P, R, Q and T are as in Proposition 6. The centre of force is the centre C of the ellipse (so change S for C in formula (3.1)). The lines CA and CB are the major and minor semi-axes. DK, the conjugate diameter to PG, is parallel to the tangent PR; PF is drawn perpendicular to DK, Qv is drawn parallel to PR.

From Apollonius's *Conics* it is well known that for an ellipse the following relation holds:

$$(Pv \cdot vG)/Qv^2 = PC^2/CD^2. \tag{3.4}$$

Since triangles QvT and PCF are similar,

$$Qv^2/QT^2 = PC^2/PF^2. \tag{3.5}$$

Eliminating Qv^2 from the two preceding equations,

$$Pv/(QT^2 \cdot PC^2) = PC^2/(PF^2 \cdot vG \cdot CD^2). \tag{3.6}$$

By construction $QR = Pv$. Furthermore, when Q tends to P, vG tends to $2PC$. One thus obtains that, when Q tends to P:

$$QR/(QT^2 \cdot PC^2) = PC/(2 \cdot PF^2 \cdot CD^2). \tag{3.7}$$

In Lemma 12, Section 2 Newton states that:

All the parallelograms described about any conjugate diameters of a given ellipse or hyperbola are equal to one another.‡

---

† *Variorum*: 114. Cf. *Principles*: 53.
‡ *Variorum*: 114. Cf. *Principles*: 53. Note that in the first edition Lemma 12 has a slightly different formulation. For an English translation see Brackenridge (1995): 255.

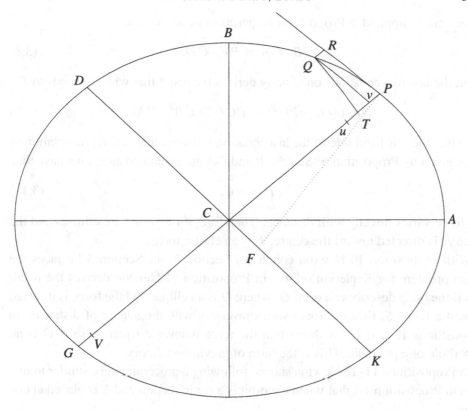

Fig. 3.6. Elliptic trajectory generated by elastic force: Prop. 10. After *Variorum*: 115

Newton simply states that this nontrivial property is derived from the writers on conic sections. He writes cryptically: 'constat ex Conicis'. This is actually Proposition 31 from the 7th book of Apollonius's *Conics*. However, it is not clear where Newton got it: whether from Giovanni Alfonso Borelli and Abraham de l'Echelle's 1661 edition (Bologna), or Christian Rau's 1667 edition (Kiel), or from the *Opus geometricum* (1647) of Grégoire de Saint-Vincent.† Or he might even have proved it for himself.‡ It is very difficult to locate Newton's sources in classic geometry.

† See Grégoire de Saint-Vincent (1647), 4: Prop. 72. Both Borelli's and Rau's editions were based on Arabic paraphrases. It was only in 1710 that Halley produced a reliable edition of the *Conics*, 1–7: he based Books 1–4 on a Greek text, 5–7 on a manuscript in Arabic. The restoration of Book 8 was based on Pappus's outline of it. As for the first four books of the *Conics*, Newton employed Barrow's 'edition'. In Harrison (1978), a book devoted to Newton's library, we find the following works: the first four books of Apollonius's *Conics* (Barrow's ed.) (London, 1675), Halley's edition of Apollonius's *de sectione rationis* (Oxford, 1706) and *Conics*, Books 1–8 (Oxford, 1710).

‡ As Whiston says in his *Memoirs* (1749): 39.

Lemma 12 applied to Proposition 10 permits one to state that

$$CB \cdot CA = PF \cdot CD. \tag{3.8}$$

From the last two equations one finally derives the result that when Q tends to P:

$$QR/(QT^2 \cdot PC^2) = PC/(2 \cdot CB^2 \cdot CA^2). \tag{3.9}$$

Note that the left-hand side of the last equation is the geometrical representation of force given by Proposition 6. Since CB and CA are given constants, we have that

$$F \propto PC, \tag{3.10}$$

the force varies directly with distance. Therefore, if the orbits are ellipses and the force F is directed toward the centre, F is an elastic force.

With Proposition 10 Newton concludes Section 2. In Section 3 he faces the direct problem for Keplerian orbits. In Proposition 11 Newton derives the result that if the body describes an orbit O, where O is an ellipse and the force is directed toward a focus S, then the force varies inversely with the square of distance. In Propositions 12 and 13 he shows that the force is inverse square also if O is an hyperbola or a parabola. This is the birth of gravitation theory.

In Propositions 11–13 Newton shows, following a procedure very similar to that seen in Proposition 10, that when the orbit is a conic section and S is placed at one focus,

$$QR/QT^2 \propto 1/L, \tag{3.11}$$

where L is a constant (the *latus rectum*), and the ratio QR/QT^2 is evaluated, as always, as the first or last ratio 'with the points Q and P coming together'.[†] Therefore the force varies inversely with the square of distance.

3.4.4 *The inverse problem: Cor. 1 to Props. 11–13*

In Corollary 1 to Propositions 11–13 Newton states that if the force is inverse square, than the trajectories are conic sections such that a focus coincides with the force centre. Corollary 1 (1687) reads as follows (see fig. 3.7):

From the last three propositions [i.e. Propositions 11–13] it follows that if any body P departs from the place P along any straight line PR with any velocity whatever and is at the same time acted upon by a centripetal force that is inversely proportional to the square of the distance[a] from the centre, this body will move in some one of the conics having a focus in the centre of forces; and conversely.[‡]

[†] 'punctis Q & P coeuntibus'. *Variorum*: 119.
[‡] [a] 'of places' added in second edition. *Variorum*: 125. Cf. *Principles*: 61.

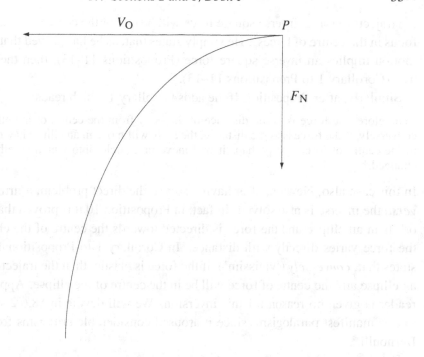

Fig. 3.7. In Corollary 1 (1687 version) Newton claims that if F is inverse square then the orbit is a conic section. See §8.6.2.1 for the proof given in second and third edition of the *Principia*. In the emended Corollary Newton makes clear the following. He considers a 'body' of given mass. Initial velocity $\vec{v}_0$ and normal component of central force F_N at P determine the tangent and the curvature of the orbit at P. There is a unique conic with a focus in the force centre which satisfies the geometric properties implied by the initial conditions. From Propositions 11–13 one knows that such a conic is a possible trajectory. The proof is completed by adding a uniqueness condition according to which for any initial condition only one trajectory is possible

As we will see, quite understandably this terse statement was subject to criticisms in the 1710s. Newton himself, while revising the second edition of the *Principia*, emended Corollary 1 adding the sketch of a proof (§8.6.2.1).†

This Corollary is an example of the inverse problem of central forces: the central force F (force law and force centre S) is given, and what is required is the trajectory (a singular, assuming uniqueness!) corresponding to any initial position and velocity of a 'body' of given mass acted upon by such force. Notice that the position in time of the body after it is shot from the given initial place with the given initial velocity is not required. What is required is the plane curve which the body will have as its trajectory. Newton does not explain here the reason why

---

† This Corollary has recently attracted the attention of several mathematicians and historians. See, e.g., Aiton (1964), Cushing (1982), Arnol'd (1990), Whiteside (1991a), Pourciau (1992a) and (1992b), Brackenridge (1992) and (1995).

the trajectory for an inverse square force will be one of the conic sections 'having a focus in the centre of forces'. He simply states that, as he has proved that Keplerian motion implies an inverse square force (Propositions 11–13), then the inverse is true (Corollary 1 to Propositions 11–13).

Similarly, after Proposition 10 he adds Corollary 1 which reads:

*a*Therefore, the force is as the distance of the body from the centre of the ellipse; and,*a* conversely, if the force is as the distance, the body will move in an ellipse having its centre in the centre of forces, or perhaps it will move in a circle, into which an ellipse can be changed.†

In this case also, Newton, after having solved the direct problem, affirms that *vice versa* the inverse is also solved. In fact, in Proposition 10 it is proved that if a body orbits in an ellipse and the force is directed towards the centre of the ellipse, then the force varies directly with distance. In Corollary 1 to Proposition 10 Newton states that, *conversely* ('vicissim'), if the force is elastic, than the trajectory will be an ellipse and the centre of force will be in the centre of the ellipse. Apparently the reader is given no reason for this inversion. We will devote in §8.6.2 some space to this manifest paralogism, since it aroused considerable criticisms from Johann Bernoulli.

For the moment some comments are in order:

- Corollary 1 to Proposition 10 and Corollary 1 to Propositions 11–13 are both true (even if unproven) statements;
- after having proved in Proposition 9 that motion along a logarithmic spiral (centre of force coinciding with the centre of the spiral) implies an inverse cube force, Newton *does not* state that an inverse cube force implies motion along a logarithmic spiral (this statement would be false);
- in Proposition 17, Section 3, Book 1, Newton presents a constructive geometrical technique for determining the unique conic trajectory which answers given initial conditions (position and velocity), when the given force is inverse square (§3.4.5);
- the inverse problem of central forces is faced by Newton in a more general way in Propositions 39, 40 and 41 of Book 1.

3.4.5 Scaling conics: Props. 16 and 17

As I said above, Proposition 17 is related to the solution of the inverse problem. Here Newton considers a body of given mass accelerated by a centripetal inverse square force directed towards S. The body is fired at P, in the direction PR with a given initial speed (see fig. 3.8). Newton *assumes* that the trajectory will be a conic section and that one of the foci will be located at the force centre S. He determines the unique conic which satisfies the given initial conditions.

† First edition lacks *aa*. *Variorum*: 117. Cf. *Principles*: 54.

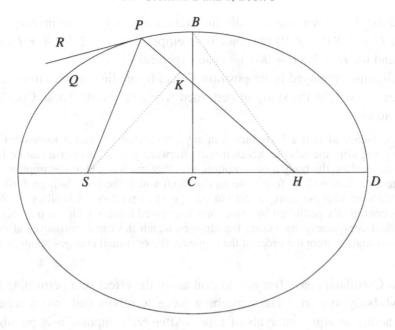

Fig. 3.8. Unique conic corresponding to given initial position and velocity. After *Variorum*: 131. Some lines have been omitted

According to Proposition 16 (again *assuming* the fact that the trajectory is a conic) the initial conditions determine the *latus rectum*, *L*, of the conic trajectory.[†] Because of the reflective property of conics, the lines joining *P* with the foci make equal angles with the tangent at *P*, therefore the direction of the line where lies the second focus *H* is also determined. Newton shows that the following geometrical property holds for any conic section:

$$(SP + PH)/PH = 2(SP + KP)/L, \qquad (3.12)$$

where SK falls orthogonally on PH. Since SP, L and KP are unequivocally determined by the initial conditions, PH also is given. And the second focus is thus found. We have now enough information to build the required conic: we have the foci S and H and the major axis $SP + PH$.[‡]

[†] In modern terms one can render Proposition 16 as follows. The semi *latus rectum*, $L/2$, of the sought conic trajectory is: $L/2 = b^2/a = (v_0^2 \cdot SY^2)/(m\kappa) = h^2/(m\kappa)$, where v_0 is the initial speed, SY falls perpendicularly on the tangent at P, a and b are the major and minor semi-axes, h is angular momentum, m the mass and κ is a constant such that $F = -\kappa/SP^2$. Thus the given initial conditions determine the *latus rectum* of the conic trajectory.

[‡] I have followed in broad outlines the analysis of Proposition 17 given by Aiton (1989b): 49, and Chandrasekhar (1995): 108.

Proposition 17 allows one to scale the conic trajectory for an inverse square force. For $L < 2(SP + KP)$ the conic is an ellipse, for $L = 2(SP + KP)$ it is a parabola and for $L > 2(SP + KP)$ it is an hyperbola.

The technique employed in Proposition 17 can be applied, as Newton states in Corollaries 3 and 4, to the study of perturbed Keplerian orbit. These Corollaries read as follows:

Corollary 3. Hence also, if a body moves in any conic whatever and is forced out of its orbit by any impulse, the orbit in which it will afterward pursue its course can be found. For by compounding the body's own motion with that motion which the impulse alone would generate, there will be found the motion with which the body will go forth from the given place of impulse along a straight line given in position. Corollary 4. And if the body is continually perturbed by some force impressed from outside, its trajectory can be determined very nearly, by noting the changes which the force introduces at certain points and estimating from the order of the sequence the continual changes at intermediate places.†

These two Corollaries pave the way to evaluating the effect of a perturbing force over a two-body system. The perturbing force is subdivided into a series of impulses acting at equal intervals of time. After each impulse it is possible to determine the velocity of the point mass and use this velocity as a new initial condition. Applying Proposition 17 one can then determine the parameters of the conic along which the body will move until the next impulse causes a successive change of parameters. This and related techniques are employed by Newton in the study of perturbations.

3.5 Quadratures applied to the general inverse problem: Sections 7 and 8, Book 1

In these sections Newton faces the general inverse problem of central forces: i.e. the problem of determining the trajectory, given initial position and velocity, of a 'body acted upon' by *any* central force.‡ He deals first with 'rectilinear ascent and descent' and then with curvilinear motion. The inverse problem for inverse square central forces was faced by Newton in Corollary 1 to Propositions 11–13 and in Proposition 17. In Proposition 41 a general solution of the inverse problem is provided. I have found it convenient to discuss Proposition 41 in Chapters 7 and 8, since it was the object of much interest and criticism.§ This proposition is based on the assumption that a method for the 'quadrature of curvilinear figures' is given. As we known, Newton had developed in his youth the 'method of fluxions and infinite series'. He was able to 'square', in Leibnizian terms 'integrate', a

---

† *Variorum*: 133. Cf. *Principles*: 67.

‡ The assumption, which was generally made, that the trajectory is unique will be discussed in §8.6.2.1 and §8.6.2.7.

§ See especially §8.6.2.

large class of curves (see §2.2.7). However, in the *Principia* he chose not to make wholly explicit his mathematical discoveries in this field: 'quadratures' are generally avoided. When Newton in the *Principia* reduces a problem to a difficult quadrature, he follows the policy of giving the solution without the demonstration. He simply shows that the solution depends upon the quadrature of a curve and leaves the reader without any hint on how to proceed. Other examples of these mysterious reductions to quadratures can be found in Newton's treatment of the attraction of extended bodies (see §3.8) and of the solid of least resistance. These parts of the *Principia* were really puzzling for his readers. They were told that a result depended upon the quadrature of a certain curve, but the method by which this very quadrature could be achieved was not revealed.

As far as Proposition 41 is concerned the following points should be noticed:

- it is entirely 'geometrical', but it can be easily translated into calculus terms by substituting infinitesimal linelets for Newtonian moments (or Leibnizian differentials);
- the final result is easily translatable into a couple of fluxional (or differential) equations;
- Newton was aware that a translation into calculus (i.e. into the analytical method of fluxions) was feasible.

The last point is supported by overwhelming evidence. Firstly, Newton says that the demonstration of Proposition 41 depends upon the 'quadrature of curvilinear figures', a reference to his 'catalogues' of curves. Secondly, in Corollary 3 he applies the general result of Proposition 41 to the case of an inverse cube force. The question to be answered is: which trajectories are described by a body accelerated by an inverse cube force? In Corollary 3 Newton gives only the solution in the form of a geometrical construction: he constructs some spiral trajectories which answer the problem. *He could have obtained this result only by application of his 'catalogues' or 'tables'* (in Leibnizian terms, integral tables). In fact, as we will see in §8.6.2.2, the result achieved in Proposition 41 applied to an inverse cube force leads to the quadrature of a curve included in the 'catalogues' of the *De methodis* (1671). In Corollary 3 Newton does not perform this quadrature explicitly, but simply states the result. He then adds:

All this follows from the foregoing proposition [41], by means of the quadrature of a certain curve, the finding of which, as being easy enough, I omit for the sake of brevity.†

Thirdly, when David Gregory, during a visit he paid to Newton in May 1694, asked about the mysterious method applied in the solution of Corollary 3, the Lucasian Professor answered by translating the basic result of Proposition 41 as a fluxional equation. He applied this equation to the case of an inverse cube force and obtained the result stated in the *Principia*. As Gregory remarked in a memorandum of this visit:

† *Variorum*: 223. Cf. *Principles*: 133. On this Corollary see Erlichson (1994b).

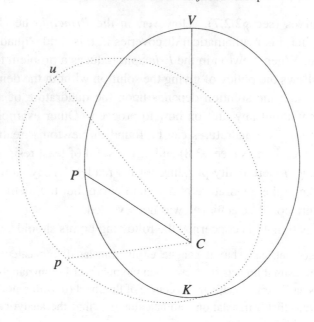

Fig. 3.9. Mobile orbit. After *Variorum*: 226

on these [quadratures] depend certain more abstruse parts in his philosophy as hitherto
published, such as Corollary 3, Proposition 41 and Corollary 2, Proposition 91.†

3.6 The geometric representation of central force (Prop. 6) and infinite series applied to 'mobile orbits': Section 9, Book 1

At the end of Section 8 Newton remarks:

And so far we have considered the motion of bodies in nonmoving orbits. It remains for
us to add a few things about the motion of bodies in orbits that revolve about a centre of
forces.‡

Section 9 is devoted to motion in 'mobile orbits'. This section is the foundation of
Newton's theory of the precessions of apsides. In Proposition 43 Newton defines a
'mobile orbit'. He considers (see fig. 3.9) a closed orbit VPK 'at rest', traversed
by a 'body' whose position is indicated by P. The point C is the centre of the
centripetal force [that I will denote with F] which accelerates the body along
VPK. The orbit that is 'at rest' has both a pericentre K and an apocentre V:
the line VK connecting the apsides is fixed in space. Furthermore (because of

---

† Memorandum July 1694, translation from Latin in *Correspondence*, **3**: 386. We will discuss Corollary 2 to
Proposition 91 in §3.8 and §7.3.
‡ *Variorum*: 225. Cf. *Principles*: 134.

Proposition 1, Book 1) CP sweeps equal areas in equal times. The 'mobile' orbit is constructed as follows. The position p of the body orbiting in the mobile orbit is such that at any time: (a) $CP = Cp$, and (b) the angles VCP and VCp are such that $VCp = (g/f)VCP$ (g and f constants). It is easy to show that Cp sweeps equal areas in equal times: therefore the body orbiting in the mobile orbit (by Proposition 2, Book 1) is accelerated by a centripetal force (which I will denote by F_m) directed towards C. Since the motion along the mobile orbit is caused by a centripetal force, we are entitled to apply Proposition 6.

By application of Proposition 6 Newton is able to prove Proposition 44 which states that $F - F_m$ is inversely proportional to the cube of distance from C:

The difference between the forces under the action of which two bodies are able to move equally – one in an orbit that is at rest and the other in an identical orbit that is revolving – is *a*inversely in the triplicate ratio*a* of their common *b*height*b*.†

In fig. 3.10 the mobile orbit and the orbit that is at rest are constructed as in Proposition 43. PK and the 'similar and equal' arc pk are infinitesimal.‡ While the body 'in the orbit that is at rest' moves from P to K in a small interval of time, the body in the mobile orbit moves from p to n. Point n is reached by the body by the composition of motion along arc pk and a precession of the line of apsides of the mobile orbit. The angles pCn and PCK (= angle pCk) are such that $pCn = (g/f)PCK$. A circle of radius $CK(= Ck)$ is drawn. The infinitesimal line kr is perpendicular to Cp at r, kr is extended so that $mr = (g/f)kr$.

Newton splits the motion of the body on the mobile orbit into a radial and a transradial component. As regards the transradial component one has the following:

the transverse component of motion of the body p will be to the transverse component of motion of the body P as the angular motion of line pC to the angular motion of line PC, that is, as the angle VCp to the angle VCP.§

Thus, since $mr = (g/f)kr$, m would be the position of the body after the small time interval, when the radial component of motion is not considered. Note that, since the force is central, there is no transradial component of force acting on the body.

But m cannot be the position of the body reached after the small time interval, since $Cm > CK$. Clearly mn is the displacement due to the difference $F - F_m$: i.e. it is proportional to $F - F_m$. Applying the geometric representation of force of Proposition 6, we find that the dynamical problem of determining $F - F_m$ is

---

† Note that Cohen and Whitman translate *aa*, 'in triplicata ratione inverse', as 'inversely as the cube'. *bb*, 'altitudo', is the distance from the force centre C. *Variorum*: 227. Cf. *Principles*: 136.

‡ '& punctorum P, K distantia intelligatur esse quam minima' is added in the second edition. *Variorum*: 227.

§ *Variorum*: 228. Cf. *Principles*: 137.

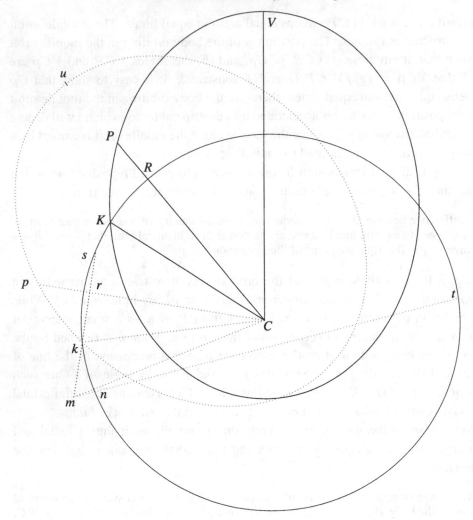

Fig. 3.10. Geometric representation of $F - F_{\mathrm{m}}$. After *Variorum*: 228

reduced to the geometric problem of studying mn in the limiting situation as K tends to P.

Produce mk and mn to s and t. By construction $mr = (g/f)kr$. Furthermore (by Euclid, 3, 36):

$$mn = \frac{mk \cdot ms}{mt}. \tag{3.13}$$

Because of Kepler's area law 'the triangles pCk [$= PCK$] and pCn are, in a given time, given in magnitude.' Therefore 'kr and mr and their difference mk

and sum *ms* are inversely as the height Cp.' Moreover *mt* in the limit is equal to $2CP = 2Cp$. Newton concludes as follows:

these are the first ratios of the nascent lines; and hence $(mk \times ms)/mt$ (that is, the nascent line-element *mn* and, proportional to it, the difference between the forces $[F - F_m]$) becomes ᵃinversely as the cubeᵃ of the height Cp.†

How the above mathematical result can be applied to the study of the motion of apsides can be briefly explained.‡ Newton first shows that the force which accelerates a body in a mobile ellipse has magnitude

$$F_m = \frac{f^2}{A^2} + \frac{\lambda(g^2 - f^2)}{A^3}, \tag{3.14}$$

where A is the 'height' $CP = Cp$ and 2λ is the *latus rectum*. Notice that, in accordance with Proposition 11, Book 1, $F = f^2/A^2$ is the magnitude of the inverse square force which makes the body move in the fixed ellipse: thus f is the absolute value of force in the fixed ellipse while g corresponds to the velocity of rotation of the moving ellipse.§ Newton then proceeds in Proposition 45 to determine the relationship between a given centripetal force and the motion of the apsides. This proposition is restricted to 'orbits that differ very little from circles'.¶ Newton illustrates his procedure with several examples. The first one will suffice.

Let us suppose that we have a constant centripetal force and a body which orbits in a nearly circular trajectory. The problem is to have an approximate estimate of the motion of apsides. Newton approximates the trajectory of the body accelerated by the constant force with a mobile ellipse. He denotes, with reference to fig. 3.10, CV by T, and $CV - CP$ by X: thus $T - X = A$. Substituting in equation (3.14) we have

$$F_m = \frac{f^2T - f^2X + \lambda g^2 - \lambda f^2}{A^3}. \tag{3.15}$$

The centripetal constant force can be expressed by A^3/A^3 or $(T - X)^3/A^3$. But 'orbits will acquire the same shape if the centripetal forces with which those orbits are described, when compared with each other, are made proportional at equal heights'. Therefore, Newton obtains, after expanding $(T - X)^3$ and collating terms in the numerators,

$$\frac{\lambda g^2 - \lambda f^2 + Tf^2}{T^3} = \frac{-f^2}{-3T^2 + 3TX - X^2}. \tag{3.16}$$

† *aa* is the translation of 'reciproce ut cubus'. *Variorum*: 230. Cf. *Principles*: 138.
‡ For a fuller account see Caldo (1929), Chandler (1975), Waff (1975) and Chandrasekhar (1995).
§ Chandler (1975): 68.
¶ *Variorum*: 234. Cf. *Principles*: 141.

But, since the orbit is nearly circular, $\lambda \approx T$ and 'X is diminished *in infinitum*'. Finally Newton achieves

$$\frac{g^2}{f^2} = \frac{1}{3}.$$ (3.17)

Therefore:

g is to f, that is the angle VCp is to the angle VCP, as 1 to $\sqrt{3}$. Therefore, since a body in an immobile ellipse, in descending from the upper apsis to the lower apsis, completes the angle VCP (so to speak) of 180 degrees, another body in the mobile ellipse [...] will in descending from the upper apsis to the lower apsis, complete the angle VCp of $180/\sqrt{3}$ degrees.†

Newton has thus given an approximate estimate of the precession of the line of apsides in the case of a constant force. He makes it clear that this approximation is accurate enough only for orbits 'that differ very little from circles'.‡

In the other examples Newton generalizes this result. For instance, following a similar procedure he shows that if the force varies as the $(n - 3)$th power of the distance, then the rate of apsidal motion is $g/f = 1/\sqrt{n}$. Hence 'if the centripetal force is as some power of the height, that power can be found from the motion of the apsides, and conversely'.§

Newton employs the results achieved in Section 9 to obtain an approximate estimate of the motion of the Moon's apogee. For reasons that will be clear below, he assumes that only the radial component (directed along the Moon–Earth distance) of the perturbing force of the Sun on the Earth–Moon system is relevant. So the effect of the Sun's gravitation is to add to the inverse square attraction of the Earth on the Moon an extra component. An estimate of this component can be derived from astronomical data.

Newton assumes that a very distant and massive body S (the Sun) perturbs the orbital motions of a two-body system $P–T$ (Moon–Earth). Assuming that the orbit of P is nearly circular, Newton shows that the radial component (directed along TP) of the perturbing force goes approximately as $a + b \cos 2\alpha$ (where we denote by α the angle STP), while the transradial component (at right angles to radius vector) goes approximately as $\sin 2\alpha$. Next he evaluates the radial and transradial components averaging over a full synodic revolution. Therefore he thinks that he can ignore the transradial component, while the radial component turns out to be independent of α. Averaging allows Newton to apply to the two-body system $P–T$ the techniques valid for central forces. The perturbed P's orbit is in fact approximated by an orbit generated by a centripetal force equal to the sum of the centripetal interaction between P and T plus the average value of the radial

† *Variorum*: 235. Cf. *Principles*: 142.
‡ *Variorum*: 234. Cf. *Principles*: 141.
§ *Variorum*: 239. Cf. *Principles*: 145.

perturbing force. Furthermore the mass of P is much less than the mass of T, therefore Newton assumes that T is a fixed centre of force. The mathematical model is thus reduced to one body plus one fixed centre of force (i.e. to the model considered in Section 9). With these approximations, Newton's theoretical value of the motion of the Moon's apogee is badly off target. As is well known, he obtains a value which is only half that measured by astronomers. It should be noted that Newton is not able to estimate the error caused by the approximations. This bad disagreement between theory and observation was considered as one of the weakest points of the *Principia*. It was only in the middle of the eighteenth century that mathematicians were able to reconcile gravitation theory with the motion of the Moon.

3.7 Qualitative geometrical modelling applied to the three-body problem: Section 11, Book 1

The Newtonian world is regulated by the dynamics of inverse square universal gravitation. Since in the Newtonian Universe all the masses are interacting, the mathematical principles of natural philosophy have to predict perturbations of elliptical Keplerian orbits. Newton's mathematical work on perturbations occupies Section 11 of Book 1 and Propositions 25–35 of Book 3. In Section 11 Newton employs qualitative geometrical methods. What he does is to deal with the very difficult three-body problem (a problem that was created by him with the introduction of universal gravitation) in terms of a graphical representation of trajectories and forces. The qualitative results obtained in Section 11 are transformed into quantitative analysis in Book 3, when the mathematical models of Book 1 are applied to real astronomical data. Newton's theoretical values often fail to approach the values computed from observation, but *they are of the right order of magnitude*. The approximate estimates given in Book 3 of the perturbations of Keplerian orbits had, notwithstanding their imprecision, an extremely important consequence: they confirmed the existence of the perturbations predicted in Section 11 and thus gave good reasons for believing that universal inverse square gravitation rules the planetary motions.†

In the great Proposition 66 with its 22 Corollaries in Section 11 (see fig. 3.11) Newton assumes that a light body located at P orbits around a body located at T because of a mutual gravitational attraction. In most of the Corollaries the orbit of the light body is assumed to be nearly circular. A massive and distant body located at S, which in most of the Corollaries is assumed to be in the plane of the P–T orbit, perturbs their Keplerian motions. Let us say that the bodies located at S, T and P are the Sun, the Earth and the Moon. Let SK be the mean distance of the

---

† For details see Chandrasekhar (1995) and Wilson (1989). In what follows I am particularly indebted to Wilson.

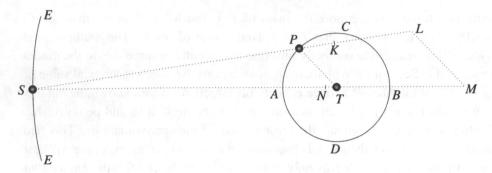

Fig. 3.11. Sun, Earth and Moon: Newton's model for the three-body problem. After *Variorum*: 279

Moon from the Sun, and let SK represent the accelerative force (the force per unit mass) exerted by the Sun on the Moon at that distance. Then the Sun's accelerative force on the Moon at P will be

$$SL = SK \left(\frac{SK}{SP} \right)^2 , \qquad (3.18)$$

since the force is inverse square. Decompose SL into the components LM, parallel to TP (radial component) and SM. So the Moon will be urged by a threefold accelerative force: (i) the force of the Earth on the Moon at P which alone would cause motion in an ellipse having its focus at the centre of mass of the Moon and Earth; (ii) a force LM, the radial component of the perturbing force of the Sun on the Moon, which causes a variation of Keplerian motion, *but* which does not cause a departure from the area law (TP sweeps equal areas in equal times); and (iii) a force SM. The third force SM, not being parallel to TP, will cause the radius TP to describe areas not proportional to the times. Let the accelerative force exerted by the Sun on the Earth be expressed by SN. If the two bodies at P and T are accelerated in the direction ST by equal forces, this will have no effect on their Keplerian motion. It is the difference NM which disturbs the proportionality of the areas and times, and the elliptical figure of the orbit.†

    With this model in mind Newton is able to deal with most of the perturbations of a Keplerian orbit. For instance, in Corollary 2 Newton proves that the radius vector TP has a greater areal velocity near the conjunction A and opposition B than near the quadratures C and D. From this result he can state in Corollary 3 that the body 'moves more swiftly in the conjunction and opposition than in the quadratures',

---

† In establishing that only NM matters, Newton deploys the very important Corollary 6 to the laws of motion. *Variorum*: 278–80. Cf. *Principles*: 174–5.

while in Corollary 4 he points out that the orbit 'is more curved at the quadratures than at the conjunction and opposition'.†

Let us see how Newton proves Corollary 4:

For the swifter bodies move, the less they deflect from a rectilinear path. And besides, the force KL, or NM, at the conjunction and opposition, is contrary to the force with which the body T attracts the body P, and therefore diminishes that force; but the body will deflect the less from a rectilinear path the less it is impelled towards the body T.‡

Here Newton employs the relation between normal component of force F_N, velocity v and radius of curvature ρ of the orbit that recurs so often in his writings:

$$F_N \propto \frac{v^2}{\rho}. \tag{3.19}$$

The radius ρ is greater at the conjunction and opposition since there v is greater and the total force acting on P is smaller.

In Corollaries 10 and 11 Newton shows that the overall movement of the line of nodes (the intersection between the plane of the Moon's orbit and the plane of the Earth's orbit) is a movement of regression. He uses this result in the succeeding Corollaries 18–22. Here he imagines many little fluid bodies orbiting at equal distances from the Earth. We have next to imagine that the little fluid moons unite in a fluid annulus. This ring will have the same behaviour as each little moon: its nodes will regress. Newton then imagines that the spherical Earth is enlarged until it touches the ring, which solidifies and is made to adhere to the Earth, as if it were an equatorial bulge. The Earth has to participate in the motion of the equatorial bulge. This is Newton's explanation, in terms of universal gravitation, of the precession of the equinoxes.§

In this section §3.7 we have briefly considered some of the Corollaries to Proposition 66. Most of the work in celestial mechanics since 1687 was based on this Proposition. Newton was able, thanks to a firm grasp of the relations between force and the geometric properties of trajectories (curvature, inclination etc.), to predict qualitatively the main perturbations of the Moon's orbit, and account for the precession of equinoxes, and he laid down an explanation of tidal motion (see §3.15). It should be stressed, however, that there is no complex mathematics involved. Newton proceeds by representing graphically the trajectory $ADBC$. He then represents the components of accelerative force LM, MN, SN, etc. Next he studies the variations of the Keplerian orbit of the Moon caused by the perturbing force exerted by the Sun. He considers the displacements of the light body at P in small intervals of time and in several conditions (at quadratures, in opposition,

---

† *Variorum*: 282. Cf. *Principles*: 177.
‡ *Principles*: 177.
§ Wilson (1989): 269.

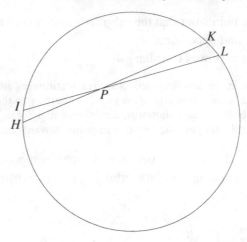

Fig. 3.12. Attraction inside a spherical shell. After *Variorum*: 299

etc.). Everything is expressed in connected prose. Sometimes the reader has to concentrate hard in order not to lose sight of the argument. But there is no difficult mathematics. The myth that Section 11 is very difficult is not well founded. Indeed, as Airy wrote in 1834, Section 11 is one of the few 'attempts at popular explanation in general use' of the theory of perturbations.†

3.8 Synthetic techniques and analytical quadratures applied to the attraction of extended bodies: Sections 12–13, Book 1

Sections 12 and 13 are devoted to the attraction of spherical and nonspherical bodies upon particles situated inside or outside these bodies. In Proposition 70 Newton shows that if a 'corpuscle' is situated inside a spherical homogeneous shell composed of particles which exert an inverse square attraction, then the corpuscle is not 'impelled' in any direction. In fact (see fig. 3.12) let P be the position of the corpuscle and construct a cone of infinitesimal solid angle with P as vertex. The cone intersects two infinitesimal areas KL and IH on the spherical surface. The areas KL and IH are as the square of the distance from P. Since the forces are 'as the particles directly and the square of the distances inversely',‡ the surface elements KL and IH exert equal and opposite forces on the corpuscle. Since this holds in any chosen direction, Newton concludes that the shell does not 'impel' in any way the corpuscle at P.

Notice that Newton in a Scholium to Proposition 73 makes it clear that:

† The other example being John F. W. Herschel's 'admirable treatise on Astronomy'. Airy (1834): ix–x.
‡ 'ut particulae directe, & quadrata distantiarum inverse'. *Variorum*: 299. Cf. *Principles*: 193.

The surfaces of which the solids are composed are here not purely mathematical, but orbs so extremely thin that their thickness is as null: namely, evanescent orbs of which the sphere ultimately consists when the number of those orbs is increased and their thickness diminished *in infinitum*[a]. Similarly, when lines, surfaces, and solids are said to be composed of points, such points are to be understood as equal particles of a magnitude so small that it can be ignored.†

As a matter of fact the demonstrations in Sections 12 and 13 are based on a partition of extended bodies into infinitesimal constituents. Newton obtains the total attraction exerted by an extended body by summing vectorially (as one would say in modern terms) the contributions of each constituent. In the above Scholium he warns us that the infinitesimal language that he adopts should be read *cum grano salis*. The underlying idea is that of first and ultimate ratios: *rigorously*, following the prescription of Section 1, Book 1, one should first partition the solid into finite elements, and then pass to the limit.

One should furthermore note that the attraction exerted by the extended bodies is evaluated by placing a corpuscle in space. The mass of such a corpuscle is negligible in comparison with the mass of the extended body, so that no reaction force of the corpuscle on the body is considered.

Proposition 71 reads as follows:

With the same conditions being supposed [as in Proposition 70], I say that a corpuscle placed outside the spherical surface is attracted to the centre of the sphere by a force [a]inversely proportional to the square of its distance from that same centre[a].‡

Newton considers two equal homogeneous spherical shells with centres S and s (see fig. 3.13) and two corpuscles P and p at different distances from the centres. Let there be drawn from the two corpuscles the lines PHK, PIL, phk, pil, such that they cut off equal arcs $HK = hk$ and $IL = il$. Construct SE perpendicular to PIL, SD and IR perpendicular to PHK, IQ perpendicular to AB (and similarly on the second sphere construct se, sd, ir and iq). When 'the angles DPE and dpe vanish' it is possible to show (by similarity of triangles in the limiting situation) that the following holds:

$$(PI^2 \cdot pf \cdot ps)/(pi^2 \cdot PF \cdot PS) = (HI \cdot IQ)/(hi \cdot iq). \qquad (3.20)$$

Furthermore, the ratio between the attraction on corpuscle P towards S caused by the annular surface obtained by rotating arc HI about the axis AB and the attraction on corpuscle p towards s caused by the annular surface obtained by

† *a* The first edition has here 'juxta Methodum sub initio in Lemmatis generalibus expositam', a phrase which is cancelled in subsequent editions. *Variorum*: 303. Cf. *Principles*: 196.

‡ *aa*, 'reciproce proportionali quadrato distantiae suae ab eodem centro'. *Variorum*: 299–300. Cf. *Principles*: 193.

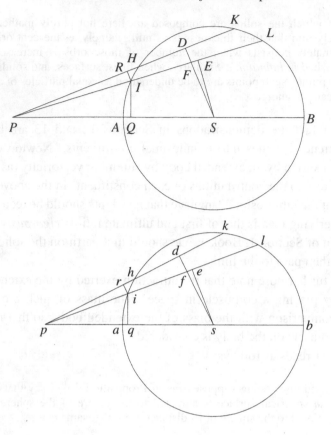

Fig. 3.13. Attraction outside a spherical shell. After *Variorum*: 300

rotating arc hi about axis ab is

$$\left(\frac{HI \cdot IQ}{PI^2} \frac{PF}{PS}\right) \bigg/ \left(\frac{hi \cdot iq}{pi^2} \frac{pf}{ps}\right). \tag{3.21}$$

In fact we have that the attraction is proportional to the surface $[2\pi \cdot HI \cdot IQ]$, inversely proportional to the square of the distance $[PI^2]$, and the cosine factor PF/PS is introduced in order to take into consideration only the horizontal component (by symmetry, only the horizontal components of the forces exerted by the particles composing the surface are relevant).

   Eliminating the ratio $(HI \cdot IQ)/(hi \cdot iq)$ from (3.20) and (3.21), we deduce that the ratio of the attractions of the two infinitesimal annular surfaces is ps^2/PS^2. "*a*By composition*a*", the forces of the total spherical surfaces exercised upon the corpuscles [P and p] will be in the same ratio'.† That is, the corpuscles are

† *aa*, 'per compositionem'. *Variorum*: 301. Cf. *Principles*: 194–5.

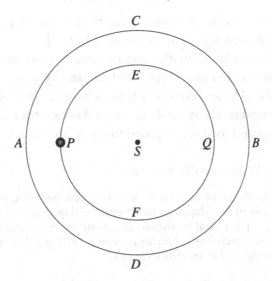

Fig. 3.14. Attraction inside a sphere. After *Variorum*: 303

attracted towards the centres of the spherical shells by forces varying inversely
as the square of the distance from the centre.

The subsequent Propositions 72–76 require no complex mathematical argument:
here Newton is at his best with regard to physical intuition. In Propositions 72
and 74 he shows that, under the hypothesis of inverse square attraction, the force
exerted by a sphere of given homogeneous density upon a point outside the sphere
is proportional to the sphere's volume and inversely proportional to the square of
the distance from the centre. In Proposition 73 it is proved that the attraction on a
point inside such a sphere is proportional to its distance from the centre.

In order to give an idea of Newton's ability in drawing consequences from his
physical models I quote Proposition 73 with its demonstration (which is essentially
that still adopted in modern textbooks):

If towards each of the separate points of any given sphere there tend equal centripetal forces
decreasing in the *a*duplicate ratio*a* of the distance from those points, I say that a corpuscle
placed inside the sphere is attracted by a force proportional to the distance of the corpuscle
from the centre of the sphere.

Let a corpuscle P be placed inside the sphere $ACBD$ [see fig. 3.14], described about the
centre S; and about the same centre S with radius SP, suppose that an inner sphere $PEQF$
is described. It is manifest (by Proposition 70) that the concentric spherical surfaces of
which the difference $AEBF$ of the spheres is composed do not act at all upon body P,
their attractions having been annulled by opposite attractions. There remains only the
attraction of the inner sphere $PEQF$. And (by Proposition 72) this is as the distance PS.†

† *aa*, 'duplicata ratione', translated as 'squared ratio' by Cohen and Whitman. *Variorum*: 303. Cf. *Principles*:
196.

The force on P is in fact directly proportional to the volume $[(4/3)\pi S P^3]$ and inversely proportional to the square of the distance $[S P^{-2}]$.

Newton extends his analysis to the mutual attraction of two homogeneous spheres (Proposition 75), and to spheres in which density is any function of the radius (Proposition 76). He next considers (Propositions 77 and 78) the case of an attractive force varying directly as the distance. In this case the attraction between two homogeneous spheres is directly proportional to the distance between their centres.

In the Scholium to Proposition 78 we read:

I have now given explanations of the two major cases of attractions, namely, when the centripetal forces decrease in the ᵃduplicate ratioᵃ of the distances or increase in the simple ratio of the distances [...] *It would be tedious to go one by one through the other cases which lead to less elegant conclusions. I prefer to comprehend and determine all the cases simultaneously under a general method as follows.*†

Lemma 29 and Propositions 79–82 are demonstrated by means of this 'general method'. The problem is to deal with the attraction exerted by homogeneous spheres when the law of attraction between the corpuscles is any function of distance. The 'general method' is in fact a reduction to quadratures of this problem. There is an interesting analogy with the solution of the inverse problem of central forces achieved in Sections 3 and 8. The cases of elastic force (Proposition 10 with its Corollary) and inverse square force (Propositions 11–13 with their Corollary) are dealt with by geometric methods. The general case of a central force (force varying as any power of distance; Propositions 39–42) is instead dealt with by reduction to quadratures. When dealing with attraction of spheres Newton proceeds similarly. He deals geometrically with the inverse square case (Propositions 70–76) and with the elastic force case (Propositions 77 and 78). He then deals with the general case (force varying as any power of distance) by reduction to quadratures: that is, he devises a geometrical construction which allows the reduction of the general problem to the calculation of an area subtended by a curve. Newton's 'general method' thus requires quadrature techniques. Of course Newton might simply have avoided the geometric treatment of the special cases, which are included in the general treatment. But he does not want to do this. This very peculiar characteristic of Newton's mathematization of natural philosophy will occupy us in the next chapters.

Lemma 29 and Propositions 79–82 deserve our attention since here Newton comes very close to disclosing the potential of his analytical method of fluxions. I will just note that Newton in Proposition 81 proceeds to perform some quadratures taking into consideration the case in which the force varies inversely as the distance

---

† *aa*, 'duplicata ratione', translated as 'squared ratio' by Cohen and Whitman. *Variorum*: 310. Cf. *Principles*: 202–3. My emphasis.

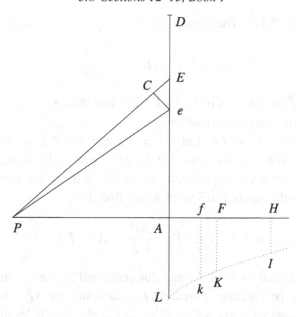

Fig. 3.15. Attraction exerted by a disk. After *Variorum*: 329

(Example 1), inversely as the cube of the distance (Example 2), and inversely as the fourth power of the distance (Example 3). As Newton remarks, the quadratures that he has to perform are achievable 'by ordinary methods', i.e. they require shared knowledge about quadrature methods.† These calculations are within the reach of a reader not acquainted with those difficult quadratures that were performed in Newton's 'catalogues' on the squaring of curves.‡

Newton's peculiar approach to quadratures can be illustrated by Propositions 90 and 91, which deal with the attraction of nonspherical bodies. These two Propositions are actually restricted to the the attraction exerted by homogeneous solids of revolution on external points situated on the prolongation of the axis of revolution.§

In Proposition 90 Newton determines the attraction exerted by a uniform circular disk with centre A and radius AD (see fig. 3.15) on a point situated in P (PA is normal to the disk). Let PH be equal to PD and let the ordinates of curve IKL be such that, if $PF = PE$, then FK is 'as the force by which the point E attracts the corpuscle P'.¶ Consider the 'small line' Ee and the annulus generated by rotation

---

† 'quarum areae per methodos vulgatas innotescunt'. *Variorum*: 317. Cf. *Principles*: 207.
‡ For details see Whiteside's notes in *Mathematical Papers*, **6**: 216–20.
§ Proposition 91 is concluded by the very important Corollary 3 (which is actually independent from Proposition 91). Newton shows by elegant geometric reasoning that the lamina situated between two similar and similarly placed ellipsoids, a 'geoid', exerts no attraction upon a point situated inside the geoid.
¶ *Variorum*: 329. Cf. *Principles*: 218.

of Ee about the axis PAH. The attraction exerted by the annulus is

$$F_{\text{annulus}} \propto FK \cdot AE \cdot Ee \cdot \frac{AP}{PE}, \tag{3.22}$$

since the surface of the annulus is $2\pi \cdot AE \cdot Ee$, and the cosine factor (AP/PE) is introduced by symmetry considerations.

Draw eC such that $PC = Pe$. Let $Pf = Pe$ and thus $CE = Ff$. When the angle ePE tends to zero, one can state that $AE/PE = CE/Ee$ (here, as it is often the case, Newton does not make explicit this limit argument, but simply relies on the fact that Ee is infinitesimal†). Therefore we find that

$$F_{\text{annulus}} \propto FK \cdot PE \cdot Ff \cdot \frac{AP}{PE} = AP \cdot FK \cdot Ff. \tag{3.23}$$

That is, the force exerted on P by the annulus generated by the revolution of Ee is proportional to 'the [infinitesimal] area $FKkf$ multiplied by AP'. In conclusion, the force directed towards A exerted on P by the circular disk is 'as the whole area $AHIKL$ multiplied by AP'.‡

The problem faced in Proposition 90 is thus reduced to the calculation of the area $AHIKL$. In the next three Corollaries Newton performs some exemplary quadratures. For instance, in Corollary 2 he writes that 'if the forces are inversely as any power D^n of the distances (that is, if FK is as $1/D^n$), then the area $AHIKL$ is as $1/PA^{n-1} - 1/PH^{n-1}$, and the attraction of the corpuscle P toward the circle is as $1/PA^{n-2} - PA/PH^{n-1}$'.§ This simple quadrature is achieved by application of results known in the pre-calculus period, and is thus rendered explicitly in the *Principia*. In fact, the area over the interval AH and under the curve whose ordinate is $FK = -k/PF^n$ is, applying a well-known law, $(k/(n-1))(1/PA^{n-1} - 1/PH^{n-1})$.¶

In Proposition 91 Newton considers the attraction exerted by a homogeneous solid of revolution on an external point situated on the axis of revolution. He partitions the solid into circular disks and obtains the total attraction by summing the component attractions of the disks. In Corollaries 1 and 2 the case in which the particles composing the solid exert an inverse square force is considered. The quadrature necessary to determine the attraction in the case of a cylinder is easily achieved and expressed explicitly by Newton in Corollary 1. In Corollary 2 he faces the nontrivial case of the attraction exerted by an oblate spheroid on a point situated

---

† 'linea quam minima'. We know, however, from Section 1 that we have to translate infinitesimal arguments according to the method of first and ultimate ratios.
‡ *Variorum*: 330. Cf. *Principles*: 219.
§ *Variorum*: 330–1. Cf. *Principles*: 219.
¶ The law, well known to Newton's predecessors, can be written as: $\int x^n = x^{n+1}/(n+1)$.

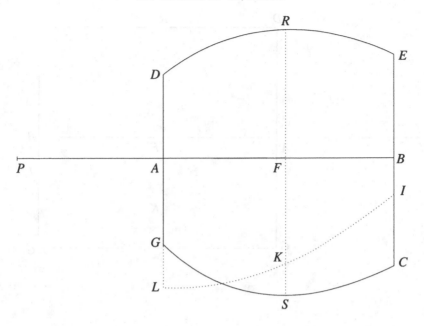

Fig. 3.16. Attraction exerted by a solid of revolution. After *Variorum*: 331

on the prolongation of the minor axis.† Here Newton gives only the solution in the form of a geometrical construction, but does not perform the quadrature explicitly. Indeed, only by application of his 'catalogues' could one solve this problem. As usual in the *Principia*, details about nontrivial quadratures are avoided. This Proposition has been analysed by Whiteside: I will follow his explanation in broad outline.‡

In fig. 3.16 the solid is generated by the rotation of the curve DRE around the axis AB. The ordinate FK represents the attraction exerted by the disk RS on the point P. Thus the total attraction exerted by the solid is given by the area $LABI$. According to Proposition 90, if the force is as any power of the distance, $-k/D^n$,

$$FK \propto \frac{1}{PF^{n-2}} - \frac{PF}{PR^{n-1}}. \qquad (3.24)$$

If the force is inverse square,

$$FK \propto 1 - \frac{PF}{PR}. \qquad (3.25)$$

† Notice that this and the subsequent corollary appear on a separate sheet bound at the end of the manuscript Book 1 held at the Royal Society. See Cohen (1971): 140. It thus seems that they were added somewhat after the completion of Book 1.
‡ For details, I refer the reader to *Mathematical Papers*, **6**: 222–7.

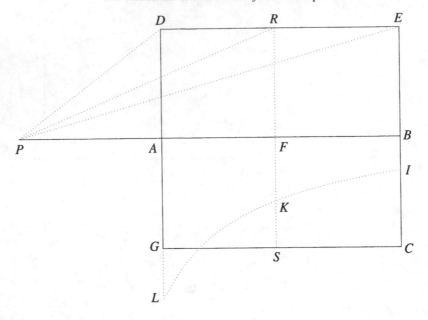

Fig. 3.17. Attraction exerted by a cylinder. After *Variorum*: 332

In Corollary 1 the solid is a cylinder and the force is inverse square (see fig. 3.17). In this case the area under $1 - PF/PR$ is easily calculated ('which can easily be shown from the quadrature of the curve LKI'†). For the reader's convenience, before quoting Newton's explanation, I will give (following Whiteside) a Leibnizian version of this quadrature.

Since the radius FR is constant and $PR^2 - PF^2 = FR^2$, we have that

$$d(PF) = \frac{PR}{PF}d(PR). \qquad (3.26)$$

This yields

$$\int_{PA}^{PB} (1 - PF/PR)d(PF) = \int_{PA}^{PB} d(PF) - \int_{PD}^{PE} d(PR) = AB + PD - PE.$$
$$(3.27)$$

In the demonstration of Corollary 1 Newton gives every detail necessary to render this quadrature understandable even to the less experienced mathematician. The master of quadrature techniques, the author of the 'catalogues' of the *De methodis*, is here behaving like a schoolmaster who wishes to avoid any passage that might be frightening to his pupils:

For the ordinate FK (by Corollary 1, Proposition 90) will be as $1 - PF/PR$. The unit part

† 'id quod ex curvae LKI quadratura facile ostendi potest'. *Variorum*: 332. Cf. *Principles*: 220.

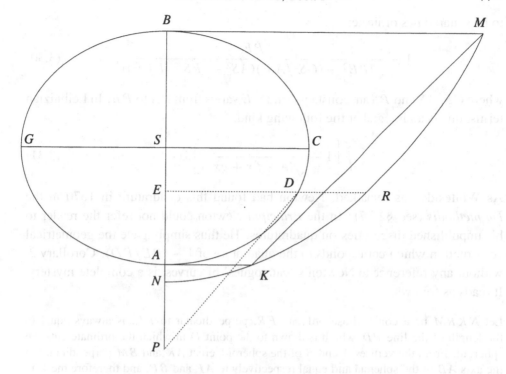

Fig. 3.18. Attraction exerted by an ellipsoid. After *Variorum*: 333

of this multiplied by the length AB describes the area $1 \times AB$, and the other part PF/PR multiplied by the length PB describes the area $1 \times (PE - AD)$, which can easily be shown from the quadrature of the curve LKI; and similarly the same part $[PF/PR]$ multiplied by the length PA describes the area $1 \times (PD - AD)$, and multiplied by AB, the difference of PB and PA, describes $1 \times (PE - PD)$, the difference of the areas. [a]From the first product[a] $1 \times AB$ take away the last product $1 \times (PE - PD)$, and there will remain the area $LABI$ equal to $1 \times (AB - PE + PD)$. Therefore the force proportional to this area is as $(AB - PE + PD)$.†

Corollary 2 concerns an oblate spheroid (see fig. 3.18). The attraction on P, under the hypothesis that the force is inverse square, is given by the area subtended by the curve whose ordinate is

$$1 - \frac{PE}{PD}. \tag{3.28}$$

For an ellipse:

$$PD = \sqrt{PE^2 + (CS^2/AS^2)(AS^2 - ES^2)}, \tag{3.29}$$

since $DE^2 = (CS^2/AS^2) \cdot AE \cdot EB$. Furthermore $ES = PS - PE$, so the curve

† *aa* translates 'De contento primo'. *Variorum*: 332. Cf. *Principles*: 220.

to be squared has ordinate:

$$1 - \frac{PE}{\sqrt{[PE^2 + (CS^2/AS^2)(AS^2 - (PS - PE)^2)]}},\qquad(3.30)$$

where CS, AS and PS are constants, and PE varies from PA to PB. In Leibnizian terms, this is an integral of the following kind:

$$\int \left[1 - \frac{x}{\sqrt{(e + fx + gx^2)}} \right] dx.\qquad(3.31)$$

As Whiteside has remarked, Newton had found this quadrature in 1670 in the *De methodis* (see §2.2.7). In the *Principia* Newton could not refer the reader to his unpublished discoveries on quadratures. He thus simply gave the geometrical construction which corresponds to the quadrature of $1 - (PE/PD)$. Corollary 2, without any reference to Newton's 'catalogues' of curves, is a complete mystery. It reads as follows:

Let $NKRM$ be a conic whose ordinate ER, perpendicular to PE, is always equal to the length of the line PD, which is drawn to the point D in which the ordinate cuts the spheroid. From the vertices A and B of the spheroid, erect AK and BM perpendicular to the axis AB of the spheroid and equal respectively to AP and BP, and therefore meeting the conic in K and M; and join KM cutting off the segment $KMRK$ from the conic. Let the centre of the spheroid be S, and its greatest semidiameter SC. Then the force by which the spheroid attracts the body P will be to the force by which a sphere described with diameter AB attracts the same body as $(AS \times CSq - PS \times KMRK)/(PSq + CSq - ASq)$ to $AScub./3PSquad..$†

Whiteside has shown how the above geometrical construction corresponds to the result achieved via Newton's 'catalogues'.‡

I conclude this section devoted to Newton's study of the attraction of extended bodies by noting that:

- Proposition 71 is demonstrated by geometrical first and ultimate ratios;
- Propositions 70 and 73 are based on ingenious physical modelling;
- Propositions 90 and 91 are based on quadrature techniques;
- these quadratures are made explicit when they are trivial;
- difficult quadratures are left hidden: only the result is revealed.

Indeed in Sections 12 and 13 Newton deploys a large amount of his demonstrative repertoire. The physicist, the geometrician and the master of integration are all represented at the highest level.

---

† Notice that CSq means CS^2, $AScub.$ means AS^3, etc. *Variorum*: 332–3. Cf. *Principles*: 220–1.
‡ See Whiteside's commentary in *Mathematical Papers*, **6**: 226.

3.9 Power series applied to optics: Sections 13–14, Book 1

At the end of Section 13 Newton considers the attraction on a corpuscle exerted by a semi-infinite plane-parallel slab. This leads him in Section 14 to consider some mathematical models of interest for corpuscular optics. The path of the corpuscle is to be understood as the path of a light ray constituted by many such corpuscles. Newton derives the law of reflection, and 'Snell's law', and deals with Descartes' ovals.

What is interesting for us is the Scholium to Proposition 93. In the first lines Newton notes that, if the force is given, in order to determine the path of a body 'attracted perpendicularly toward a given plane' one has to apply Proposition 39. That is, one has to decompose motion into an horizontal inertial component and a vertical accelerated component. Proposition 39 can be applied to the vertical component. Newton next notes that, if the path is given, in order to determine the force 'the procedure can be shortened by resolving the ordinates into converging series'.†

Newton thus assumes that we know the trajectory of the corpuscle and he poses the problem of determining the force, directed normally to the plane, which accelerates the corpuscle along the given trajectory. He denotes with B the ordinate, i.e. the distance of the corpuscle from the plane, and assumes that $B = A^{m/n}$, where A is the abscissa. He writes:

I suppose the base to be increased by a minimally small part O, and I resolve the ordinate $(A + O)^{m/n}$ into the infinite series

$$A^{m/n} + \frac{m}{n} O A^{(m-n)/n} + \frac{mm - mn}{2nn} O O A^{(m-2n)/n} \&c, \qquad (3.32)$$

and I suppose the force to be proportional to the term of this series in which O is of two dimensions, that is, to the term

$$\frac{mm - mn}{2nn} O O A^{(m-2n)/n}.$$

From this it follows that the force sought is as

$$\frac{mm - mn}{nn} B^{(m-2n)/m}. \qquad (3.33)$$

Therefore:

If the ordinate traces out a parabola, where $m = 2$ and $n = 1$, the force will become as the given quantity $2B^0$, and thus will be given. Therefore with a given [i.e. constant] force the body will move in a parabola, as Galileo demonstrated. But if the ordinate traces out a hyperbola, where $m = 0 - 1$ and $n = 1$, the force will become as $2A^{-3}$ or $2B^3$; and therefore with a force that is as the cube of the ordinate, the body will move in a hyperbola.‡

---

† *Variorum*: 337. Cf. *Principles*: 225.

‡ Note that here Newton is using the algebraic symbols for raising powers. *Variorum*: 338. Cf. *Principles*: 225.

We note that Newton is here rather insecure in distinguishing the direct from the inverse problem. After having shown that if the trajectory is a parabola then the force is constant, he adds that *therefore* with a constant force the body will move in a parabola. He repeats the same reasoning for the hyperbola. As we will see in Chapter 8 he will be criticized for not distinguishing the two problems from one another.

It is remarkable to see how Newton attributes a dymanical meaning to the terms of a power series expansion. He does so also in Proposition 10, Book 2, as we will see in §8.6.3.

In the 1690s Newton introduced a new notation for fluxions and higher-order fluxions. He wrote $\dot{x}$, $\ddot{x}$, etc., for first, second, etc., fluxions. He also had the notation $\acute{x}$ for the fluent of x. Dots and accents could be repeated to generate higher-order fluxions and higher-order fluents. Newton also used the notation $\overset{n}{x}$ for the nth fluxion of x.

In discussing higher-order fluxions Newton stated that every ordinate y of a curve in the x–y plane can be expressed, assuming $\dot{x} = 1$, as a power series whose nth term is equal to the nth fluxion of y, i.e. $\overset{n}{y}$, divided by $n!$.† In fact, Newton assumed that y is expressible as a power series such as

$$y = a + bx + cx^2 + dx^3 + ex^4 + \cdots. \tag{3.34}$$

One gets immediately that $y(0) = a$, $\dot{y}(0) = b$, $\ddot{y}(0) = 2c$, etc.

Newton was thus expressing a series, nowadays called the Taylor series, which was to play an important role in the development of eighteenth century calculus. We are used in modern mathematics to include a treatment of the rest of the series and the study of the interval of convergence. This was not done either by Newton or by any mathematician until the end of the eighteenth century.

The statement of the Taylor series and the understanding of the dynamical meaning of the successive terms is already in the *Principia*. This is relevant, since it reveals Newton's awareness of the possibility of expressing his science of motion in terms of series and fluxions. In fact, if we put together the result expressed in the 1690s with the Scholium to Proposition 93, Book 1, and Proposition 10, Book 2, we arrive at the statement that the force is proportional to the second fluxion of displacement. As we will see in §4.5, Newton in the 1690s drew this conclusion and attempted to give a fluxional representation of central force.

† *Mathematical Papers*, 7: 96–8.

3.10 Book 2: an overview

Book 2 is devoted to the motion of bodies in resisting media. It contains many pages devoted to experimental results (the general Scholium in Section 6, Proposition 40 and its Scholium in Section 7 on air resistance), pages related to the physics of fluids (hydrostatic, the density of the atmosphere in Section 5, the efflux of water from a hole at the bottom of a vessel in Section 7) and a refutation of the vortex theory of planetary motions (Section 9). In dealing with the above-mentioned topics, Newton does not employ advanced mathematical methods. Book 2 is thus much less mathematical and more experimental than Book 1. The mathematical parts of Book 2 are very problematic. Compared with the mathematical methods of the first book, those of the second were considered, during Newton's lifetime, the less satisfactory, and in some cases just mistaken.

Propositions 1–4 deal with projectile motion in a medium whose resistance is proportional to velocity [$\mathcal{R} = k_1 v$], Propositions 5–10 with projectile motion in a medium whose resistance is proportional to the square of velocity [$\mathcal{R} = k_2 v^2$], while in Propositions 11–14 $\mathcal{R} = k_1 v + k_2 v^2$. Lemma 2, Section 2, concerns the calculation of moments of variable quantities. Propositions 15–17 are devoted to spiral motion in resisting media, Propositions 25–31 to pendular motion in resisting media, Proposition 34 and its Scholium to the solid of least resistance.†

3.11 Geometrical representation of logarithms applied to projectiles in resisting media: Props. 1, 2 and 5, Book 2

The motion of projectiles in resisting media aroused the interest of many seventeenth-century natural philosophers.‡ Huygens had studied this topic in the years 1668–69 both theoretically and experimentally. James Gregory in 1672 and Robert Anderson in 1674 had considered this matter, which was of fundamental importance for ballistics. In 1674, after reading Anderson's work, Newton had made some calculations on trajectories in resisting media. During the composition of *De motu* in 1684 he considered the topic afresh. Motion of projectiles in resisting media might seem a subject unrelated to the theory of gravitation: however, in Descartes' cosmology planets are seen as bodies which move in (and are moved by) an ethereal medium. A treatment of motion of point masses and extended bodies in resisting media is preliminary to the discussion (and would-be refutation) of the vortex theory of planetary motions which occupies the last section of Book 2.

As we will see, most of the Propositions dealing with motion in resisting media lead to solutions in which magnitudes are related by logarithmic functional relations. In order to represent these relations, Newton employs the properties

---

† Proposition 34 in the second and third editions is Proposition 35 in the first.
‡ Hall (1952). On Newton's contributions see Blay (1987).

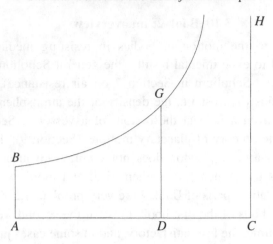

Fig. 3.19. Representation of logarithmic relations. After *Variorum*: 350

of the equilateral hyperbola. This is what he did also in his fluxional writings. Newton generally avoided a notation for transcendental functions. He expressed the logarithm as $\boxed{1/x}$: i.e. as the area bounded by the hyperbola.†

Newton's mathematical model consists in conceiving a body which moves in a resisting medium as accelerated by impulsive forces which act at equal finite intervals of time. He then passes to the limit in order to obtain a resistance which acts continuously on the body. In his words:

> Divide the time into equal particles; and [...] at the very beginning of the particles a force of resistance [...] acts with a single impulse [...] let those equal particles of times be diminished, and their number increased *a in infinitum*, so that the impulse of the resistance becomes continual.‡

Section 1 deals with the 'motion of bodies that are resisted in proportion to their velocity' [$\mathcal{R} = k_1 v$]. In Proposition 1 Newton shows that in this case 'the motion lost as a result of the resistance is as the space described in moving'. In fact 'the motion lost in each of the equal particles of time' ('motion' is equal by definition to mass times velocity) is by hypothesis as the velocity at the beginning of the particle of time, i.e. 'as a particle of the path described' and then 'by composition, the motion lost in the whole time will be as the whole path'.§

In Lemma 1 Newton states that if $A : A - B = B : B - C = C : C - D$ etc. then $A : B = B : C = C : D$ etc.

---

† In Leibnizian terms $\ln(x) = \int_1^x (1/x)\mathrm{d}x$.

‡ Note that Cohen and Whitman translate *aa* as 'indefinitely'. *Variorum*: 349. Cf. *Principles*: 236.

§ *Variorum*: 348. Cf. *Principles*: 235. Following Blay (1992): 257ff. and Erlichson (1991a): 281 one can translate this into symbols as follows. By hypothesis $\mathrm{d}v \propto v$. Since during each equal small interval of time motion is uniform, $\mathrm{d}x \propto v$. Therefore, $\mathrm{d}x \propto \mathrm{d}v$. And, integrating, $\Delta x \propto \Delta v$.

Proposition 2 states that:

If a body is resisted in proportion to its velocity and moves through a homogeneous medium by its inherent force alone and if the times are taken as equal, the velocities at the beginning of the individual times are in geometric progression, and the spaces described in the individual times are as the velocities.†

The proof is as follows. Newton subdivides time into equal finite intervals. By hypothesis the decrement of the velocity in each of the intervals of time will be proportional to the velocity. Following Blay (1987) we can translate in symbols as:

$$v_{i+1} = v_i - kv_i, \tag{3.35}$$

where v_i is the velocity during the ith interval of time and kv_i is the resistance exerted by the medium at the beginning of the ith interval. Thus, $v_i/(v_i - v_{i+1}) = k^{-1}$, and therefore (by Lemma 1, Book 2) the velocities are continually proportional. That is, in symbols not employed by Newton:

$$v_1/v_2 = v_2/v_3 = v_3/v_4 = \cdots. \tag{3.36}$$

At this stage Newton considers equal portions of time, each constituted by an equal number of intervals of time. It is easy to show that the velocities at the beginning of each equal portion of time are in a geometrical progression. For instance, taking portions constituted by ten intervals, we have

$$v_1/v_{11} = v_{11}/v_{21} = v_{21}/v_{31} = \cdots. \tag{3.37}$$

Now increase the number of intervals to twenty. The velocities at the end of the twenty equal intervals will be altered slightly, but a relation such as (3.37) will still hold. The velocity still decreases in a stepwise fashion, but we are approaching the real smoothly decreasing velocity. In fact, Newton passes to the limit, allowing the intervals of time to shrink to zero, while keeping fixed the equal portions of time. The English translation by Motte and Cajori reads as follows:

Let those equal intervals of time be diminished, and their number increased *in infinitum*, so that the impulse of resistance may become continual; and the velocities at the beginning of equal [portions of] times, always continually proportional, will be also in this case continually proportional.‡

In the Corollary to Proposition 2 Newton gives a mathematical representation of the results achieved in the previous two Propositions. He considers (see fig. 3.19) the equilateral hyperbola with asymptotes AC and CH. The lines DG and AB are two ordinates. It is a property of the hyperbola, known for example to Grégoire

---

† *Variorum*: 349. Cf. *Principles*: 236.

‡ In the Motte–Cajori translation a useful distinction between 'portions' and 'intervals' of time is made. For this reason, I have preferred to adopt here the Motte–Cajori translation. *Principles*: 236. Cf. *Variorum*: 349–50. I must thank Herman Erlichson for his useful advice on Proposition 2, Book 2. See Erlichson (1991a).

de Saint-Vincent (1647), that if the abscissae grow in a geometric progression then the areas under the hyperbola delimited by the corresponding ordinates grow in an arithmetic progression. This is exactly the relation between velocity and time obtained in Proposition 2. Furthermore, by Proposition 1, the distance described is as the velocity lost. Summing up, if DC represents velocity 'after some time is elapsed', the time may be expressed by the area $ABGD$, and the distance described by the line AD. In fact:

If the area $[ABGD]$ is increased uniformly by the motion of point D, in the same manner as the time, the straight line DC will decrease in a geometric ratio in the same way as the velocity, and the parts of the straight line AC described in equal times will decrease in the same ratio.†

Of course AC expresses the initial velocity.

Newton's procedure can be summarized as follows. He deploys the impulsive model for finite intervals of time in order to deduce a relationship between time, velocity and space. He passes to the limit in order to state this relation for a continuous resistance. Then he refers to the geometric properties of the hyperbola in order to offer a mathematical representation of the relations achieved. This mathematical representation allows one to draw several conclusions about the relations between time, space, velocity and force.

Newton follows this procedure also in Proposition 5 which deals with a body which moves 'by its inherent force alone' in a medium which resists as the square of the velocity. By definition, 'if the time is divided into innumerable equal particles, the squares of the velocities at each of the beginnings of the times will be proportional to the differences of those same velocities'.‡ In the notation proposed by Blay:

$$v_i - v_{i+1} = k v_i^2. \tag{3.38}$$

In order to relate this property to the hyperbola Newton shows by simple geometry that (see fig. 3.20) if the axis AD is divided into equal finite intervals AK, KL, LM, ... then:

$$(AB - Kk)/AK = (AB \cdot Kk)/(AB \cdot CA). \tag{3.39}$$

He then passes to the limit as follows:

Hence, since AK and $AB \times CA$ are given, $AB - Kk$ will be as $AB \times Kk$; and ultimately, when AB and Kk come together, as ABq. And, by a similar argument, $Kk - Ll$, $Ll - Mm$, &c., will be as $Kkquad.$, $Llquad.$, &c.§

† *Variorum*: 350. Cf. *Principles*: 237.
‡ *Variorum*: 360. Cf. *Principles*: 245.
§ Note that ABq means AB^2, $Kkquad.$ means Kk^2, etc. *Variorum*: 360. Cf. *Principles*: 245.

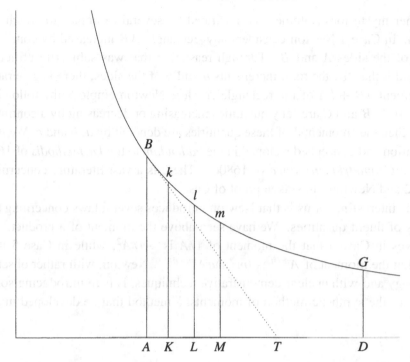

Fig. 3.20. Representation of logarithmic relations. After *Variorum*: 360

Thus the abscissae can represent time and the ordinates velocity. In fact the relation 'if the time is divided into innumerable equal particles, the squares of the velocities at each of the beginnings of the times will be proportional to the differences of those same velocities' is mirrored in the hyperbola. Of course, since fig. 3.20 is a time–velocity diagram, the distance covered in time AM will be represented by the area $AMmB$.

3.12 Synthetic method of moments and fluxions: Lemma 2, Section 2, Book 2

Before Proposition 8 Newton has to introduce with Lemma 2 some techniques concerning the calculation of moments. In the style of *Geometria curvilinea* he states:

The moment of a generated quantity is equal to the moments of each of the generating ^{a}roots^a multiplied continually by the exponents of the powers of those ^{a}roots^a and by their coefficients.†

† aa, which reads 'laterum' in second and third editions, reads 'terminorum' in first edition. Cohen and Whitman translate both 'latus' and 'terminus' as 'root'. *Variorum*: 364. Cf. *Principles*: 249.

This rather mysterious statement is illustrated by several examples to which we now turn. In Case 1 Newton considers any 'rectangle AB increased by continual motion' of the sides A and B. Through reasoning that was subject to criticism he concludes that 'by the total increments a and b of the sides, there is generated the increment $aB + bA$ of the rectangle'.† Here Newton employs the following notation. If A, B and C are 'any quantities increasing or decreasing by a continual motion', then the 'moments' of these quantities are denoted by a, b and c. We find here notation and concepts developed in the *addendum* to the *De methodis* of 1671 and in the *Geometria curvilinea* of 1680ca. There is a vast literature concerning Lemma 2 and Newton's mistaken proof of Case 1.

What is interesting for us is that Newton introduces several laws concerning the moments of fluent quantities. We have seen above the moment of a product. He also proves in Case 4 that the moment of $1/A$ is $-a/A^2$, while in Case 5 it is proved that the moment of $A^{m/n}$ is $(m/n)aA^{(m-n)/n}$. Newton, with rather obscure terminology and with unclear demonstrative techniques, is here introducing some elements of the synthetic method of moments, a method that he developed in the 1670s.

3.13 Application of Lemma 2 to motion in resisting media: Prop. 8, Book 2

Lemma 2 is applied by Newton in Propositions 8, 9, 11 and 14 on resisted motion and Proposition 29 on resisted pendular motion. It should be said that in Proposition 39, Book 1, Newton had already calculated the moment of v^2. Let us consider in some detail Proposition 8.‡

Newton considers a body that ascends or descends in a straight line perpendicular to the ground. On the body two forces act (both in the direction of motion): the constant force of gravity and the resistance. The resistance is proportional to the square of velocity. Let the 'given line' AC (see fig. 3.21) represent the force of gravity, the 'indefinite line' AK the resistance. During fall the 'absolute force' acting on the body will be $AC - AK = KC$. Let AP be such that

$$AP^2 = AK \cdot AC. \tag{3.40}$$

AP represents velocity: in fact AC is constant and AK (the resistance) varies as the square of AP.

Let the linelet KL be 'the increment of the resistance occurring in a given particle of time'. Let the linelet PQ be the 'simultaneous increment of the velocity'. By Lemma 2 we have that the 'moment KL' will be 'as the moment

---

† *Variorum*: 366. Cf. *Principles*: 250.
‡ *Variorum*: 368–71. Cf. *Principles*: 252–3. On this Proposition see Galuzzi (1991).

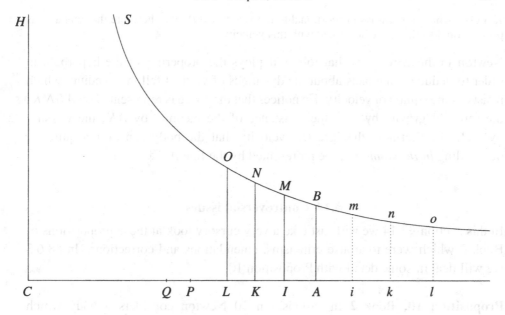

Fig. 3.21. Representation of logarithmic relations. After *Variorum*: 370

$2AP \cdot PQ$':

$$KL \propto 2AP \cdot PQ. \tag{3.41}$$

This is why Newton needs Lemma 2: he has to determine the moment of (3.40) (in Leibnizian symbols $AC\mathrm{d}(AK) = 2AP\mathrm{d}(AP)$).

Now Newton draws the equilateral hyperbola as in fig. 3.21. The centre is at C, the asymptotes are CH and CA, and the ordinates AB, KN and LO are drawn.

By the second law of motion the increment PQ of the velocity *acquired in a given time* is 'proportional to the generating force KC'. Substituting in (3.41) KC for PQ and multiplying by KN:

$$KL \cdot KN \propto AP \cdot KC \cdot KN. \tag{3.42}$$

Since $KC \cdot KN$ is given (by definition of the hyperbola):

$$KL \cdot KN \propto AP. \tag{3.43}$$

Next Newton passes to the limit in his characteristic style:

But the ultimate ratio of the hyperbolic area $KNOL$ to the rectangle $KL \times KN$, when points K and L come together, is the ratio of equality.   Therefore that evanescent hyperbolic area is as AP. Hence the total hyperbolic area $ABOL$ is composed of particles

$KNOL$, which are always proportional to the velocity AP, and therefore the total area is proportional to the space described with this velocity.†

Newton in the corollaries that follow deploys the properties of the hyperbola in order to deduce some facts about the dynamics of vertical fall in a medium which resists as the square of velocity. He notices that the space is represented by $ABNK$, the force of gravity by AC, the resistance of the medium by AK, the velocity by AP. Furthermore the 'greatest velocity that the body can ever acquire by descending *in infinitum*' will be represented by the line AC.‡

3.14 Controversial issues

In this section §3.14 we will just take a very cursory look at those propositions of Book 2 which were to arouse criticisms, emendations and corrections. In §8.6.3 we will deal in some detail with Proposition 10.

Proposition 10, Book 2 In Proposition 10 Newton considers a body which moves under the influence of a uniform constant gravity [g] and a resistance [$\mathcal{R}$] proportional to the product of the density [ς] of the medium and the square of the velocity [$\mathcal{R} \propto \varsigma v^2$]. Given the trajectory, it is required to find the density and the velocity at each point of the trajectory.

Notice that Newton does not address the general inverse problem, which would be: given initial position and velocity, and given g and ς, determine the trajectory of a body under the assumption that $\mathcal{R} \propto \varsigma v^2$. In Proposition 10 Newton does not address the problem that one would expect at this point of the *Principia*. The inverse ballistic problem proved to be very difficult for early-eighteenth-century mathematicians.

Proposition 10 is followed by four 'examples' handled by application of power series ('Taylor series'). The given trajectory is supposed to be a half circle (Example 1), a parabola (Example 2), an hyperbola (Example 3), and a generalized hyperbola (Example 4).

As we will see, after Johann and Niklaus Bernoulli's criticisms, Newton changed this proposition in the second edition of *Principia*.

Propositions 15–18, Book 2 In these propositions Newton considers a body which moves in a resisting medium and which is accelerated by a central force. In Proposition 15 he states that if the density is inversely proportional to the distance from the force centre, and if the centripetal force is as the square of the

† *Variorum*: 370. Cf. *Principles*: 253. Denoting the area by $\mathcal{A}$ and the space by s, we have $d\mathcal{A}/dt \propto v$, hence $\mathcal{A} \propto s$.

‡ *Variorum*: 371. Cf. *Principles*: 253.

density, then a logarithmic spiral is a possible orbit. In Proposition 16 the centripetal force is assumed to be proportional to the inverse of any power of the distance. In this case too Newton states that the trajectory can be a logarithmic spiral. These propositions were criticized by Roger Cotes and Johann Bernoulli. Basically these criticisms were a direct consequence of the criticisms advanced against Proposition 10.

Propositions 25–31, Book 2 These propositions are devoted to resisted motion in a cycloidal pendulum. Free motion in a cycloidal pendulum had already been treated by Newton in Propositions 50–2, Book 1. Propositions 25–31 were to be criticized by James Stirling, Jacopo Riccati, and Niklaus and Johann Bernoulli.†

Proposition 35/34, Book 2 In Proposition 35 (34 in second and third editions) Newton considers a medium composed of 'equal' particles 'arranged freely at equal distances from one another'. He assumes that a cylinder (circular base with diameter D) and a sphere of radius $D/2$ move 'with equal velocity along the direction of the axis of the cylinder'. Newton proves that the resistance acting on the sphere is half that acting on the cylinder. In the Scholium he considers a frustrum of a right-circular cone with a given altitude and base which moves with constant velocity in the direction of its axis. He identifies the shape of the frustrum that minimizes resistance. Newton concludes the Scholium by addressing the problem of determining the shape of the solid of revolution that moves in the direction of its axis with the least possible resistance.

In the Scholium no proof is given. Newton simply states, without demonstration, the geometrical properties of the frustrum and of the general solid of revolution of least resistance. Manuscripts dating from 1685 containing Newton's fluxional analysis of the Scholium are extant. They are the only surviving preparatory manuscripts of the *Principia* in which the fluxional algorithm occurs. Newton solves the frustrum problem by application of simple maxima and minima fluxional methods. The general problem is much more difficult. It is one of the first complex variational problems considered in the history of mathematics. Newton's fluxional calculation, which leads to the geometrical property enunciated in the Scholium, has been discussed at length by Goldstine and Whiteside.‡

The Scholium to Proposition 35/34 aroused the interest of most of the mathematicians during the period taken into consideration in this book. We may here list Huygens,§ David Gregory, Fatio de Duillier, Leibniz and Johann Bernoulli. In fact, in the late 1690s, variational problems (such as the brachistochrone problem)

---

† *Mathematical Papers*, **6**: 441–2.
‡ *Mathematical Papers*, **6**: 456–80. Goldstine (1980): 7–29. See also Forsyth (1927).
§ *Oeuvres*, **22**: 335–41.

received great attention by those mathematicians who were practising the new calculi of Newton and Leibniz. We will not deal in our book with the solid of least resistance. It should however be noted that this is another case in which Newton in the *Principia* includes a result achieved via advanced techniques of the fluxional analytical method but withholds all the details of the algorithmic demonstration. The fluxional analysis of this problem was in fact circulated. In 1694 David Gregory got the solution from Newton: John Keill and William Jones transcribed it. Newton's fluxional solution was published by Fatio de Duillier in (1699), by Hayes in (1704) and in 1729 in an appendix to Motte's English translation of the *Principia*. Newton's manuscripts relating to the solid of least resistance were found in the Portsmouth papers and published in the *Catalogue of the Portsmouth Collection* at the end of the nineteenth century, and were reproduced by Ball and by Cajori.†

3.15 Book 3: some remarks

In Book 3 Newton abandons the predominantly mathematical level of the first two books in order to face the intricacies of the real 'System of the World'. It is in the third book that he develops the theory of universal gravitation: every two masses attract each other with a force proportional to the product of the masses, and inversely proportional to the square of the distance between the masses. This attractive force, responsible for long-range interaction between Sun, planets and stars, is the very force which attracts bodies towards the centre of the Earth; it is the force which we commonly experience as weight. The argument in favour of universal gravitation contained in Book 3 is extremely complex: it is a masterpiece of mathematical insight, physical intuition and experimental accuracy. Newton has recourse to the Moon test, to astronomical observations on the motion of planets and comets, to experiments with pendula aimed at proving the equivalence between gravitational and inertial mass, and to data concerning the shape of planets and the flow and ebb of the sea.‡

Book 3 immediately attracted the interest of the natural philosophers and astronomers. Reactions ranged from enthusiasm to scepticism. Newton's physical concepts, his methodology, and the experimental and observational data were carefully analysed. The mathematical methods of Book 3 attracted much less attention. It is only at the very end of the period considered in the present work that the mathematical methods of the third Book began to be analysed by competent mathematicians. Until the third decade of the eighteenth century the debate on the

---

† Hayes (1704): 146–50; Ball (1893): 101–3; *Principles*: 657–9. See also *Correspondence*, **3**: 323, 375–7, 380–2.

‡ For a careful analysis of Newton's Propositions on universal gravitation see Densmore (1995): 239–406.

mathematics of the *Principia* concerned some specific and fundamental questions related to the first two books. It is not difficult to understand the reason for this. In the first place Book 3 is the least 'mathematical'. As Cohen puts it, Newton abandons the pure study of 'mathematical constructs', rather he applies some mathematical results to the real world. Thus mathematicians directed their attention to the first two books, where the preliminary mathematical results were achieved. In the second place, the mathematical parts of Book 3 are difficult in the extreme. Eager to achieve a result related to astronomy, geodesy or tides, Newton neglects to complete the proofs, which are generally just sketches given in broad outline. Even Johann Bernoulli, one of the best mathematical minds of the period, had to admit in correspondence with Maupertuis that he found relevant parts of Book 3 (those related to the shape of planets) 'obscure and impenetrable'.† Even though Book 3 was not touched on in the early-eighteenth-century debate on the *Principia*'s mathematical methods, it is worthwhile to offer some succinct observations.

As I stated before, most of the results in Book 3 are applications of 'mathematical constructs' developed in the first books. Most notably the existence of an inverse square force directed towards the Sun receives three separate proofs. The first proof assumes circular planetary orbits and Kepler's third law. In Proposition 4, Book 1, Newton had shown that the centripetal acceleration for uniform circular motion is proportional to the square of velocity and inversely proportional to the radius. It is a matter of simple algebra to deduce, from Kepler's third law, that acceleration is inverse square. But, of course, planetary and cometary orbits are (approximately) conic sections, having the Sun at a focus. Propositions 11–13, Book 1 (§3.4.3), are thus invoked in order to demonstrate that the Sun exerts an inverse square centripetal force. Finally, the results achieved in Section 9, Book 1, are deployed by Newton. The apsides of planets are approximately fixed. In Section 9 Newton had shown that departure from the inverse square law would make the lines of apsides move (§3.6).

In order to push his mathematization of the System of the World ahead Newton could not, however, rely only on results achieved in the first books. He had to develop new mathematics in Book 3, in order to deal with comets (where interpolation techniques that he had discovered in the 1670s were deployed), with the shape of planets, with the motion of the Moon and with tides.

According to Newtonian theory the planets are flattened at the poles.‡ In the third book Newton bases his discussion of planetary shape mostly on Propositions 72 and 91, Book 1. He assumes that the Earth is an ellipsoid of revolution. This assumption is based on the unproven statement according to which the form

† Johann Bernoulli to Maupertuis (1 April 1731) Basel MS LIa 662 no. 8. Quoted in Greenberg (1995): 12.
‡ Greenberg (1987) and (1995) contain a great deal of information on Newton's theory of the shape of the Earth.

of equilibrium of a homogeneous rotating fluid is an ellipsoid. His problem is to determine whether it is prolate or oblong and, subsequently, to estimate the ellipticity. He does so by applying his principle of balancing columns: fluid polar and equatorial columns of infinitesimal breadth must balance each other at the Earth's centre. Here the result achieved in Corollary 2 to Proposition 91, Book 1, turned to be useful (§3.8). Because of the effect of rotation this is possible only if the Earth is flattened at the poles. It has been noted by Greenberg that the choice of ellipsoids was determined by the mathematical knowledge available to Newton. The geometry of the ellipse, and of surfaces generated by rotating an ellipse around one of its axes, was a well-known topic. The properties of the ellipse could be, for instance, used in Corollary 3 to Proposition 91, where it is shown that a body placed within a 'geoid' feels no net force. Furthermore, Newton's approximations, based on Taylor series expansions, were all valid for infinitesimal ellipticity [$\epsilon^2 \ll \epsilon$] (they hold good only to terms of first order).† Newton's theory of the Earth's shape attracted attention and much criticism. However, during the first three decades of the eighteenth century these analyses, carried on by, for example, Dortous de Marain, Desaguliers and Maupertuis, were mostly phenomenological. The first mathematical breakthrough came in the late 1730s with Clairaut. Employing results that he had achieved in the study of orthogonal trajectories to families of plane curves and other topics in the geometry of curves, he was able to develop a new mathematical approach based on multivariate functions ['fonctions de x et de y'] and complete differential [différentielle complette'].‡ Later on, the development of potential theory allowed the theory of planetary shapes to be placed on a new mathematical level. It seems clear that, from the middle of the eighteenth century, the development of the concept of multivariate function and of partial differential calculus allowed Continental mathematicians to deal in new mathematical terms with the problems of planetary shapes left open in the third book of the *Principia*.

Some of the most complex parts of the third book are the Propositions devoted to the theory of the Moon. The tools developed in the first book proved to be insufficient, since they did not allow quantitative accurate predictions of the Moon's motion. Most notably, the approach followed in Section 9 did not allow an accurate prediction of the motion of the Moon's apogee (§3.6). In fact that approach was based on the idea that it is legitimate to ignore the transverse component of the Sun's disturbing force. Newton must have realized that a new perturbation method was needed. He provides a new one in Propositions 25–35 of Book 3. This method too could not be considered as successful. Unpublished manuscripts, elaborated during the composition of the *Principia*, related to a

---

† Greenberg (1995): 5, 10.
‡ For a brief discussion see Guicciardini (1989): 79–81. A detailed study is Greenberg (1995).

third perturbation method are extant.† Michael Nauenberg has recently analysed Newton's perturbation methods for the three-body problem and their application to lunar motion.‡ He has established that Newton made recourse to several analytical techniques. Binomial expansions were used for approximations (a technique that was used also in the Scholium to Proposition 93 and in Proposition 45, Book 1, and in Corollary 3 and in the four Examples of Proposition 10, Book 2). The determination of curvature of the trajectory at given points was related to the normal component of force, following a method sketched in Corollaries 3 and 4 to Proposition 17, Book 1 (§3.4.5).§ Furthermore, some numerical results obtained by Newton render it extremely probable that in his lost worksheets he must have integrated sinusoidal functions.

Curtis Wilson has recently compared Newton's perturbation methods with those elaborated in the 1740s by Continental mathematicians such as Clairaut, Euler and d'Alembert.¶ From his analysis it emerges that Newton's analytical techniques were always intertwined with geometrical procedures. Newton began with some simplifying assumptions concerning the shape of the Moon's orbit and tried to establish the perturbation caused by the Solar force. As Wilson states, this approach led to several problems.

- The partitioning of the perturbing forces into components was problematic. Newton stated that certain components could be ignored since their effect was negligible when compared to other components, but these unproven statements are often wrong.
- The level of approximation was not under control. There was no internal check on the level of accuracy of the various approximations introduced (see §3.6 and §8.6.3.5).

Newton had recourse to analytical techniques only in certain crucial passages of his demonstrations. His demonstrations were based mainly on those geometrical, qualitative, reasonings on the shape of the trajectory that we have briefly discussed in our analysis of Section 11, Book 1 (§3.7). Wilson observes that, from Newton to d'Alembert, the essential theoretical advance in lunar theory consisted in the decision to start from a set of differential equations, while relinquishing the demand for direct geometrical insight into the particularities of the lunar motions. The mid-eighteenth-century Continentals approached the three-body problem from a highly algorithmic and iterative point of view, which is extraneous to Newton's procedures. The algorithmic, iterative approach followed by the

---

† *Mathematical Papers*, **6**: 508–35. On Newton's lunar theory see Whiteside (1975).

‡ Nauenberg in (1994) and in a forthcoming paper entitled 'Newton's perturbation methods for the 3-body problem and its application to lunar motion' devoted to Newton's theory of the Moon has maintained that Newton was able to deal in calculus terms with some quite advanced topics in perturbation theory. I am grateful to Prof. Nauenberg for sharing this information with me. See Nauenberg (1995), (1998a) and (1998b).

§ The curvature method is particularly evident in Proposition 28, Book 3.

¶ I am grateful to Prof. Wilson for having sent me his forthcoming essay 'Newton on the Moon's variation and apsidal motion: the need for a newer "new analysis".' Wilson (1995). On Newton's Moon theory see also Chandler (1975), Waff (1975) and (1976), Aoki (1992).

Continentals allowed them to avoid dangerously intuitive approximations: the level of approximation was regulated by analytical tools. Wilson, furthermore, makes the very interesting point that only after Euler's systematization of integration techniques of sinusoidal functions was it possible to implement the algorithmic approach to the three-body problem. Indeed, as he puts it, in order to deal in calculus terms with lunar motion there was a 'need for a newer "new analysis".'

Nowhere is the lack of a newer new analysis felt more conspicuously than in the theory of tides. In the *Principia* Newton sketched two separate theories. In a prototype of the so-called 'dynamic theory' he considered a canal ('a fluid annulus or ring') encircling the Earth along the equator. The motion of the particles of water composing the canal could be, he suggested in Corollaries 19 and 20 to Proposition 66, Book 1, studied by employing the results deduced for the motion of a satellite orbiting around the Earth and disturbed by a third body (Moon or Sun). Newton, however, did not develop quantitatively the dynamic theory; instead, in Proposition 24, Book 3, he addressed the problem from the point of view of the so-called 'equilibrium theory'. Here one first studies the shape of equilibrium of a fluid nonrotating Earth attracted by the Moon or Sun. Gravitation theory predicts the formation of two bulges, one having its maximum situated at the point closest to the disturbing body, the other situated at the antipodes. The Earth's movement of rotation around the polar axis is then superimposed: the two bulges sweep the surface of the ocean causing two high tides every 24 hours. It should be noted that the tide-generating force acting upon a water particle is defined by Newton as the difference between the force with which the Moon (or the Sun) attracts the fluid particle and that with which it would attract the same particle if it were placed at the centre of the Earth. Newton, furthermore, considers only the component of this force normal to the ocean's surface as responsible for the formation of the two water bulges: his intuitive idea was that the Moon's and Sun's normal-to-surface tidal forces pull the waters up at syzygies, and press them down at quadratures. In this case Corollary 2 to Proposition 91, Book 1, again turned out to be useful (§3.8). In the modern equilibrium theory one instead considers the tangential-to-surface component to be the main cause of tide generation: it is this component which makes the water particles slide until an equilibrium surface is obtained. The equilibrium theory is valid when the Moon's revolution around the Earth and the Earth's rotation around the polar axis are infinitely slow. In the real world one cannot forget that tides are a dynamic phenomenon. Newton's equilibrium theory in fact was in bad disagreement with experimental data. The improvements of the equilibrium theory achieved around 1740 by Colin Maclaurin, Daniel Bernoulli and Leonhard Euler did not bring theory much closer to observation. The first mathematical treatment of the dynamic theory is due to Laplace, who in the 1770s applied partial differential equations to tidal oscillations.†

From this brief survey of Book 3 it emerges that some of the most advanced topics in gravitation theory (planetary shapes, lunar motion and tides) were mathematized by Newton thanks to a mixture of physical intuition, geometric and analytical tools. Newton simplified his models having recourse to often unproven physical simplifications. In order to mathematize these models he could only rely in part on results achieved in the first two books. He had to develop new mathematical tools, which were a typical mixture of geometry and analysis. Calculation of curvature, power series and integration were deployed from time to time, but an algorithmic approach was by no means followed coherently and exclusively. In a rather pragmatic way, Newton used the tools which were at his disposal, whether geometrical or analytical. His only coherent policy was that of not disclosing in print the analytical parts of his demonstrations. Newton's procedures in Book 3 left most of his competent readers perplexed. His physical assumptions were questionable, his mathematical demonstrations obscure. It is only from the late 1730s that mathematicians began to work on the advanced problems in gravitation theory that Newton had attacked in his pioneering, almost visionary, Book 3. This was rendered possible by the creation of a 'newer new analysis' which included: the calculus of trigonometric functions, the concept of multivariate function, the calculus of partial derivatives, the theory of partial differential equations and the calculus of variations. It was only after the creations of mathematicians such as Euler, Clairaut and d'Alembert, that Book 3 ceased to be an object of wonder or frustration and became a motivation for further advances in the higher calculus.

3.16 Conclusion: quadrature avoidance

I hope that the effort of reading this long chapter has not been useless. We have in fact gone through a whole variety of methods and of different argumentative strategies. The plurality of the *principia mathematica* employed by Newton should now have some solid meaning for the reader!

Even though my purpose is to read the *Principia* through the eyes of Newton's contemporaries, I think that there is some justification in attempting at this stage of the book a classification of these methods. Before leaving the word to the seventeenth- and eighteenth-century geometers, before listening to their reactions

---

† Aiton (1955). In the above characterization of Newton's theory of tides I owe much to my student Paolo Palmieri, who is writing a thesis on the theory of tides in Galileo. In his thesis he also has a great deal to say on several drawbacks of the Newtonian theory. See also Proudman (1927).

to Newton's mathematical methods for natural philosophy, let us try to recapitulate for ourselves what we have seen in the *Principia*.

We have encountered a plurality of methods which have their origin in different periods of Newton's intellectual development. We can broadly classify them into three groups: (i) geometrical constructs independent of the method of fluxions, (ii) geometrical limit procedures based on the synthetic method of fluxions, and (iii) symbolic methods based on the analytical method of fluxions.

(i) Some of Newton's demonstrative geometrical techniques are based on physical insights rather than mathematical procedures. We can refer (see §3.8) to the physical modelling at work in Propositions 70 and 73, Book 1, dealing with attraction inside spherical shells and spheres. Or we can refer to the reasoning underlying Corollary 1 (1713 version) to Propositions 11–13, Book 1, on the inverse problem for inverse square forces (see §3.4.4 and §8.6.2.1). Also, the modelling in the study of perturbations in Section 11 (see §3.7) is mostly independent from fluxions. Here we find at work Newton the physicist rather than the mathematician.

(ii) Most of the demonstrations in the *Principia* are based on geometrical limits: the so-called 'method of first and ultimate ratios' that Newton had developed in the late 1670s and later termed the synthetic method of fluxions. The most important application of this method is the geometrical representation of force (Proposition 6, Book 1), a representation which is based on the determination of the limit to which tends the ratio QR/QT^2, when P and Q 'come together' (§3.4.2). Newton was familiar with limits from his fluxional analytical works. But, from the 1670s (typically in *Geometria curvilinea*), he was familiar also with 'geometrical' limits, which he conceived as an alternative to the former. In order to determine first and ultimate ratios Newton asks the reader to refer to geometrical figures and to 'visualize' the limit to which the ratio between vanishing geometrical quantities tends. As a typical example, in handling the direct problem of central forces, Newton considers the ratio QR/QT^2, for a *finite* arc PQ. He then shows, deploying known geometrical properties of the trajectory, that this ratio is equal to a ratio of some finite geometrical elements of the trajectory. He then asks the reader to consider what happens to these finite elements as P tends to Q. The reader does not have to 'calculate' a limit: he is asked to visualize what is the limit to which the geometric elements of the trajectory tend (consider the demonstration of Proposition 10 in §3.4.3). Geometric limits belong to the method of fluxions: they belong, however, to the 'synthetic method of fluxions' not to the 'analytical' one (see Chapter 2).

De Gandt defines Newton's synthetic method as a geometry which is neither classic geometry nor calculus – a mathematical style which he aptly terms a 'geometry of the ultimate'.[†]   Newton's geometrical limit procedures are not classical: they could be conceived only by a man well acquainted with the analytical method of fluxions. A translation of these limit procedures into the analytical method is of course possible,[‡] but often unnatural. This translation often spoils the evidence and informative content of Newton's geometric proofs. It happens very frequently that a solution by geometry becomes very complicated when translated into algebraic terms. And *vice versa*, what is solved in a few lines by algebra, becomes impracticable when translated into geometric terms. The issue of the translatability of Newton's geometric limit procedures into analytical form, and of the advantages and disadvantages of such a translation, was central to the debate to which my book is devoted.

(iii*a*) The analytical method of fluxions, which Newton had developed in his early years, enters into other parts of the *Principia*. Infinite series occur in several places: the binomial expansions of $(T - X)^n$ in Proposition 45, Book 1 (§3.6), and of $(A + O)^{m/n}$ in the Scholium to Proposition 93, Book 1 (§3.9); the Taylor series $Qo + Ro^2 + So^3 + \cdots$ in Proposition 10, Book 2 (§8.6.3). Newton shows an understanding of (and makes use of) the kinematical meaning of the successive terms of these power series. (iii*b*) Another 'analytical' technique is reduction to quadratures. Most notably in Propositions 39–42, 53–56, 79–83 and 90–91 in Book 1 and 29–31 in Book 2, Newton reduces the dynamical problems to the calculation of an area subtended by a curve. These reductions to quadratures are performed deploying a geometry of infinitesimally small quantities (e.g. one considers the infinitesimal arc of the trajectory, the infinitesimal increment of velocity acquired in the moment of time necessary to traverse such an arc, etc.). Translation into calculus notation is in these cases quite straightforward. In some cases (e.g. in Propositions 79–83, 90, and Corollary 1 to Proposition 91) Newton performs explicitly the quadratures employing symbolic methods (see §3.8). He does so when the quadratures require standard simple techniques (e.g. techniques known to a reader of Wallis's *Arithmetica infinitorum*). In other, more advanced, cases he simply says that the problem is solved 'granting the quadratures of curvilinear figures'. He does so, for instance, in Corollary 3 to Proposition 41, Book 1, on the inverse problem for inverse cube forces (see §8.6.2.1). He does so in Corollary 2 to Proposition 91 on the attraction exerted by an homogeneous ellipsoid on the prolongation of the axis of revolution (see §3.8). These 'difficult' quadratures can be performed using the 'catalogues' of curves that Newton had developed (notably in the *De methodis*), but not published yet. In the Scholium

---

[†] De Gandt (1986).

[‡] For example, the ratio QR/QT^2 has been given analytical fluxional form by Routh and Brougham, Whiteside and Brackenridge. See Brougham & Routh (1855): 43–7; *Mathematical Papers*, 6: 42; Whiteside (1991a): 736–7; Brackenridge (1988): 454–5; Brackenridge (1995): 213–15.

to Proposition 35/34 Newton states the geometric properties of the solid of least resistance without offering any proof to the reader of the *Principia*. We know, however, that he had performed in 1685 the necessary calculation leading to a fluxional equation.

One of the most striking aspects of the mathematical style of the *Principia* is what we may call 'quadrature avoidance'. In several places in Newton's arguments there are gaps that can be filled only by a method for 'squaring curves'. Why did Newton avoid making this method explicit? We will answer this question in the next chapter.† Here we can briefly point out four motivations for Newton's quadrature avoidance:

- the competence of his readers could not include the inverse algorithm of the method of fluxions (§2.2.7), which was still unpublished;
- the Grecian façade of the *Principia* was in harmony with Newton's methodological turn of the 1670s (§2.3.1);
- it was common in the early seventeenth century to give priority and publicity to the geometrical construction which solves the problem, rather than to the analysis necessary to achieve such a construction (an analysis which was often kept hidden by mathematicians);‡
- the analytical method did not provide a unifying tool for Newton's natural philosophy. Newton made recourse to quadratures and series only in a few isolated cases, therefore it was not compulsory to give a prominent role to it.

The *Principia* is thus a repertoire of mathematical techniques which have their roots in different periods of Newton's development as a mathematician. When confronted with this rich mathematical arsenal, the readers of the *Principia* had differing reactions, from enthusiasm to scepticism. The rest of this book is devoted to them. The first reader that I will consider is Newton himself.

† On this topic see also Erlichson (1996).
‡ Henk Bos pointed out this aspect of early-seventeenth-century mathematics to me.

Part two

Three readers

4

Newton: between tradition and innovation

4.1 Purpose of this chapter

The second and third editions of the *Principia* appeared in 1713 and 1726 respectively. The emendation and variations between these editions are remarkable and have been studied in detail by scholars such as Hall and Cohen.† However, in broad outline, the structure of the first edition remained unaltered. The number and order of the propositions, as well as the methods of proof, remained almost unchanged.

Thanks to the recent edition of Newton's *Mathematical Papers*, we now know that more radical restructurings were considered from the early 1690s up to the late 1710s. Despite the fact that nothing of these projects appeared in print during Newton's lifetime, it is interesting to consider them since they reveal Newton's evaluations of his own mathematical methods for natural philosophy. For instance we know of projects of gathering all the mathematical Lemmas in a separate introductory section.‡ From David Gregory's retrospective memorandum of a visit he paid to Newton in May 1694 we learn about projects of expanding the geometrical Sections 4 and 5, Book 1, into a separate appendix on the 'Geometria Veterum', and of adding a treatise on the quadrature of curves as a second mathematical appendix in order to show the method whereby 'curves can be

---

† Hall (1958), Cohen (1971). All the variants can be found in Newton *Variorum*. As we will see in Part 3, Newton revised some propositions (most notably Corollary 1 to Proposition 13, Book 1, and Proposition 10, Book 2) and added an alternative geometrical representation of central force (see §4.5).

‡ *Mathematical Papers*, **6**: 600–9. Cohen (1971): 171–2.

squared'.† In 1712 Newton was still thinking of adding a treatise on series and quadratures as an appendix to the *Principia*.‡ From a draft letter of Cotes to Newton written in 1712 we read: 'I am glad to understand by Dr. Bentley that You have some thoughts of adding to this Book [the *Principia*] a small Treatise of Infinite Series & the Method of Fluxions'.§ In general, after the publication of the *Principia*, Newton showed a concern for two problems. The first was to relate his mathematical methods for natural philosophy to the tradition of ancient geometry. The second was to relate them to the new analytical method of fluxions. Both problems became urgent and indeed an obsession for Newton after the inception of the priority quarrel with Leibniz.

In fact, during the priority dispute, Newton found himself in a double trap. From one point of view, he wished to use the *Principia* as proof of his knowledge of calculus prior to the publication of Leibniz's *Nova methodus* (1684). This led him to state that all the propositions of the *Principia* had been found by means of the 'new analysis', even though they had been published in a different, 'synthetic', form. It was easy, he claimed, to revert the synthetic demonstrations into analytical form. On the other hand, Newton did not want to define the synthetic methods of the *Principia* as equivalent to the analytical method of fluxions. In his opinion, the choice between the analytical method and the synthetic method of fluxions was not merely a question of language or of presentation. Newton was convinced that his geometrical mathematical way was superior to Leibniz's reliance on algorithm.

In this chapter I will consider Newton's evaluations and justifications of his own mathematical methods for natural philosophy. The historical evidence comes mainly from manuscripts and letters written in the 1690s and 1710s. We have to distinguish Newton as the writer from Newton as the reader of the *Principia*. When he wrote it his concern was to achieve a scientific result and communicate it to his peers. In the moment of composition he used the methods that appeared to him more natural and efficient. During the 1690s and 1710s Newton's approach to the *Principia* was that of an author who is called to defend his work from critics, who wishes to propose it as a model to his disciples, and who attempts to attribute to it a place in the historical development of mathematics. As we will see, this process of justification led Newton to re-read his work with the purpose of achieving a coherent mathematical methodology. Newton was thus compelled to

---

† *Mathematical Papers*, **6**: 601; Cohen (1971): 193–4 and 345–9. In May 1694 Gregory copied part of a manuscript on quadratures related to the *De quadratura* (1704): *Mathematical Papers*, **7**: 508ff. Gregory noted that this tract 'post *Principia* editurus est'. *Correspondence*, **3**: 334–6. Whiteside conjectures that in 1694 Newton was thinking of adding as an appendix to the *Principia* a *Geometria* in two parts: the former being a reworking of the *Geometria curvilinea*, the latter the *De quadratura*. See *Mathematical Papers*, **7**: 402ff.

‡ A copy of the *De quadratura* recast by Newton for publication as an appendix to the *Principia* is in *Mathematical Papers*, **8**: 258ff.

§ Edleston (1850): 119. Cohen (1971): 238–9.

face the intricacies, the errors, the complexities and the inconsistencies of his own *magnum opus*.

4.2 *Prisca geometria*

As we already know (see §2.3.1), Newton in the 1670s distanced himself from the 'new analysis' and began looking in Pappus's *Collections* for the lost geometrical 'analysis of the Ancients'. After the publication of the *Principia* his mathematical classicism was reinforced. The result of Newton's research on lost books of the Greek tradition, such as the book on *Porisms*, was a long treatise on geometry which has been published only recently.†

It is to be noted that during the 1690s Newton began conceiving the idea of publishing some scholia, in which it was stated that his natural philosophy was a 'rediscovery' of an ancient wisdom, in a new edition of the *Principia*. For instance, as regards atomism he wrote:

That all matter consist of atoms was a very ancient opinion. This was the teaching of the multitude of philosophers who preceded Aristotle, namely Epicurus, Democritus, Ecphantos, Empedocles, Zenocrates, Heraclides, Asclepiades, Diodorus, Metrodorus of Chios, Pythagoras, and previous to these Moschus the Phoenician whom Strabo declares older than the Trojan war. For I think that same opinion obtained in that mystic philosophy which flowed down to the Greeks from Egypt and Phoenicia, since atoms are sometimes found to be designated by the mystics as monads.‡

Although the so-called 'classical scholia' were not published in the second edition of the *Principia*, they appeared in the introduction to David Gregory's treatise on astronomy (1702). That the myth of the rediscovery of an ancient wisdom was deeply rooted in Newton's mind appears from the very opening lines of the *De mundi systemate*, written in 1686. Here Newton states that the Ancients held that the Earth moves as a planet around the Sun. According to Newton, the Copernican theory was taught by 'Philolaus, Aristarchus, Plato in his riper years [...] the whole sect of the Pythagoreans [...] Anaximander [...] and Numa Pompilius'. It was only after Eudoxus, Callippus and Aristotle that 'the ancient philosophy began to decline, and to give place to the new prevailing fictions of the Greeks'.§ A page below, Newton cautiously states that it is unclear how the ancient philosophers explained the motions of the planets in void space. However, in an intended preface to the *Principia* written in the late 1710s Newton attributed to the 'Chaldeans', to the 'Ancients', to the 'Pythagoreans' and to the 'Greeks and Romans' a knowledge

---

† *Mathematical Papers*, 7: 185–561.
‡ Add. 3965.6: 270^r (University Library, Cambridge). Quoted in McGuire and Rattansi (1966): 115.
§ *Principles*: 549–50.

of universal gravitation.† Newton considered himself a rediscoverer of an ancient knowledge and came to attribute to the Ancients the doctrines of atoms, of the void, of the planetary nature of the Earth and of universal gravitation.

Newton not only believed that his *philosophia naturalis* was a rediscovery of ancient philosophy, he also stated that his *principia mathematica* were a rediscovery of ancient geometrical methods. The role attributed by Newton to Pythagoras deserves to be underlined here. According to Newton, Pythagoras had derived from Egypt and Phoenicia, via Moschus, knowledge about the pristine religion and natural philosophy of Noah. The Pythagoreans not only maintained atomism and heliocentrism, but also held a correct view about the supremacy of a unique God, and had expressed in musical-harmonic terms the cosmological truths about gravitation and the planetary system. Of course the Pythagorean school was also distinguished by a supreme mathematical knowledge. The myth of a *prisca sapientia*, and in particular the role of Pythagoras in transmitting it to the Greeks, was shared for a long time by a variety of thinkers, most notably by the Florentine and Cambridge Platonists.‡ We can cite, among many possible examples of the pervasiveness of this myth, Joseph Glanville who, in 1665, maintained that the four Sages who brought the learning of the Egyptians among the Greeks were 'Orpheus bringing in Theology, Thales the Mathematicks, our Democritus, natural Philosophy, and Pythagoras all Three'. Newton was repeating, for his own purposes, what was stated again and again in the theological, chronological and alchemical books which composed his library.§

Evidence that a relationship between Newton's philosophical and mathematical classicisms can be drawn is to be found in the intended preface to the *Principia* to which I have just referred. This preface, devoted to the *prisca sapientia*, begins with reference to the 'ancient geometers', thus linking the two classicisms in a continuous way:

The ancient geometers investigated things sought through analysis, demonstrated them when found out through synthesis, and published them when demonstrated so that they might be received into geometry. Once analysed they were not straightaway received into geometry: there was need of their solution through composition of their demonstrations. For the force of geometry and its every merit lay in the utter certainty of its matters, and

---

† 'Planetas in orbibus fere concentricis & Cometas in orbibus valde excentricis circum Solem revolvi, Chaldaei [Veteres] olim crediderunt. Et hanc philosophiam Pythagoraei in Graeciam invexerunt. Sed et Lunam gravem esse in Terram & stellas graves esse in se mutuo, et corpora omnia in vacuo aequali cum velocitate in Terram cadere, adeoque gravia esse pro quantitate materiae in singulis notum fuit Veteribus [*sc.* Graecis et Romanis]. Defectu demonstrationum haec philosophia intermissa fuit [ad nos propagata non fuit & opinioni vulgari de orbibus solidis cessit]. *Eandem non inveni sed vi demonstrationum in lucem tantum revocare conatus sum.*' Add. 3968.9: 109^r/109^v (University Library, Cambridge) and variants from a manuscript in private possession inserted in square brackets. Quoted from *Mathematical Papers*, **8**: 459. My emphasis.

‡ A useful outline of the Pythagorean myth from Copernicus to Newton is provided by Casini (1996). See also Yates (1977).

§ Glanville's citation is taken from Gaukroger (1991): xi. On the *prisca* see: McGuire & Rattansi (1966), Casini (1981), Gouk (1988), Gascoigne (1991), Trompf (1991).

that certainty in its splendidly composed demonstrations. In this science regard must be paid not only to the conciseness of writing but also to the certainty of things. And on that account I in the following treatise synthetically demonstrated the propositions found out through analysis.[†]

Here Newton justifies the geometric style of the *Principia* stating that he wants to adhere to the Ancients' way of presenting their theorems by synthesis.

Newton next comes to a question of possible discontinuity between his geometry and classic geometry, a discontinuity that from the point of view of a twentieth-century historian is evident. Newton's geometry in the *Principia* appears to us extremely innovative since it is applied to motion and force, velocity and acceleration. We view it in relation to to seventeenth-century kinematic methods, such as those of Roberval and Barrow. Rather than stressing this element of modernity, Newton reinforces the argument of continuity:

> The geometry of the ancients had, of course, primarily to do with magnitudes, but propositions on magnitudes were from time to time demonstrated by means of local motion: as, for instance, when the equality of triangles in Proposition 4 of Book 1 of Euclid's *Elements* were demonstrated by transporting either one of the triangles into the other's place. Also the genesis of magnitudes through continuous motion was received in geometry: when for instance, a straight line were drawn into a straight line to generate an area, and an area were drawn into a straight line to generate a solid. If the straight line which is drawn into another be of a given length, there will be generated a parallelogram area. If its length be continuously changed according to some fixed law, a curvilinear area will be generated. [...] If times, forces, motions and speeds of motion be expressed by means of lines, areas, solids or angles, then these quantities too can be treated in geometry. Quantities increasing by continuous flow we call fluents, the speeds of flowing we call fluxions and the momentary increments we call moments.[‡]

Newton's statement of continuity between his synthetic geometric methods and ancient geometry, as well as between his natural philosophy and the ancient wisdom, played an important rhetorical role in the years following the publication of the *Principia*. As Casini has shown this rhetoric was not an exception during Newton's lifetime.[§] We should not regard it as something unexpected in a late-seventeenth-century natural philosopher. Whiteside has furthermore stressed that Newton's interest in the rediscovery of the lost analysis of the Ancients was not an antiquarian affair. In fact, Newton was researching, stimulated by the reading of Pappus, into what nowadays we would call projective geometry. Part of his research in this field carried on in the 1680s appeared in Sections 4 and 5 of Book 1. We know from one of Gregory's memoranda of his 1694 visit to Cambridge that Newton was thinking of reformulating these sections as an independent treatise

[†] *Mathematical Papers*, **8**: 453–5. Translation by Whiteside.
[‡] *Mathematical Papers*, **8**: 455. Translation by Whiteside.
[§] Casini (1981).

on ancient geometrical analysis.† Gregory further noted that this treatise on the geometry of the Ancients was to be published in an appendix to the *Principia* together with another treatise which would contain 'his Method of Quadratures [i.e.] the methods of infinite series, of tangents, of maxima and minima, of curvature and of the rectification of curves'.‡ The tension between the new and the old methods, inherent in the *Principia*, would have thus found an explicit form.

Devoting attention to Newton's 'classicism' can induce us to speculate on the implications that it might have had for the strategies of presentation and publication of the *Principia*. Indeed, in contrast with Leibniz, who promoted the acceptance and use of his 'new method', Newton kept a rather secretive attitude. As is well known, he was almost forced to give publicity to fluxional analysis. He employed it in natural philosophy in correspondence with his closest friends, or in manuscripts that he showed to deferential and devout pilgrims who paid visits to him (e.g. Craig, Halley, Fatio de Duillier, De Moivre, Cotes and Gregory). Having access to the fluxional analysis of central forces was a privilege accorded to few. Newton gave publicity to a book which had the appearance of a classic and behaved as a new Pythagoras. He surrounded himself with a circle of initiates. William Stukeley's *Memoirs* allow us a glimpse into this aspect of the attitude of Newton's disciples towards their master:

We always took care on Sunday to place ourselves before him, as he sat with the heads of the Colleges; we gaz'd on him, never enough satisfy'd, as on somewhat divine [...] drawn forth into light before, as to his person, from his beloved privacy in the walls of a college.§

Was the *Principia* the work of a man who made a point of being inaccessible?

4.3 Newton and geometry in the *Principia*

In the seventeenth century the idea that the language for natural philosophy had to be geometrical was deeply rooted. Since Galileo's times the 'Book of Nature' had been thought to be written in 'circles and triangles, and other geometrical figures': it was not written in algebraic symbols. In writing about motion, velocity and trajectories in terms of geometry, Newton, rather than trying to use an esoteric language, was just inscribing himself in a school of natural philosophy that

---

† According to Gregory's recollections: 'In Editione nova Philosophiae [...] Sectiones IV et V [Lib. I] eximuntur et tractatus separati [de Veterum Geometria] Partes fiunt. [...] In hoc tractatu genuinum Veterum institutum explicatur, [...] Euclid[is] Porismatum liber, Apollonij libri deperditi reliquique Veterum explicabuntur ex illis quae a Pappo alijsque de ijs dicuntur quaeque a nemine hactenus intellecta [et] pars erit de inventione orbium ex dato umbilico vel etiam neutro umbilico dato, quae Sect: IV et V constituunt'. Gregory Codex C42 (University Library, Edinburgh). See *Correspondence*, **3**: 384–5 and 335, and *Mathematical Papers*, **6**: 601.

‡ *Correspondence*, **3**: 386.

§ William Stukeley, *Memoirs of Sir Isaac Newton's Life*, ed. A. Hastings White, 1936: 10. Quoted in Iliffe (1995): 159.

reckoned Galileo and Huygens as its principal exponents. If we turn our attention to astronomers, an important (perhaps the most important) category amongst the readers of the *Principia*, we find again that the language of geometry dominated their works. The works of Streete and Wing, which Newton might have had as sources, have the geometrical representation of trajectories as their object of study.

In the seventeenth century the language of geometry proved to be extremely useful in the study of kinematics. In fact, geometry allowed the modelling of the basic kinematic magnitudes. Displacements and velocities could be represented by geometrical continuous quantities, and thus kinematics could be studied in terms of the theory of proportions. But Newton's research programme was wider: he had to mathematize force, not only kinematical magnitudes. One of his aims in Section 2 of Book 1 was to find a geometrical representation of force. He employed the proportionality of force to displacement from inertial motion acquired in an infinitesimal interval of time (§3.4.2).

Around 1705, in an attempt to classify the objects, methods and principles of arithmetic, geometry and mechanics, Newton wrote:

Arithmetic has to do with numbers and numerable things; geometry with measures and all things long, wide and deep which come needing to be measured; mechanics with forces and motions from place to place. And arithmetical solutions of questions are ones which are achieved through the operations of arithmetic alone, viz.: adding, subtracting, multiplying, dividing and extracting roots; geometrical ones are those accomplished by the mechanical operations of drawing lines and constructing figures accurately by dint of postulates; *and mechanical are those effected through other mechanical operations of applying forces and directing bodies in motions following assigned lines.* Of course, geometry is a particular species of mechanics, and is rendered accurate by force of postulates. Geometrical and mechanical quantities, however, and all others are, insofar as they can be numbered or expressed by numbers, treated in arithmetic; and likewise those not geometrical are yet, insofar as they can be expressed by geometrical quantities, treated also in geometry.†

The possibility of establishing a proportionality between force and displacement allows Newton to geometrize force. The language of geometry is thus what permits the modelling of the world of forces and accelerations, once forces and accelerations are expressed in terms of displacements. In fact Newton wrote:

In mechanics it is lawful to postulate: To cut a given body by a given knife-edge carried straight through; To cut a given body by a given knife-edge rotating round a given axis; and *To move a given body by a given force in a given direction.*‡

The possibility of modelling force in terms of geometrical displacements allowed Newton to obtain a representation which proved to be advantageous. The traditional tools of classic geometry and the kinematic results of Galileo could be applied to the physics of centripetal forces. For instance, the parallelogram rule and

---

† *Mathematical Papers*, **8**: 173. Translation by Whiteside. My emphasis.
‡ *Mathematical Papers*, **8**: 177. Translation by Whiteside. My emphasis.

Euclidean theorems on similar triangles could be employed in the demonstration of the law of areas (Propositions 1 and 2 of Book 1 in §3.4.1). Furthermore, the trajectory of a body accelerated by a central force could be locally represented as a small Galilean fall (Proposition 6, Book 1 in §3.4.2). Therefore, Galileo's results on free fall could be locally applied to the trajectories considered in Sections 2 and 3 of Book 1 (§3.4.3). In this way a completely new theory about forces and accelerations was reduced within the framework of established mathematical theories, such as the geometrical theory of proportions, the geometry of Euclid's *Elements* and Apollonius's *Conics*, Galileo's kinematic of falling bodies and Huygens's kinematic of pendula: a framework which was familiar to late-seventeenth-century natural philosophers. As Newton wrote in a manuscript which dates from the late 1710s:

The forces and speeds of movable bodies do not properly pertain to geometry, but they can be expressed by means of lines, surface-areas, solids and angles, and to that extent reduced to geometry.†

As I have stressed in §2.3.2, Newton valued his method of fluxions highly as much for its ontological content as for its continuity with ancient tradition. Similarly, recourse to geometry in natural philosophy was motivated not only by classicism, but also by the desire to represent actually existing motion and force.

4.4 Newton and his readers

Despite the fact that reference to the 'Ancients' played an important role in Newton's justification of the geometrical structure of the mathematical methods for natural philosophy, the *Principia* was written to be read by his contemporaries. However, between the first (1687) and the second (1713) editions the competence of the possible readers of Newton's *magnum opus* changed dramatically. Whereas in 1687 the method of series and fluxions and the differential and integral calculus were almost unpublished discoveries, in the first decade of the eighteenth century the group of 'geometers' who mastered fluxions and differentials began to be numerous. In 1713 there were a number of 'schools' (e.g. in Basel, Paris, London) and several treatises where the enthusiast could find instructions about the new analysis.

In 1687 only Leibniz's *Nova methodus* (1684) and *De geometria recondita* (1686) were in print, while Newton had communicated to a happy few something of his discoveries.‡ The difficulties that the brothers Jacob and Johann Bernoulli had in understanding Leibniz's *Nova methodus* tell of the conceptual leap necessary in

---

† *Mathematical Papers*, **8**: 453. Translation by Whiteside.
‡ *LMS*, **5**: 220–6 and 226–33.

order to enter into the new mathematical world disclosed by the new calculus.†
Newton's *Principia* did not require the effort of acquiring knowledge in new
mathematical methods: it was written in a language comprehensible for his
readers. Classical geometry, a knowledge of Euclid's *Elements* and some results
by Archimedes‡ as well as some elementary properties of conic sections would
suffice. Furthermore, essential to Newton's mathematical methods for natural
philosophy was a use of geometrical infinitesimals (justified in terms of first and
ultimate ratios) that readers acquainted with seventeenth-century mathematical
literature had learnt to digest. A mixture of elementary classical geometry
and geometrical infinitesimals was fashionable in 1687. Astronomers, natural
philosophers and geometers were well schooled in this mathematical language.
Of course, the *Principia* had its own mysteries, where, in a few isolated cases,
reference to a method whereby 'curvilinear figures may be squared' occurred, or
where Taylor series were employed.

Newton's awareness that the competence of his readers had changed is evident
in the following lines, which were written in the late 1710s:

> To the mathematicians of the present century, however, versed almost wholly in algebra
> as they are, this [i.e. the *Principia*'s] synthetic style of writing is less pleasing, whether
> because it may seem too prolix and too akin to the method of the ancients, or because it is
> less revealing of the manner of discovery. And certainly I could have written analytically
> what I had found out analytically with less effort than it took me to compose it. I
> was writing for Philosophers steeped in the elements of geometry, and putting down
> geometrically demonstrated bases for physical science. And the geometrical findings
> which did not regard astronomy and physics I either completely passed by or merely
> touched lightly upon.§

While the 'philosophers' in 1687 were steeped in geometry, the generation of
mathematicians, formed at the Bernoulli's school in Paris and Basel, who began
their studies in higher mathematics reading l'Hospital's *Analyse des infiniment
petits* (1696) found the *Principia* obscure. Their first question concerned the
translatability of Newton's mathematical methods into the familiar new language
of calculus. Newton repeated many times that 'any man who understands Analysis'
could 'reduce the Demonstrations of the Propositions from their composition back
into Analysis'.¶ This process would have easily yielded, he claimed, the fluxional
analysis thanks to which the propositions had been discovered. However, the fact
that this translation was not a trivial exercise is revealed by a passage by Newton

---

† See the beautiful and colourful recollections of Johann Bernoulli's first encounter with the differential calculus
 quoted in Dupont & Roero (1991): 139–41.
‡ A 'bare reference (in Book 1, Proposition 79) to the main result of Archimedes's *Sphere and cylinder*'.
 Whiteside (1970): 117.
§ *Mathematical Papers*, **8**: 451. I have slightly altered Whiteside's translation from the Latin.
¶ *Mathematical Papers*, **8**: 259 and Cohen (1971): 294.

himself contained in a draft letter to Pierre des Maizeaux, who was collecting information on the priority dispute:

I wrote the Book of Principles in the years 1684, 1685, 1686, & in writing it made much use of the method of fluxions direct & inverse, but did not set down the calculations in the Book it self because the Book was written by the Method of Composition as all Geometry ought to be. *And ever since I wrote that Book I have been forgetting the Methods by which I wrote it.*†

The question naturally arises: was Newton able to apply the analytical method of fluxions to his natural philosophy of force, mass and acceleration? In Chapter 3 we have already found an answer to this question. In some places of the *Principia* there are gaps which can be filled only by employing the methods of quadratures that Newton had developed in his early *anni mirabiles*. That is, Newton states results (e.g. trajectories in an inverse cube force field, attraction of spheroids, solid of least resistance) which can be achieved only by rather sophisticated integration techniques. In the manuscripts and letters written after 1687 we find further evidence of Newton's competence in dealing in calculus terms with some Propositions of the *Principia*.‡

4.5 Newton's analytical approach to central force

In 1960 Clifford Truesdell wrote:

Except for certain simple if important special problems, Newton gives no evidence of being able to set up differential equations of motion for mechanical systems.§

Seven years later Pierre Costabel reinforced this judgment:

One would be wrong in thinking that Newton knew [...] how to formulate and resolve problems through the integration of differential equations.¶

After the publication of Volumes 6, 7 and 8 of Newton's *Mathematical Papers*, we know that both Truesdell and Costabel were underestimating Newton's ability to tackle dynamical problems in terms of the fluxional calculus.

In the 1691 version of the *De quadratura* Newton fully employed an intuition that he had from his youth.‖ In December 1664 Newton noted that the normal

---

† *Mathematical Papers*, **8**: 523 and Cohen (1971): 296. My emphasis.

‡ See Chapter 7 and Guicciardini (1998a).

§ Truesdell (1960): 9.

¶ Here Costabel is referring to Corollary 1 to Proposition 13, Book 1. He further states that Varignon and Johann Bernoulli were able to provide the demonstration in 1710–11 of the inverse problem of central forces 'envisaged in the Newtonian corollary'. Costabel (1967): 125–6. Similar conclusions were drawn by Fleckenstein (1946). See §8.6.2 for a refutation of Costabel's and Fleckenstein's thesis on Corollary 1.

‖ Whiteside (1991b): 14. On the role of curvature in Newton's representation of central force see Brackenridge (1992) and Nauenberg (1994). According to Nauenberg the curvature representation was adopted by Newton well before the writing of the *Principia*. As he notes Newton employed this representation in the 1687 edition of the *Principia* where he applied it to a sophisticated perturbative solution of the lunar motion in Proposition 28, Book 3.

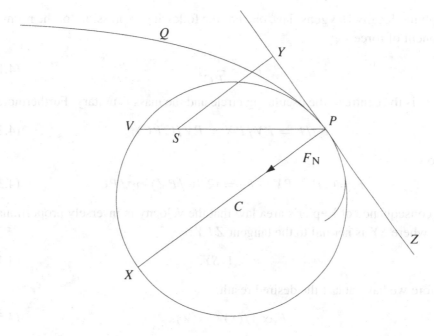

Fig. 4.1. Huygens' law applied to central forces. After Guicciardini (1995): 548

component of a 'centrifugal' force is directly as the square of speed and inversely as the radius of curvature:

Where bodies revolve in curves, to find their centrifugal forces. Solution. They are directly as the squares of speed and reciprocally as the radii of curvature.†

This observation led Newton to formulate an alternative geometrical representation of central force. It was published in the second edition (1713) of the *Principia* as a Corollary to Proposition 6.

In this representation Newton makes use of a different local approximation to the trajectory from that employed in the first edition. He uses the *osculating circle*: i.e. the circle which touches the orbit at P and has a radius equal to the radius of curvature of the orbit at P (see fig. 4.1). The 'vanishing' arc PQ can (in the limit, when Q tends to P) be considered equal to the arc of the osculating circle. The 'body', accelerated by the centripetal force F (S is as usual the force centre) can be thought of as moving during an infinitesimal interval of time with uniform circular motion.

† *Mathematical Papers*, **8**: 100.

Applying locally Huygens' law, one has the following expression for the normal component of force F_N:

$$F_N = \frac{v^2}{PC}, \tag{4.1}$$

where C is the centre of the osculating circle and the mass is unitary. Furthermore,

$$F_N/F = PV/PX = PV/(2PC), \tag{4.2}$$

therefore

$$F = (2PC/PV) \cdot F_N = (2PC/PV) \cdot v^2/PC. \tag{4.3}$$

It is a consequence of Kepler's area law that the velocity is inversely proportional to SY, where SY is normal to the tangent ZPY:†

$$v \propto 1/SY. \tag{4.4}$$

Therefore we have at last the desired result:

$$F \propto 1/(SY^2 \cdot PV). \tag{4.5}$$

This alternative formulation for the geometrical representation of force lends itself to easy translation into the symbolism of the analytical fluxional method, since it implies the calculation of the radius of curvature.

The fluxional calculation of the radius of curvature was of course well within Newton's reach. Around 1670 he had developed several formulas, e.g. in Cartesian and polar coordinates, for the determination of the radius of curvature.‡ In early winter of 1691, in a version of *De quadratura* which was not published until the recent edition of the *Mathematical Papers*, he showed how to apply the method of fluxions to some problems 'which shall have some connection with the physical world'.§ After an elaborate fluxional treatment of curvature, he presented a variant of formula (4.5) and applied it to the solution of the direct problem of central forces. He also gave some hints on how the inverse problem could be tackled. The dynamical problems of centripetal forces were thus reduced to the analytical problem of determining the radius of curvature of the orbit.

Another variant of (4.5) occurs in a manuscript which dates from the mid-1690s. Here Newton states that

$$F \propto SP/(SY^3 \cdot PC), \tag{4.6}$$

where PC is the radius of curvature at P. This follows easily from (4.5) and from the similarity of triangles SYP and PVX. This manuscript is of great historical

† Nowadays we would understand (4.4) in terms of conservation of angular momentum h: $v = h/(m \cdot SY)$.
‡ See §2.2.6.1, formula (2.10).
§ *Mathematical Papers*, 7: 122–3.

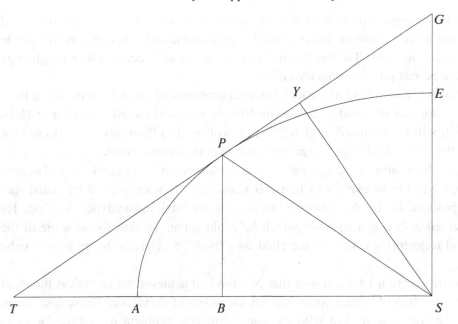

Fig. 4.2. Newton's diagram for formula (4.9). I have added the line SY normal to TG and altered the lettering slightly. After *Mathematical Papers*, **6**: 598

significance because it is one of the few extant instances in which Newton applies the analytical method of fluxions to central force motion.

Let S be the centre of force, APE the trajectory (see fig. 4.2). Newton writes:

> If it is to be investigated what the law of centripetal force is which will move a body along an orbit APE and around a centre S: let $AB = x$ be the abscissa and $BP = y$ the ordinate taken at right angles. Take BT in the same ratio to BP which $\dot{x}$ has to $\dot{y}$ and TP will be tangent to the curve in P. Draw SG, parallel to the ordinate BP, and let it meet the tangent in G, and the force which will move the body along the orbit APE and around the centre S will be as $(\ddot{y}SP)/(\dot{x}\dot{x}SG^3)$.†

In fact, since triangles SYG and TBP are similar,

$$GY/SY = BP/BT = \dot{y}/\dot{x}. \tag{4.7}$$

Newton knows that the radius of curvature is given by (see §2.2.6.1, formula (2.10))

$$PC = \frac{\left(1 + (\dot{y}/\dot{x})^2\right)^{3/2}}{\ddot{y}/\dot{x}^2} = \frac{\left(1 + (GY/SY)^2\right)^{3/2}}{\ddot{y}/\dot{x}^2} = \frac{SG^3/SY^3}{\ddot{y}/\dot{x}^2}. \tag{4.8}$$

Hence, substituting in formula (4.6) this value for PC, he obtains

$$F \propto \frac{SP\ddot{y}}{SG^3\dot{x}^2}. \tag{4.9}$$

† *Mathematical Papers*, **6**: 598. The translation from Latin is mine.

The manuscript ends with a few applications to instances of the direct problem of central forces. Newton makes several computational mistakes, apparently due to haste or oversight. The manuscript is rather messy and seems to be a rough copy where he just jots down his formulas.

Another example of analytical fluxional treatment of central forces dating from the 1690s can be found in a letter by Newton to David Gregory, written in 1694, which will be discussed in §8.6.2.2. In Corollary 3 to Proposition 41, Book 1, of the *Principia* Newton gives a partial solution of the inverse problem of inverse cube forces 'by means of the quadrature of a certain curve'.† In 1694 David Gregory asked Newton to explain to him this Corollary. Newton replied by translating Proposition 41, Book 1, into the language of the fluxional analytical method. He obtained a fluxional equation which he could solve: he thus found some of the spiral trajectories which are travelled by a 'body' accelerated by an inverse cube force.

In this section I have shown that Newton had achieved an analytical fluxional representation of central force through the radius of curvature. I have also stated that he was able to deal with the general inverse problem of central forces in analytical fluxional terms (this matter will be discussed in Chapter 8). This clearly refutes Costabel's pessimistic judgment on Newton's ability to formulate his mathematical natural philosophy in terms of calculus.‡ However, Newton's own evaluation of the role played by the method of fluxions in the *Principia* is still to be discussed.

4.6 Newton and analysis and synthesis in the *Principia*

In the late 1710s Newton often repeated in published and unpublished writings that he had employed the method of fluxions in discovering the propositions of the *Principia*. He did so during the priority dispute with Leibniz. His intention was clearly that of establishing his knowledge of fluxions prior to Leibniz's *Nova methodus* (1684). Actually, explicit reference in the *Principia* to the 'method of fluxions' was inserted in the second edition (1713) in Lemma 2 to Proposition 38, Book 3.§ Newton's reconstructions of the process of discovery of his mathematical natural philosophy are thus far from being objective; they do not lead us to historical truth. It is nonetheless interesting to read them since they allow us to appreciate Newton's evaluations of the role played by fluxions in the *Principia*.

---

† See §3.5. *Variorum*: 223. Cf. *Principles*: 133.

‡ Nauenberg's recent researches on Newton's treatment of the three-body problem should be mentioned here. Nauenberg in (1995), (1998a) and (1998b) has shown that Newton employed several analytical techniques in dealing with advanced topics of perturbation theory. Brackenridge has studied in depth Newton's curvature measure of force.

§ *Variorum*: 680.

In the following lines, written in 1719, Newton states:

About three years after the writing of these Letters [1679?] by the help of this Method of Quadratures I found the Demonstration of Kepler's Propositions that the Planets revolve in Ellipses describing with a Radius drawn to the sun in the lower focus of the Ellipsis, areas proportional to the times. And in the year 1686 I set down the Elements of the Method of Fluxions & Moments, & demonstrated them synthetically in the second Lemma of the second Book of Principles in order to make use of the Lemma in demonstrating some following Propositions.†

Here Newton might be referring to the 1679 exchange of letters with Hooke. It seems very unlikely that he had already broached Keplerian motion in terms of quadratures: this would mean that he had in about 1682 faced the inverse problem of central forces in terms of fluxional equations. However, it is true that in Lemma 2 in the second book of the *Principia*, i.e. by 1686, he had set down some rules about the direct synthetic method of fluxions (§3.12). In subsequent propositions on resisted motion he employed the rule for the moment of a product of two flowing quantities (§3.13).‡

A statement more adherent to the truth is to be found in an intended preface to a projected joint edition of the *Principia* and the *De quadratura* written in 1719:

...and because several Problems are proposed in the Book of *Principles* to be solved *concessis Figurarum Quadraturis*, I have added to the end of the Book of Quadratures some Propositions taken from my Letters already published for reducing quantities into converging series & thereby squaring figures.§

As a matter of fact several propositions, most notably those concerning the general inverse problem of central forces (§3.5) and those concerning the attraction of extended bodies (§3.8), consist in reduction to quadratures. In some simple cases Newton published the quadratures in the *Principia*. In other cases he referred to a method for 'squaring curvilinear figures': clearly his 1671 'catalogues'. There is no doubt that he used his 'catalogues' (i.e. 'integral tables') during the composition of the *Principia* in solving Corollary 3 to Proposition 41 and Corollary 2 to Proposition 91. In no other way could he have reached the solution. In the 1690s Newton had already considered the opportunity of publishing his analytical method of quadratures as an appendix to the *Principia*. Furthermore, as we have noted above in §4.5, in the unpublished version of *De quadratura*, in several manuscripts and in a letter to Gregory of 1694, Newton shows his awareness of the possibility of applying the inverse analytical method of fluxions to those propositions of the *Principia* whose solution is based on the squaring of curves.

† *Mathematical Papers*, **8**: 659.
‡ Cf. also 'The elements of the method of fluxions and moments are given synthetic proof in Lemma II of Book 2, and a number of the propositions following are synthetically demonstrated by means of this lemma'. *Mathematical Papers*, **8**: 447–9.
§ *Mathematical Papers*, **8**: 663.

In the following lines Newton draws a distinction between the occurrence in the *Principia* of the analytical and the synthetic methods of fluxions:

The synthetic method of fluxions occurs widespread in the following treatise [the *Principia*], and I have set its elements in the first eleven Lemmas of the first book and in Lemma II of the second. Specimens of the analytical method occur in Proposition XLV and the Scholium to Proposition XCIII of Book 1, and in Propositions X and XIV of Book 2. It is, furthermore, described in the scholium to lemma II of Book 2. And from their composed demonstrations, also, the analysis by which the propositions were found out can be learnt by going backwards.†

The first eleven Lemmas are those in which Newton presents the method of first and ultimate ratios (§3.3): quite rightly Newton defines them as the foundation of the use of the synthetic method of fluxions. In Proposition 45 (§3.6), in the Scholium to Proposition 93 (§3.9), and in Proposition 10 (§8.6.3) power series are employed: certainly a constituent of the analytical method. In the Scholium (which was radically varied in the third edition) to Lemma 2, Book 2, one finds a description of some letters that Newton sent concerning the analytical method of fluxions. This Lemma, defined as belonging to the synthetic method, is said in its Scholium to be 'the foundation' of the analytical method.‡   Lemma 2, we remind the reader, contains some rules for determining the 'moment' of (i.e. differentiating) expressions obtained by sums, products and powers (§3.12). These rules are applied in subsequent demonstrations, e.g. in Proposition 14, Book 2 (§3.13).

Summing up, we have seen that Newton identifies as examples of the synthetic method of fluxions Section 1, Book 1, on first and ultimate ratios, Lemma 2 on moments and the subsequent propositions in Book 2. He describes as examples of the analytical method three propositions (45, Book 1; Scholium to 93, Book 1; and 10, Book 2) where series are employed. In several places Newton refers to those propositions which lead to quadrature problems as examples of the analytic method (see §3.5 and §3.8). Even though Newton was writing during the quarrel with Leibniz, it would be wrong to say that his evaluations are completely unreliable.§ It seems correct to say that these parts of the *Principia* are examples of application of the analytical and synthetic methods of fluxions.

---

† *Mathematical Papers*, **8**: 455–7. Another interesting document is a loose catalogue entitled 'De methodo fluxionum' inserted in a copy of the *Principia* which was in Newton's possession. 'De methodo fluxionum: Lib. I. Sect. I. est de methodo rationum primarum et ultimarum [...] haec est methodus momentorum synthetica. Eadem si Analytice tractetur evadit methodus momentorum Analytica, quam etiam methodum fluxionum voco. Lib. I. sect XIII Prop. 93. Schol. pag. 202. Methodus solvendi Problemata per serie & momenta conjunctim exponitur. Lib. II. Lem. II. pag. 224. ostendo quomodo fluentium ex lateribus per multiplicatonem divisionem vel extractionem radicum genitarum momenta et fluxiones inveniri possunt. Lib. II. Prop. 19 [for 14]. pag. 251 argumentum procedit per differentiam momentorum, ideoque ideam tunc habui momentorum secondorum, et primus hanc ideam in lucem edidi.' *Variorum*: 793–4.

‡ *Variorum*: 368.

§ 'The fable of fluxions' according to Hall (1992): 212–3.

But from this we cannot infer, with Newton, that it is easy to discover the hidden fluxional analysis in *all* the pages of the *Principia*. Even though examples of application of fluxions are present in the *Principia*, one has to admit several facts which disprove Newton's famous statement that 'any man who understands Analysis' can 'reduce the Demonstrations of the Propositions from their composition back into Analysis'.† We may note how painful the process of translation into analysis of the *Principia* was even for the best mathematicians in Europe. As a matter of fact, the propositions indicated by Newton in the above catalogues created considerable embarassment. We can separate these propositions, following Newton, into examples of the synthetic and of the analytical methods.

The synthetic propositions (most notably Propositions 6–13, Book 1, discussed in §3.4.2 and §3.4.3) based on first and ultimate ratios do not lend themselves, as I have already noted, to an easy translation into symbolic terms. The geometric limits make appeal to geometric properties and to a visualization of limits: they are thus framed in terms which rarely lead to easy algebraical formulas. As we will see below (§8.4), Varignon showed a considerable ingenuity in performing such a translation. Far from being a mere exercise for a 'skillful man', the treatment of Propositions 6–13, Book 1, in terms of the differential calculus was considered by Leibniz and Varignon a major research subject. Another example we can think of is Proposition 1, Book 1, which is proved in the *Principia* thanks to a geometric limit procedure. This fundamental proposition was proved in terms of the integral calculus by Hermann and by other mid-eighteenth-century mathematicians, such as Euler (§8.5). These calculus demonstrations are considerably different from the geometric procedures employed by Newton.

On the other hand, the analytical propositions (e.g. Proposition 41, Book 1) are indeed easily translatable in terms of calculus – it is just a matter of substituting geometrical linelets with differentials – but they often require integration techniques which were not elementary even for a mathematician of the level of Johann Bernoulli (§8.6).

4.7 Conclusion: private discoveries and public style

In this chapter I have shown that:

● Newton knew how to apply the analytical method of fluxions to some simple problems concerning motion and force;

† *Mathematical Papers*, **8**: 259. See also: 'By the help of this new Analysis Mr Newton found out most of the Propositions in his *Principia Philosophiae*. But because the Ancients for making things certain admitted nothing into Geometry before it was demonstrated synthetically, he demonstrated the Propositions synthetically that the systeme of the heavens might be founded upon good Geometry. And this makes it now difficult for unskillful men to see the Analysis by wch those Propositions were found out.' *Mathematical Papers*, **8**: 598–9.

- he did not publish these results;
- the analytical method had a limited role in the *Principia*.

Newton's 'quadrature avoidance' in the first edition of the *Principia* could be justified by the need to take into consideration the competence of the readers. I also observed in §2.4 that seventeenth-century publication practice did not require one to reveal in printed form the methods of discovery. Such justifications, however, do not hold for the successive editions. In the early 1710s – with the *De quadratura* at last in print – the time was ripe for revealing the 'hidden analysis' by which several propositions had been demonstrated. As we will see in more detail in Chapter 7, Newton chose to reveal this analysis in private exchanges with his close associates: Fatio de Duillier, Gregory, Cotes, De Moivre and Keill. The reasons why Newton did not publish are related to the values that directed his research in his mature life, values which led him to distance himself from calculus in favour of geometry. We have already devoted some space to this topic. Let me just add an episode which is illustrative of Newton's reticence in publishing calculus solutions. When Johann Bernoulli proposed the brachistochrone problem as a challenge 'to the sharpest mathematicians in the whole world', Newton's solution appeared anonymously in the *Philosophical Transactions*. Newton had probably achieved this solution through a fluxional equation similar to that (unpublished†) employed for the solid of least resistance. In Newton's paper one finds a geometrical construction of the curve required (a cycloid), but not the fluxional demonstration.

The limited role of the analytical method in the *Principia* is apparent from the fact that Newton could include very few propositions as examples of the use of the 'hidden' fluxional analysis in his retrospective catalogues. This is, I believe, due to the relative weakness of calculus in 1687. As we have seen in §4.5, Newton knew how to apply the analytical method of fluxions to the simplest dynamical problems. Manuscripts in which Newton writes differential equations (or better, fluxional equations) of motion for the direct and inverse one-body problems are extant, and, as Nauenberg has shown, Newton was able to intertwine calculus and geometry in dealing with perturbations.‡ However, universal gravitation is an assumption which implies very difficult mathematical problems. The possibility of mathematizing the theory of planetary shapes or the theory of tides was crucial for Newton and his followers. The analytical method of fluxions was not powerful enough to allow such dynamical studies. Geometry on the other hand offered a means to tackle these problems, at least at a qualitative level (§3.15). Employing geometrical methods was not therefore a defensive, a backward move motivated

---

† First published in the Appendix to Motte's English translation of the *Principia* (1729).
‡ Nauenberg (1995), (1998a) and (1998b).

uniquely by classicism, but was rather seen by Newton as a progressive move, a choice of a more powerful and unifying method.

5

Huygens: the *Principia* and proportion theory

5.1 Purpose of this chapter

At the time that Newton's *Principia* was published, Christiaan Huygens was universally recognized as the greatest authority in natural philosophy and a leading geometer of his age. His work in mathematics, mechanics, technology, astronomy and optics was outstanding. One can enumerate his main contributions: the mathematical treatment of probability, the study of impact laws, the observations of the planetary system, the mathematical treatment of simple and compound pendula, the discovery of the tautochronism of the cycloid, the understanding of conservation laws in dynamics nowadays seen as 'equivalent' to energy conservation, the mathematical study of centrifugal force, the wave theory of light and the study of double refraction in a crystal of Iceland spar. This is indeed an impressive list: both Newton and Leibniz declared their indebtedness to the Dutch natural philosopher.

What is perhaps more relevant with Huygens is that he showed how far-reaching could be the use of mathematics in the understanding of natural phenomena. In particular, Huygens' work in mechanics and optics was framed in a highly sophisticated mathematical language. While Galileo's 'two new sciences' required relatively simple mathematical tools, while Descartes' 'natural philosophy' was mainly qualitative and divorced from mathematics, Huygens' work was heavily mathematical. One of his masterpieces, significantly entitled *Horologium oscillatorium, sive de motu pendulorum ad horologia aptato demonstrationes geometricae*, published in Paris in 1673, required very complex mathematical techniques. Newton's *Principia* and Leibniz's dynamical works were written after the example of mathematization of mechanics given in the *Horologium*, and, despite the numerous differences, both men owed a great deal to this work. It is thus interesting to compare the methods employed by Newton and by Huygens in the mathematization of natural philosophy. In section §5.5 I will concentrate on Huygens' criticisms to Proposition 6, Book 1 of the *Principia*, since they reveal differences in the use of the theory of proportions. Huygens employed this theory in the *Horologium*, a work that Newton never ceased to praise. However, the use of proportion theory in the *Principia* was not acceptable to Huygens.

5.2 Huygens as a mathematician

Huygens' early mathematical education took place at home until he was sixteen years old. His father, a diplomat and a well-known poet, guaranteed to Christiaan the liberal education typical of an aristocratic Dutch family. From 1645 to 1647 Huygens studied law and mathematics at the University of Leiden. Mathematics was taught by Frans van Schooten, one of the most influential promoters of Descartes' new methods. In fact van Schooten was the editor of the Latin edition of Descartes' *Géométrie*: there could not be a better guide to the new mathematics of Viète, Fermat and Descartes. Huygens also acquired, however, a good knowledge of the geometrical writings of Euclid, Archimedes and Apollonius. This combination of new and old mathematical methods was to characterize Huygens as a creative mathematician.†

Huygens' early mathematical researches carried on in the 1650s relate to probability and to that variety of problems which were faced in the so-called pre-calculus period (quadratures, rectification of curves, tangents, etc.). In 1656–58 he discovered the tautochronism of the cycloid and he developed his theory of evolutes and centres of oscillation: these results were incorporated in the *Horologium oscillatorium* (1673). From the early 1660s to 1681 Huygens was in Paris as the leading member of the newly established Royal Academy of Sciences. There he had contacts with all the members of that great circle of French *savants*, which included Roberval, Pascal and Desargues. It was in Paris that he introduced Leibniz to higher mathematics.

It is difficult to define Huygens' mathematical style. In his published works, most notably in the *Horologium*, he gave a geometric presentation of his results, following almost exclusively Archimedean *ad absurdum* methods. However, his manuscripts and letters reveal that he practised new methods such as Descartes' application of algebra to geometry and made use of geometrical infinitesimal techniques (especially in quadrature problems). He also employed the characteristic triangle (§6.2.3) independently, and actually before Pascal, in order to transform difficult quadratures into easier ones.‡ He was thus aware of the new techniques adopted by his contemporaries. Huygens' mathematical style is typical of the transitional period in which he lived. In the period before the introduction of Newton's and Leibniz's calculi the objects of interest for mathematicians were

---

† The best study on Huygens' mathematics is Yoder (1988). For biographical information see Bos (1972).
‡ Bos (1982): 116.

geometrical ones: curves, surfaces, volumes. Some of these objects could be expressed, following Descartes, by equations. In areas where Cartesian algebra did not reach one had to use geometry. But here it was admissible to use geometrical infinitesimals: figures could be partitioned into infinitesimal components.

Leibniz's and Newton's calculi (the 'differential' and the 'fluxional') began to circulate in the late 1680s. Huygens was then an old man who had already established himself as one of the leading mathematicians in Europe. His creative period was far away. He was clearly interested in the new methods. His reaction was at first scepticism. Only around 1694 did he begin to admit the superiority of the new algorithms. He corresponded on the new calculi with his 'pupil' Leibniz, with l'Hospital, and with the Newtonians, Fatio de Duillier and David Gregory. With Fatio he discussed in the 1680s 'inverse tangent problems': i.e. the problem of determining a curve, given the properties of its tangent. These problems were to be faced by the new generation with differential or fluxional equations. In the 1690s Huygens also considered such classic test cases for the new calculus as the isochrone, the catenary, the *velaria* and the tractrice. He attempted, with partial success, geometric solutions of these problems and proposed these as alternatives to the calculus solutions developed by the Leibnizian mathematicians. The correspondence with Leibniz on the new calculus began in January 1688 with a letter by Leibniz on the isochrone curve. In subsequent years Leibniz tried to convince his mentor about the usefulness of the differential and integral calculus. Leibniz was a great champion of his *nova methodus*, he actively opposed critics such as Clüver and Nieuwentijt and tried, aided by the Bernoullis, to extend the number of supporters. For the Leibnizians it was clearly important to gain the approbation of the distinguished Dutch mathematician. It was not an easy task. Huygens did not like the abstract and mechanical nature of the calculus. His scepticism towards the new method is revealed in a letter sent to l'Hospital concerning Leibniz's calculus solution of the tractrice:

I find nothing poorer or more useless, in view of the embarrassed, and in fact impracticable, descriptions he [Leibniz] proposes. For it is almost impossible to construct this simple tractrice with any exactitude.†

Faced with a specific problem Huygens, contrary to Leibniz, was not interested in the generality of methods, but rather in the geometric constructions which allow the solution of that particular problem. Typically, Huygens could affirm that his methods were preferable to Leibniz's because they needed 'neither extraordinary calculations nor new symbols'.‡   Notwithstanding his aversion towards dry

---

† 'je ne trouve rien de plus pauvre ni de plus inutile, vu les descriptions embarassées et tout à fait impracticables qu'il apporte. Car à peine pourroit on construire avec quelque exactitude cette simple Tractoria'. *Oeuvres*, **10**: 578–9.

‡ 'ni de calcul extraordinaire ni de nouveaux signes'. Quoted in Heinekamp (1982): 104.

symbolism, Huygens followed with interest some new algorithmic developments in quadrature problems. He was much intrigued by Newton's general rule (see §2.2.7) for the quadrature of fluents of the form $y = x^\theta (e + fx^\eta)^\lambda$ which had appeared in 1693 in Wallis's *Opera*. He corresponded with David Gregory on this subject.†

In Huygens' replies to Leibniz one often finds the statement that the calculus was not new, that he already had 'equivalent' methods. Leibniz reacted always tactfully and diplomatically: it was too important to win Huygens to the cause of the new mathematics. Leibniz insisted especially on the fact that his calculus could deal with transcendental quantities.‡ As a matter of fact, in the very last years of his life, Huygens admitted to having been much impressed both by the solution of the catenary, given by Leibniz and Johann Bernoulli, and by l'Hospital's rectification of the exponential curve (a curve which Huygens had defined and studied in his youth).§ Thus in 1691 he wrote to Leibniz:

Furthermore, I have considered the reason why I missed many of your discoveries, and I think that this must be the effect of your new way of calculating.¶

5.3 Huygens and Newton's *Principia*

Huygens received a copy of the *Principia* late in 1687.‖ Fatio had already revealed to Huygens something of Newton's work. He remarked that it contained subjects which Huygens had already dealt with. Fatio mentioned the treatment of cycloids and epicycloids in Section 10, Book 1, and the theory of the ebb and flow of the sea.†† When Huygens received the book he found many other subjects overlapping with his own research. Most notably his attention was caught by the first three sections of Book 2 on motion in resisting media.

Huygens was clearly impressed by Newton's achievements, but at the same time he found difficulties in understanding the mathematics and the physics. Huygens expressed his admiration for the *Principia* ('I very much admire the beautiful inventions that I have found in the work that he sent me.'‡‡), but he had reservations about the theory of gravitation and the mathematical methods. For instance in a letter to Hudde written in 1688 he remarked:

† For details see Vermij & van Maanen (1992).
‡ e.g. *Oeuvres*, **9**: 450, 472, 533. On this subject see Breger (1986) and Bos (1996).
§ *Oeuvres*, **10**: 129, 139, 305, 307, 312–15, 325–35. For further details see Yoder (1988): 176.
¶ 'Je consideray en suite pourquoy plusieurs de vos decouvertes m' estoient echappées, et je jugeay que ce devoit estre un effet de vostre nouvelle façon de calculer.' Quoted in Heinekamp (1982): 112.
‖ Snelders (1989): 209. For a comparison between the *Principia* and the *Horologium* see Speiser (1988).
†† *Oeuvres*, **9**: 167–71.
‡‡ 'j'admire extremement les belles inventions que je trouve dans l'ouvrage qu'il m'a envoyé.' Letter to Constantijn Huygens in *Oeuvres*, **9**: 305.

Professor Newton in his Book entitled Philosophiae Naturalis Principia Mathematica has stated several hypotheses of which I cannot approve.†

In a letter sent in 1690 to Leibniz, Huygens wrote:

As for the cause of the ebb and flow given by Newton, I am not at all satisfied. I am dissatisfied also with all the other theories which he bases on the principle of attraction, which seems to me absurd [...] And I was often surprised to see how he could make such an effort to carry on so many researches and difficult calculations which have as foundation this very principle.‡

Huygens adhered to a mechanistic philosophy, Cartesian in character, according to which the idea of a force acting instantaneously at a distance could not be acceptable. Until his death he often repeated that Newton's principle of universal gravitation had to be rejected. He wrote to l'Hospital:

I have great esteem for his [Newton's] knowledge and subtlety, but, in my opinion, he has made a poor use of them in most of this work [the *Principia*], when the author researches things which have little utility, or when he builds on such an unlikely principle as that of attraction.§

In 1690 Huygens published, as an appendix to the *Traité de la lumière*, the *Discours de la cause de la pesanteur* where he developed his own theory of gravitation based on vortices of subtle matter.¶ The *Traité* and the *Discours*, where Huygens systematized researches he had been carrying on for many years, were welcomed as his mature reply to Newton's optics and celestial mechanics. In the former he opposed corpuscular optics and favoured a wave theory of light. In the latter he took a position against Newtonian attraction. It should be noted that, while Huygens was sceptical about the existence of a universal gravitational force acting at a distance, he did recognize that Newton had given a mathematically correct demonstration of the existence of an inverse square force acting among the planets, thus disproving Descartes' version of the vortex theory of planetary motions. In 1688 he remarked:

The most eminent Newton has brushed aside all difficulties [concerning Keplerian laws] together with the Cartesian vortices; he has shown that the planets are retained in their

---

† 'Professor Newton in sijn boeck genaent Philosophiae Naturalis Principia Mathematica, stellende verscheyde hypotheses die ick niet en kan approberen.' *Oeuvres*, **9**: 267.

‡ 'Pour ce qui est de la Cause du Reflus que donne Mr. Newton, je ne m'en contente nullement, ni de toutes ses autres Theories qu'il bastit sur son Principe d'attraction, qui me paroit absurde [...] Et je me suis souvent etonné, comment il s'est pu donner la peine de faire tant de recherches et de calculs difficiles, qui n'ont pour fondement que ce mesme principe.' *Oeuvres*, **9**: 538.

§ 'J'estime beaucoup son scavoir et sa subtilité, mais il y en a bien de mal emploié à mon avis, dans une grande partie de cet ouvrage lors que l'autheur recherche des choses peu utiles, ou qu'il batit sur le principe peu vraisemblable de l'attraction.' *Oeuvres*, **10**: 354.

¶ For further details see Snelders (1989) and Martins (1993).

orbits by their gravitation towards the Sun. And that the excentrics necessarily become elliptical.†

So Huygens accepted the mathematical correctness of Newton's derivation of an inverse square force. But, in his opinion, a mechanical cause of this force should be provided according to the principles of the 'vraye & saine Philosophie.' He could not accept the physical theory of universal gravitation.‡

In the *Discours* Huygens not only criticized attraction at a distance, but he also dealt with motion in resisting media and with the shape of the Earth. In sending the *Discours* to Leibniz, he wrote that there were some theorems concerning motion in resisting media 'which you have also treated, and Mr. Newton more amply than both of us'.§ In fact Huygens from 1668 to 1669 had made experiments and performed calculations on motion in resisting media. He reproduced these on pages 168–80 of the *Discours*. In these researches he had successfully dealt with the case in which the resistance is proportional to velocity, but he had encountered difficulties with the case, closer (for high velocity) to experimental evidence, in which resistance is proportional to the square of velocity.¶ Newton in Sections 2 and 3, Book 2, had also dealt with the more difficult case (§3.11 and §3.13). Huygens recognized that Newton had been able to surpass him on this point. However, he could see that in the simpler linear case the *Principia* was basically in agreement with his 1668 procedure.

In the *Discours* Huygens also dealt with the shape of the Earth. He agreed with Newton that the Earth is flattened at the poles, but his analysis and estimate of its ellipticity differed from the one given in the *Principia*. He was probably led to think about this subject upon reading Mariotte's reports, published in 1686, on the observations by Richer on the length of a one-second pendulum at the Cayenne. Richer had found that a pendulum at Cayenne had to be shorter than in Paris in order to beat seconds. Huygens attributed this effect to the centrifugal force due to the Earth's rotation. He also speculated that the Earth had to have a flattened shape.‖ After reading Newton's *Principia*, Huygens adopted Newton's principle of balancing columns in order to calculate the ellipticity of the Earth (§3.15). He published his results in the *Discours* assuming that the Earth is a homogeneous fluid rotating mass. The main point is that he assumed that gravity in the interior of the Earth is constant, while Newton had used a theorem according to which the attraction along the polar and equatorial axes varied linearly with distance from the

† 'Hasce omnes difficultates abstulit Clar. vir. Newtonus, simul cum vorticibus Cartesianis; docuitque planetas retineri in orbitis suis gravitationem versus solem. Et excentricos necessarie fieri figurae Ellipticae.' *Oeuvres*, **21**: 143. Quoted in Snelders (1989): 209.
‡ *Oeuvres*, **21**: 446.
§ 'de quoy vous avez traité aussi, et Mr. Newton plus amplement que pas un de nous deux'. *Oeuvres*, **9**: 367.
¶ *Oeuvres*, **19**: 102–18 and 144–57; **21**: 478–93.
‖ *Oeuvres*, **9**: 130–1.

centre. Huygens' assumption of the constancy of gravity led him to a much easier mathematical problem.†

Other mathematical subjects contained in Newton's *Principia* which attracted Huygens' attention were Lemma 28, Book 1, on 'ovals' (he corresponded on this topic with Leibniz) and Proposition 35, Book 2, on the solid of least resistance.‡ Huygens was able to find the cone's frustrum of minimal resistance. He confirmed the result stated by Newton without demonstration in the *Principia*. Huygens' calculation, based on a 'rule of maxima and minima', is entirely analytical and agrees with Proposition 35. He finds the minimum of an algebraic expression, a result which was simple for any student of van Schooten.

Huygens maintained a great interest in Newton's masterpiece until his death. When he visited England in 1689 he met Newton several times, and they discussed topics connected with motion in resisting media. The two men had a great esteem for each other. Huygens hoped to see a second edition of the *Principia* printed. In 1691 he wrote to Fatio:

I believe I wrote to you that it is to be hoped that this illustrious author should have a second edition of his Book done, where all these errors could be corrected, and many obscurities explained§

and in 1692 l'Hospital was writing to Huygens:

I have been assured that Mr. Newton is going to publish a second edition of his mathematical principles in a style which will be more within the reach of all.¶

The fact that Newton was working towards a 'restructuring' of the *Principia* had thus reached France and Holland. Huygens often complained about the obscurity of Newton's work.‖

In the early 1690s Fatio attracted Huygens' interest by proposing himself as the editor of this much awaited second edition.†† Later in 1694 David Gregory did the same.‡‡ The exchange of letters with Fatio, Gregory, l'Hospital and Leibniz reveals Huygens' impatience for having an edition of Newton's work that was 'more accessible' and less 'obscure'.

We thus have ample evidence of Huygens' interest in the *Principia*. The Dutch natural philosopher could not accept universal gravitation – as he explained in the

---

† For further details see Mignard (1988).
‡ *Oeuvres*, **10**: 49–52; 22: 335–41.
§ 'Je crois vous avoir escrit alors qu'il est à souhaiter que cet Illustre autheur fist faire une seconde Edition de son Livre, où tous ces Errata pourroient estre corrigez, et beaucoup de choses obscures, eclaircies'. *Oeuvres*, **10**: 209.
¶ 'l'on m'a assuré que Mr. Neuton faisoit jmprimer puor la 2e fois ses principes mathematiques d'une maniere qui estoit plus à la portée de tout le monde'. *Oeuvres*, **10**: 346.
‖ In 1690 he wrote to Römer: 'Newtonii librum cui titulos Phiolos.ae Principia Mathematica non dubito quin videris, in quo obscuritas magna. Attamen multa acute inventa.' *Oeuvres*, **9**: 490.
†† *Oeuvres*, **10**: 213, 239, 567.
‡‡ *Oeuvres*, **10**: 614.

Discours – but was extremely interested in Newton's mathematical methods. As we have seen he met and corresponded with Newton, he corresponded with several mathematicians on a project for a second edition of the *Principia*, and he devoted attention to several technical aspects of Newton's work. Few people in Europe had the competence to read Newton's mathematical natural philosophy in such depth.

On the other hand, Newton declared his admiration for Huygens as a mathematician. Newton was aware of having surpassed the Hollander, but described Huygens as a model to be followed. In particular, he was impressed by the *Horologium oscillatorium*. A list of background readings prepared in 1691 for Bentley who was having difficulty reading the *Principia* is often quoted. After having mentioned several works, including those of van Schooten and Barrow, Newton added:

These are sufficient for understanding my book: but if you can procure Hugenius' *Horologium oscillatorium*, the perusal of that will make you much more ready.†

According to Pemberton, Newton often praised Huygens' mathematical style.‡

We now turn to take a closer look at Huygens' reactions to the mathematical methods of the *Principia*. In the following pages we will devote some attention to Huygens' criticisms of Newton's geometrical representation of force. As we will see, despite the similarities between Newton's and Huygens' geometrical methods for natural philosophy, the two differed significantly.

5.4 Proportion theory

In order to appreciate Huygens' criticisms a few words on the theory of proportions are in order. These criticisms of Newton, to which we will devote the next section, depend upon Huygens' understanding of this theory and of its role in the mathematization of nature.

The theory of proportions was the main mathematical tool employed in the mathematization of natural phenomena during the first half of the seventeenth century. Most notably, Galileo and his Italian pupils had used this tool in the study of kinematics.

The theory of proportions had been developed by Euclid in the fifth book of the *Elements*. It was the only mathematical theory of antiquity which dealt with 'magnitudes' in general. In particular the theory of proportions was a powerful mean for handling continuous magnitudes. This theory operates on 'magnitudes' *A*, *B*, *C*, where a 'magnitude' is a general concept covering all instances in which the operations of adding and comparing have a meaning. For instance, a magnitude

† *Correspondence*, **3**: 155–6.
‡ Pemberton (1728): *Preface*.

can be a length, a volume, or a weight. Ratios can be formed between two magnitudes of the same 'kind' (e.g. between two areas, or between two volumes), and a proportion is a 'similitude' between two ratios. So one can say that two volumes V_1 and V_2 have the same ratio as two weights W_1 and W_2. Several notations were employed. We use:

$$V_1/V_2 = W_1/W_2. \tag{5.1}$$

Two quantities are 'homogeneous', i.e. they are of the same kind, when 'they are capable, when multiplied, of exceeding one another.'† Notice that the theory of proportions does not allow the formation of a ratio between two heterogeneous magnitudes. This is particularly important for kinematics since it is not possible, for instance, to define velocity as a ratio between space and time.

It should be stressed that ratios are not magnitudes. We tend to read ratios as numbers and proportions as equations, but in the old Euclidean tradition this was not the case. The discovery of incommensurable ratios (such as the ratio between the side and the diagonal of a square) was interpreted as implying that the concept of number was unable to cover the extension of the concept of ratio. Nonetheless operations on ratios are possible. An operation on ratios that will concern us was called 'compounding'. When A, B and C are homogeneous quantities, A/C was said to be the 'compound ratio' of A/B and B/C. Only 'continuous' ratios could be compounded: i.e. ratios such that the last term of the former is equal to the first of the latter. In order to compound two ratios which are not continuous, A/B and C/D, one has to determine the 'fourth proportional' to C, D and B, i.e. a magnitude F such that $C/D = B/F$. So the compound ratio of A/B and C/D is equal to the compound ratio of A/B and B/F, i.e. A/F.

The theory of proportions had been applied to the mathematization of natural phenomena since ancient times. In the Galilean school it was employed, most notably, in the mathematization of statics, kinematics and hydrostatics. The simple example of uniform motion will allow us to understand some characteristics of the mathematization of natural philosophy in the Galilean school. The relationship between time, space and velocity in uniform motion might be expressed as

$$s = v \cdot t. \tag{5.2}$$

This was not possible within the framework of the theory of proportions, as we would have a magnitude, velocity, equal to a ratio between two heterogeneous magnitudes, space and time. One has instead to state a series of proportions allowing only two of the three magnitudes involved (space, time and velocity) to vary. So one can state, 'when the velocity is the same and we compare two uniform

† Euclid, *The Elements*, **5**, definition 4. In modern terms, a ratio A/B (when $A < B$) can be formed if and only if there is a positive integer n such that $nA > B$.

motions the distances are as the times':

$$s_1/s_2 = t_1/t_2, \tag{5.3}$$

where s_1 (s_2) is the distance covered in time t_1 (t_2). Or, 'when the distance covered is the same, the velocities are as the times inversely':

$$v_1/v_2 = t_2/t_1. \tag{5.4}$$

When more variables are involved, one has to fix everything but two magnitudes and form proportions between a ratio of the first magnitude and a ratio of one power of the second one.

A more elaborate mathematical formulation of uniform motion can be achieved through the operation of compounding. The aim is to state that the distances are in the compound ratio of the velocities and the times. This can be achieved as follows. Let us assume that distance s_1 is covered with velocity v_1 in time t_1 and that distance s_2 is covered with velocity v_2 in time t_2. How can we 'compare' the two motions? The trick consists in considering a third uniform motion which lasts for a time $t_3 = t_1$ and with velocity $v_3 = v_2$. We can state:

$$s_1/s_3 = v_1/v_3, \tag{5.5}$$

and

$$s_3/s_2 = t_3/t_2. \tag{5.6}$$

The operation of compounding allows one to state that s_1/s_2 is the compound ratio of $v_1/v_2(= v_3)$ and $t_1(= t_3)/t_2$. This fact was expressed by statements such as 'the distance is proportional to the velocity and to the time conjointly'. However, in order to be purely Euclidean, to compound the right-hand ratios of (5.5) and (5.6) we have first to determine three homogeneous quantities (let us say segments) K, L and M such that $v_1/v_2 = K/L$ and $t_1/t_2 = L/M$. So at last we can state that

$$s_1/s_2 = K/M. \tag{5.7}$$

The theory of proportions was not sufficiently flexible to cope with the problems of seventeenth-century natural philosophy. Several proposals aimed at relaxing the rigidity of the Euclidean scheme were advanced.† Finally, Descartes' geometry overtook the theory of proportions. From this new viewpoint ratios are considered as quotients, proportions as equations. Consequently, direct multiplication of ratios, even noncontinuous ratios, is allowed. Descartes' approach won great success. Note that in the *Géométrie* multiplication of two segments is not an area, as it should be in the Euclidean tradition, but another segment: multiplication is

† Giusti (1993) is one of the best studies on this matter.

thus a closed operation for segments. Homogeneity of geometrical dimensions is not a constraint for the formation of ratios. As far as physical magnitudes are concerned, once a unit is specified for each magnitude, one can choose to represent them through numbers or segments. Since on numbers and segments the four operations are closed, the constraints on homogeneity of physical dimensions characteristic of the theory of proportions are overcome. So multiplication of time and velocity or division of weight and length is possible. While Huygens in his unpublished mathematical manuscripts clearly shows the ability to practise the new analytic art initiated by Descartes, he chose the language of proportion theory in his writings on natural philosophy, most notably in the *Horologium oscillatorium*.

Bos characterizes Huygens' mathematical style in the *Horologium* as 'geometrical physics', a style in which new results in mechanics are presented in the language of classic geometry. In the *Horologium* the theory of proportions is employed with such rigour that, with one exception, no multiplication of dimensionally different magnitudes occurs in the whole work.† The fact that relations between physical quantities are not expressed by equations but by proportions has the consequence that no proportionality factors occur. Furthermore, as Bos notes, in those cases where the modern theory uses time derivatives (as dv/dt), we find proportions between increments of related physical magnitudes acquired in equal time intervals.‡ These time intervals can be either finite or infinitesimal. In order to evaluate the time dependence of a magnitude which varies in time, Huygens splits time into finite or infinitesimal equal intervals and evaluates the changes occurring in these time intervals.

As I said above (in §2.3.1), Newton's *Principia* was probably written with the *Horologium* as a model. As a matter of fact also Newton's synthetical methods of the *Principia* can be defined as 'geometrical physics'. At first sight a parallel between Newton and Huygens can be drawn. As Huygens employed Cartesian algebra in the process of discovery, but chose to avoid it in his published work, so Newton knew the analytical method of fluxions, but did not publish it in the *Principia*. Both Huygens and Newton made recourse to geometrical methods, both showed a dissatisfaction with the mechanical algorithmic character of the calculus. Furthermore the language of proportion theory pervades the *Principia*.§ However, a closer reading reveals that the geometrical methods of Newton's natural philosophy differ from those of Huygens. Huygens' criticisms of Proposition 6, Book 1, of the *Principia* make us aware of the distance that separates the Dutch from the British natural philosopher.

† Bos (1986): xxv.
‡ Bos (1986): xxiii.
§ Sylla (1984), Grosholtz (1987), Jesseph (1989).

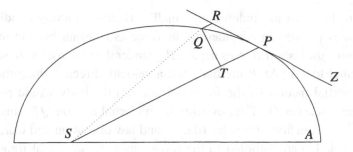

Fig. 5.1. *Principia*, Prop. 6, Book 1. After *Variorum*: 103

5.5 Huygens on Newton's geometrical representation of force, Prop. 6, Book 1

5.5.1 Proposition 6, Book 1 (1687)

5.5.1.1 Text

In Proposition 6, Book 1, Newton provides a geometrical representation for central forces.† Proposition 6 reads as follows (see fig. 5.1):

If a body P, revolving about the centre S, describes the curved line APQ, while the straight line ZPR touches the curve at any point P; and QR, parallel to distance SP, is drawn to the tangent from any other point Q of the curve, and QT is drawn perpendicular to the distance SP; then the centripetal force will be inversely as the *ᵃsolidᵃ* $SPquad \times QTquad/QR$, provided that the magnitude of that solid is always taken as that which it has ultimately when the points P and Q come together.

For in the *ᵇindefinitely smallᵇ* figure $QRPT$ the nascent line-element QR, if the time is given, is as the centripetal force (by Law 2) and, if the force is given, is as the square of the time (by Lemma 10), and thus, if neither is given, is as the centripetal force and the square of the time jointly, and thus the centripetal force is as the line-element QR directly and the square of the time inversely. But the time is as the area SPQ, or its double $SP \times QT$, that is, as SP and QT jointly, and thus the centripetal force is as QR directly and $SPquad$ into $QTquad$ inversely, that is, as $SPquad \times QTquad/QR$ inversely. Q.E.D.‡

5.5.1.2 Commentary

The idea underlying Newton's demonstration of Proposition 6 is the following. The central force acting on the orbiting 'body' changes in both magnitude and direction. However, the smaller the arc PQ, the smaller are these changes. Newton makes it clear that one has to consider – according to the synthetic method of limits premised in Section 1 on first and ultimate ratios – a limiting situation: the 'figure

---

† See §3.4.2. In the second edition (1713) of the *Principia* Proposition 6 was considerably expanded and modified.

‡ *aa*, 'solid', refers to the geometric dimension of $SPquad \times QTquad/QR$. *bb*, 'indefinitely small', translates 'indefinite parva'. *Variorum*: 103–4.

$QRPT'$ is to be taken as 'indefinitely small'. If these limiting conditions are accepted, one is justified in considering the force as constant both in magnitude and in direction: the body traverses the arc PQ under the action of a constant force directed parallel to PS. At P the body has a velocity directed along the tangent ZPR. By inertial motion (if the force did not act) the body would reach R in the same time it reaches Q. The deviation from inertial motion QR, this 'nascent line-element', is from first principles (the second law of motion and Corollary 10, Section 1, Book 1†) proportional to the force times the square of time. In fact point Q is reached by composition of inertial motion along PR and uniformly accelerated motion along RQ. The deviation from inertial motion for uniformly accelerated motion is proportional to the square of time (Corollary 10), it is proportional to the force (Law 2) and is in the direction PS (Law 2). One can thus write:

$$QR \propto F \cdot \Delta t^2. \tag{5.8}$$

Newton makes it clear that this formula, valid for constant forces, can be extended to variable central forces, provided that the limit, when 'P and Q come together', is taken.‡

In order to complete the demonstration of Proposition 6 one has just to consider that, since the force is central, Kepler's area law holds. Thus the area SPQ (a triangular area since the limit of the ratio between the vanishing chord PQ and arc PQ is 1; cf. Lemma 7, Section 1, Book 1, in §3.3) is proportional to time. In the limiting situation considered, the area SPQ is equal to $\frac{1}{2}(SP \cdot QT)$. And at last:

$$F \propto QR/(SP \cdot QT)^2. \tag{5.9}$$

5.5.2 *Huygens' notes on Proposition 6: Manuscript G, f. 15r*

A set of Huygens' notes to Newton's *Principia* written around 1689 still survives. They have been published in the *Oeuvres complètes*. These notes have to do with several points of the *Principia*.§ Here we will be concerned with Manuscript G, f. 15r. In the next subsection, §5.5.2.1, I present my English translation.

---

† Law 2 states: 'The change of motion is proportional to the motive force impressed; and is made in the direction of the right line in which that force is impressed'. *Principles*: 13. Note that 'motion' is defined as mass times velocity. Corollary 10 says that $s - s_0 - v_0 t \propto at^2 + t^3(\cdots)$. On Corollary 10 see §3.3.

‡ Formula (5.8) can be rendered in modern terms as $2(\vec{s} - \vec{s}_0 - \vec{v}_0 \Delta t) = \vec{a} \Delta t^2$, for constant $\vec{a}$. Note that Newton, as always, considers a 'body' of given mass and, writing proportions, does not make explicit the constant mass term. Force is thus equated to acceleration.

§ i.e. Corollary 4 to the laws of motion on inertial motion; Proposition 6, Book 1; Proposition 5, Book 2, on motion in resisting media; Proposition 9, Book 1, on motion in an equiangular spiral; Lemma 2, Book 2 on moments; Proposition 20, Book 3, on pendular motion. *Oeuvres*, **21**: 415–26. Manuscript G, f. 15r is on pages 417–18. It should be noted that Book G of the *Oeuvres* is now shelved as Codex Hugeniorum 7 in the University Library in Leyden.

(a)

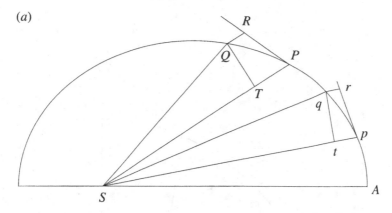

Fig. 5.2. Huygens' diagrams for Proposition 6, Book 1, of Newton's *Principia*. In (a) *S* is the force centre, PQ and pq two infinitesimal arcs such that the areas QSP and qSp are equal. In (b) the infinitesimal arcs AB and EF are such that the areas SBA, SFE are equal; *s* is the radius vector SA, *n* the radius vector SE; lines r, x and y are the deviations from tangent corresponding to the arcs AB, EX and EF; line t is perpendicular to SA, and u and $FG = t \cdot s/n$ are perpendicular to SE. After Huygens *Oeuvres*, **21**: 417–18

5.5.2.1 Text

On Proposition 6, Book 1 of Newton.

He says that the centripetal force in P is *reciprocally* as the solid SP^2 in QT^2/QR.†

Commentary. In order to say reciprocal, it is necessary to give or to conceive another point p, in which the centripetal force can be compared to the centripetal force which is at P [see fig. 5.2(a)]. Furthermore, in order to compare those forces, it is necessary that the spaces QSP and qSp be equal: that is, $SP \cdot QT$ is equal to $Sp \cdot qt$.‡ Then the centripetal forces will be as the minimal straight lines§ RQ and rq. I do not see what else this proposition could mean: in fact if it says that the centripetal force in P is to the centripetal force in p as Sp^2 in qt^2/rq to SP^2 in QT^2/RQ, that is as RQ in Sp^2 in qt^2 to rq in SP^2 in QT^2, this ratio is obviously the same as RQ to rq, because $Sp \cdot qt$ is equal to $SP \cdot QT$, and moreover $Sp^2 \cdot qt^2$ is equal to $SP^2 \cdot QT^2$. Why then say that the centrifugal¶ forces in P and p are as RQ to rq, and not rather reciprocally as $SP^2 \cdot QT^2/QR$ to $Sp^2 \cdot qt^2/qr$, or rather reciprocally as $SP \cdot QT/QR$ to $Sp \cdot qt/qr$, or rather reciprocally as $SP^3 \cdot QT^3/QR$ to $Sp^3 \cdot qt^3/qr$.‖

If one wishes to compare the centripetal forces in P and p when the spaces QSP and qSp are made unequal, this should be done.

---

† See our discussion of Proposition 6 above. 'Solid' refers to the geometric dimension of $SP^2 \cdot QT^2/QR$.

‡ Because of Kepler's area law, the areas QSP and qSp are proportional to the times. They must be equal, since one has to form a proportion between forces at P and p and displacements from tangent QR and qr acquired after equal intervals of time. Furthermore these areas, since the arcs PQ and pq are infinitesimals, are as the product SP times QT, and the product Sp times qt. Huygens' notation for $A \cdot B$ is $\square A, B$: it means the 'rectangle' with sides A and B.

§ 'minimal straight lines' translates 'rectae minimae'.

¶ Huygens here should write centripetal.

‖ Since only the case in which the areas QSP and qSp are equal can be considered, according to Huygens, any power of the product $SP \cdot QT$ cancels against the same power of $Sp \cdot qt$.

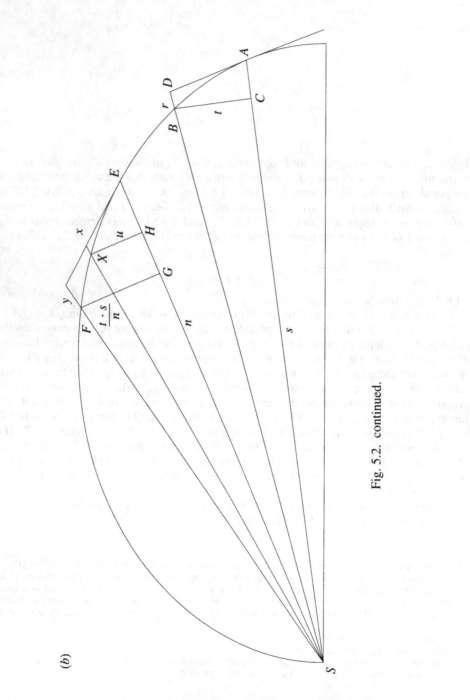

Fig. 5.2. continued.

(b)

Let the spaces SBA, SFE be set equal [see fig. 5.2(b)].

$n\top s - t/(ts/n)$.† The ratio of r to y, that is the ratio of the centripetal forces in A and E (given that the spaces SBA and SFE are put equal),‡ is composed from the ratio of r to x and the ratio of x to y.§ But x is to y as uu is to $ttss/nn$.¶ [...] Therefore $nnuu/x \top ttss/r$ is the same ratio as r to y, that is, the same ratio of the centripetal force in A to the centripetal force in E.‖ And this is what proposition 6 Book 1 states. Therefore let us say simply that the centripetal force is as $ttss/r$.††

But in the specific problems this proposition is not useful, when the value of QR can be discovered and $QT \cdot SP$ is put as given.

5.5.2.2 *Commentary*

- Huygens considers two points of the trajectory. In order to establish a proportionality Huygens needs four terms. Thus the force must be evaluated at two different points p and P.‡‡ In fact he makes it clear that another point must be considered in order to establish a proportionality between force and displacement from tangent. He aims at writing something like:

$$F(SP)/F(Sp) = QR/qr, \tag{5.10}$$

where $F(SP)$ and $F(Sp)$ denote the force at P and p respectively (see fig. 5.2(a)).

- Huygens considers equal infinitesimal intervals of time. Magnitudes which vary in time can be evaluated by establishing a proportion between changes acquired in *equal* intervals of time. This is Huygens' main criticism of Newton. He writes that the force in P and the force in p can be compared only when the deviations from tangent QR and qr are acquired after equal intervals of time, i.e. only when the areas QSP and qSp are equal.

- When different intervals of time are considered (see fig. 5.2(b)) we are in a situation in which a magnitude, force, is functionally related to two other magnitudes, displacement from tangent and time. In order to apply the theory of proportions one has to follow a process somewhat similar to that adopted in the Galilean school. In this case one has firstly to fix time (areas SBA and SFE are equal), and establish a proportion between force and displacement (i.e. $F(SA)/F(SE) = r/y$). Secondly one has to fix force (i.e. force $F(SE)$ at point E is considered), and establish a proportion between time and displacement (i.e. $x/y = (n^2 \cdot u^2)/(s^2 \cdot t^2)$, where $n \cdot u$ is area SXE and $s \cdot t$ is

---

† '$n\top s$' is the ratio of n to s. This can be interpreted as meaning that 'n is to s as t is to $t \cdot s/n$. Since the areas SBA, SFE are equal, $s \cdot t = n \cdot FG$, thus $FG = t \cdot s/n$.

‡ Huygens has shown above that, when the areas SBA, SFE are equal, the forces are as the displacements from tangent, that is as r to y.

§ Here Huygens composes continuous ratios in the style of the Euclidean theory of proportions.

¶ This is because, for infinitesimal angle θ, the versed sine, $1 - \cos\theta$, is proportional to the square of the sine, $\sin\theta$. Thus $x/y = u^2/FG^2 = u^2/(t^2 \cdot s^2/n^2)$.

‖ We can rewrite: (centripetal force in A)/(centripetal force in E) $= r/y = r/x \cdot x/y = r/x \cdot u^2/FG^2 = r/x \cdot u^2/(t^2 \cdot s^2/n^2) = (n^2 \cdot u^2/x)/(t^2 \cdot s^2/r)$. Note that $x/y = (n \cdot u)^2/(s \cdot t)^2$ where $n \cdot u$ is area SXE and $s \cdot t$ is area $SBA = SFE$, and these areas are proportional to the times. As Newton would write: 'given the force' the displacements x and y are as the squares of the times.

†† Huygens should write 'reciprocally as $ttss/r$'. Note that $s^2 \cdot t^2/r$ is Newton's ratio $SP^2 \cdot QT^2/QR$.

‡‡ Compare Newton's figure, fig. 5.1, with Huygens' figure, fig. 5.2(a).

area $SBA = SFE$, and these areas are proportional to the times). Thirdly Huygens compounds ratios and achieves Newton's result.

5.6 Conclusion: Huygens' vs. Newton's classicism

Both Huygens and Newton understood and practised the new Cartesian analytical style which had superseded the theory of proportions. However, they both published their results in natural philosophy in 'geometric style'. As one can see from Huygens' annotations to Newton's Proposition 6, their geometric 'public' styles are quite different. The difficulty that Huygens had with Proposition 6 are revealing.

It should be noted that:

- Newton evaluates force at a point; he does not need four terms. This reveals the fact that the proportions in the *Principia* are understood as equations, that is, objects not acceptable to a purist Euclidean.
- Newton employs limit arguments: he handles variable flowing quantities and takes limits according to the method of first and ultimate ratios. Newton's technique for evaluating force at point P consists in determining the limit of QR/QT^2, as 'P and Q come together'. This allows him to determine the local properties of the trajectory relevant for a study of central force.
- For Newton it is essential to study the functional relationship between several different physical magnitudes, leaving them to vary simultaneously. So in Proposition 6 a relationship is established between force, time and displacement from tangent.

Newton reached this innovative geometrical treatment of central force after a long gestation in which he often employed proportion theory along the lines indicated by Huygens. In fact, as Erlichson has convincingly argued, in his early work on central forces Newton dealt with ratios of forces and he compared forces by comparing the motion changes they produced for equal time intervals. Only at a later stage (basically during the composition of *Principia*) did Newton definitely turn his attention to the determination of force at a point: this new viewpoint was a development of his comparison of forces at two points for equal times.†

In conclusion we have seen that Huygens in his published work adheres to proportion theory. He studies variation in time of physical magnitudes by comparing changes acquired by two related magnitudes after equal infinitesimal intervals of time. On the other hand Newton in the *Principia* writes relations between a greater number of physical magnitudes (leaving them all to vary) and evaluates, through limit arguments, rates of change. Notwithstanding its classic façade, Newton's *Principia* is written in a style which is deeply innovative compared with that of

† Erlichson (1991b): 843–4.

Huygens. Newton's technique for evaluating central force at a point is geometrical and is expressed in the language of proportion theory, but it is modelled on new concepts characteristic of the synthetic method of fluxions.

6

Leibniz: not equivalent in practice

6.1 Purpose of this chapter

Leibniz's response to the *Principia* is one of the most complex intellectual events in the history of science. The *homo universalis* took a critical attitude towards a number of issues, spanning absolute time and space, the gravitational force, the basic laws of motion, and the role of God in the Universe. In this chapter I will confine myself to Leibniz's reaction to Newton's mathematical methods. Notwithstanding the priority dispute, Leibniz's evaluation of Newton's mathematics was a positive one. As a matter of fact, in their mathematical works Newton and Leibniz shared many techniques and concepts. The algorithms of their calculi (the analytical method of fluxions and the differential and integral calculus) were translatable, *and* translated, one into the other. Furthermore, contrary to what is generally believed, their ideas on the interpretation of these algorithms were strikingly similar. However, the equivalence between the two breaks down when we move from the abstract level of algorithmic techniques and foundational matters, and we consider the mathematical practices. In fact, as I will show, Newton and Leibniz oriented their research along different lines, since they held different values and different expectations for future research. The idea that I would like to convey is that the two mathematicians shared a common mathematical tool, but used it for different purposes.

6.2 The differential and integral calculus

6.2.1 *In search of a* characteristica universalis

Gottfried Wilhelm Leibniz was born in Leipzig in 1646 from a Protestant family of distant Slavonic origins. His father, a Professor at Leipzig University, died in 1652 leaving a rich library, where the young Gottfried began his scholarly life. He studied philosophy and law in the Universities of Leipzig, Jena and Altdorf.

He also received some elementary education in arithmetic and algebra. Very soon he formulated a project for the construction of a mathematical language with which deductive reasoning could be conducted. His manuscripts related to symbolical reasoning have revealed anticipations of nineteenth century algebra of logic. The programme of devising a *characteristica universalis* was never abandoned by Leibniz. As we will see, he conceived his mathematical research as part of this ambitious project. More specifically, his interest in number sequences played a relevant role in the discovery of the differential and integral calculus. After receiving his doctorate in 1666 from the University of Altdorf, he entered in the service of the Elector of Mainz. From 1672 to 1676 he was in Paris for a diplomatic mission. Here he met several distinguished scholars, most notably Christiaan Huygens, who belonged to the recently established *Académie Royale des Sciences*. It was in Paris, following Huygens' counsel, that Leibniz learned higher mathematics. In a few months he had digested all the relevant contemporary literature and was able to contribute original research. The discovery of the calculus dates from the years 1675–77. He published the rules of the differential calculus in 1684 in the *Acta eruditorum*, a scientific journal that he had helped to found in 1682. In 1676 his seminal period of study in Paris came to an end. After 1676 Leibniz worked in the service of the Court of Hanover. He embarked on political projects: the most ambitious was the reunification of the Christian Churches. Leibniz was very good at propagating his ideas, in particular his mathematical discoveries, through scientific journals and learned correspondence. While Newton kept his method secret, Leibniz made great efforts to promote the calculus. In Basel, Paris and Italy several mathematicians, such as the brothers Bernoulli, l'Hospital, Varignon, Gabriele Manfredi, and Jacopo Riccati, began to use the new calculus of sums and differences, and to defend it from its critics. Most notably at the turn of the century Johann and Jacob Bernoulli contributed to the extension of the integral calculus and to its application to dynamics.

6.2.2 Infinite series, 1672–73

Leibniz's interests in combinatorics led him to consider finite numerical sequences of differences such as

$$b_1 = a_1 - a_2, \ b_2 = a_2 - a_3, \ b_3 = a_3 - a_4, \cdots. \tag{6.1}$$

He noted that it is possible to obtain the sum $b_1 + b_2 + \cdots + b_n$ as a difference, $a_1 - a_{n+1}$. This simple law, extrapolated to the infinite, led to interesting results with infinite series. For instance, he was able to find the sum of the series of reciprocals of triangular numbers, which, expressed using modern $\sum$-notation, is as follows:

$$\sum_{n=1}^{\infty} \frac{2}{n(n+1)} = \sum_{n=1}^{\infty} b_n. \tag{6.2}$$

Leibniz noted that the terms of this series may be expressed using a difference sequence such that

$$b_n = \frac{2}{n} - \frac{2}{n+1} = a_n - a_{n+1}. \tag{6.3}$$

Therefore

$$\sum_{n=1}^{s} b_n = a_1 - a_{s+1} = 2 - \frac{2}{s+1}. \tag{6.4}$$

So, if we 'sum' all the terms, we obtain 2.

Leibniz applied this successful procedure to several other examples. For instance he considered the 'harmonic triangle' (see fig. 6.1(a)). In the harmonic triangle the nth oblique row is the difference sequence of the $(n+1)$th oblique row. It follows, for instance, that

$$\frac{1}{4} + \frac{1}{20} + \frac{1}{60} + \frac{1}{140} + \cdots = \frac{1}{3}. \tag{6.5}$$

These researches on infinite series imply an idea that played a central role in Leibnizian calculus.† The sum of an infinite number of terms b_n can be achieved via the difference sequence a_n. In fact, as he wrote to Hermann in 1716,

I arrived at the method of inassignable differences through the method of increments in the series of numbers, as the nature of things requires.‡

6.2.3 The geometry of infinitesimals, 1673–74

In 1673 Leibniz met with the idea of the so-called 'characteristic triangle'. He was reading Pascal's *Lettres de 'A. Dettonville'* (1659). Pascal, in dealing with quadrature problems, had associated with a point on a circumference a triangle with infinitesimal sides. Leibniz generalized this idea.

Given any curve (see fig. 6.1(b)) he associated with an arbitrary point P an infinitesimal triangle. One can think of the curve as a polygonal constituted by infinitely many infinitesimal sides. The prolongation of one of the sides gives the tangent to the curve. A line at right angles to one of the sides is the normal. Let us call t and n the lengths of the tangent and normal, respectively, extending from P to the x-axis.

† Bos (1980a): 61.
‡ 'Ego per Methodum incrementorum in seriebus numerorum perveni ad Methodum differentiarum inassignabilium, ut postulat natura rerum.' Leibniz to Hermann in *LMS*, **4**: 413.

(*a*)

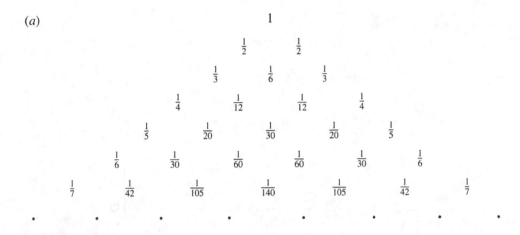

(*b*)

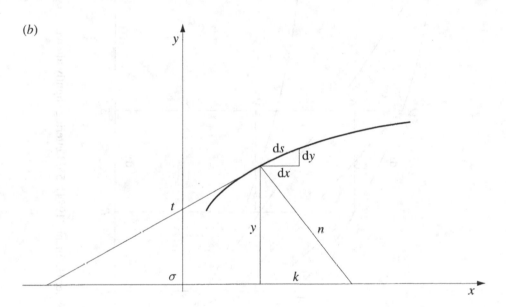

Fig. 6.1. The harmonic (*a*) and characteristic (*b*) triangles

From the similarity of the three triangles shown in fig. 6.1(*b*), Leibniz obtained several geometrical transformations which allowed him to transform a problem of quadrature into another problem. In a notation that Leibniz had not yet introduced he stated in verbal form equivalences such as $\int k\mathrm{d}x = \int y\mathrm{d}y$, $\int y\mathrm{d}x = \int \sigma\mathrm{d}y$, $\int y\mathrm{d}s = \int t\mathrm{d}y$, $\int y\mathrm{d}s = \int n\mathrm{d}x$, where n is the normal, t the tangent, k the subnormal and σ the subtangent.

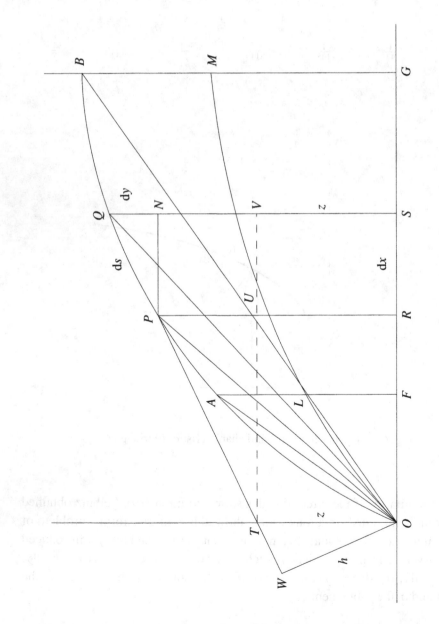

Fig. 6.2. Transmutation theorem. After Hofmann (1974): 55. Lettering slightly altered

The most useful transformation obtained by Leibniz in 1673–4, i.e. the years immediately preceding the invention of the algorithm of the calculus, is the 'transmutation theorem'.† Leibniz considered a smooth convex curve OAB (see fig. 6.2). The problem is to determine the area $OABG$. Let PQN be the characteristic triangle associated with the point P. The area $OABG$ can be seen either as the sum of infinitely many strips $RPQS$ or as the sum of triangle OBG plus the sum of infinitely many triangles OPQ. We can write

$$OABG = \Sigma RPQS = \tfrac{1}{2}OG \cdot GB + \Sigma OPQ. \tag{6.6}$$

Let the prolongation of PQ (i.e. the tangent at P) meet the y-axis in T and let OW be normal to the tangent. Triangle OTW is thus similar to the characteristic triangle PQN. Therefore

$$\frac{PN}{OW} = \frac{PQ}{OT}. \tag{6.7}$$

The area of the infinitesimal triangle OPQ is

$$OPQ = \tfrac{1}{2}OW \cdot PQ = \tfrac{1}{2}OT \cdot PN. \tag{6.8}$$

Leibniz defined a new curve OLM, related to the curve OAB through the process of taking the tangent. The new curve has ordinate at R equal to OT. Geometrically the construction is obtained by drawing the tangent in P and determining the intersection T between the tangent and y-axis. In symbols not yet available to Leibniz the ordinate z of the new curve OLM is $z = y - x\mathrm{d}y/\mathrm{d}x$, where y is the ordinate of the curve OAB.

In conclusion, Leibniz has derived that:

$$OABG = \tfrac{1}{2}OG \cdot GB + \tfrac{1}{2}\Sigma OT \cdot PN = \tfrac{1}{2}OG \cdot GB + \tfrac{1}{2}OLMG, \tag{6.9}$$

where $OLMG$ is the area subtended by the new curve. In modern symbols:

$$\int_0^{x_0} y\mathrm{d}x = \tfrac{1}{2}x_0y_0 + \tfrac{1}{2}\int_0^{x_0} z\mathrm{d}x = \tfrac{1}{2}x_0y_0 + \tfrac{1}{2}\int_0^{x_0} y\mathrm{d}x - \tfrac{1}{2}\int_0^{x_0} x\frac{\mathrm{d}y}{\mathrm{d}x}\mathrm{d}x. \tag{6.10}$$

Therefore Leibniz's geometrical 'transmutation' is equivalent to integration by parts.‡ He was later to express it as§

$$\int y\mathrm{d}x = xy - \int x\mathrm{d}y. \tag{6.11}$$

Leibniz thus achieved, through the geometry of the infinitesimal characteristic triangle, a reduction formula for quadrature. The quadrature of the curve OAB is reduced to the quadrature of an auxiliary curve OLM related to OAB through

---

† Hofmann (1974): 46–62; Bos (1980a): 62–4.
‡ Bos (1980a): 65.
§ *LMS*, **5**: 408.

the process of taking the tangent. The relation of tangent and quadrature problem thus began to emerge in Leibniz's mind. This work with the characteristic triangle made also him aware of the fruitfulness of dealing with infinitesimal quantities.

6.2.4 The calculus of infinitesimals, 1675–86

During 1675 Leibniz made the crucial steps which led him to forge the algorithm which, in revised form and in a different conceptual context, is still utilized. He began considering two geometric constructions which had played a relevant role in seventeenth-century infinitesimal techniques: viz., the characteristic triangle and the area subtended by a curve as the sum of infinitesimal strips. Let us consider a curve C (see fig. 6.3) in a Cartesian coordinate system. Leibniz imagined subdividing the x-axis into infinitely many infinitesimal intervals with extremes x_1, x_2, x_3, etc. He further defined the differential $dx = x_{n+1} - x_n$. On the curve and on the y-axis one has the corresponding successions s_1, s_2, s_3, etc. and y_1, y_2, y_3, etc. Therefore $ds = s_{n+1} - s_n$ and $dy = y_{n+1} - y_n$. The characteristic triangle has sides dx, ds, dy. The tangent to the curve C forms an angle γ with the x-axis such that $\tan \gamma = dy/dx$. The area subtended by the curve is equal to the sum of infinitely many strips ydx. Leibniz initially employed Cavalieri's symbol 'omn.', but soon replaced this notation with the now familiar $\int ydx$, where $\int$ is a long 's' for 'sum of'. The first published occurrence of the d-sign was in Leibniz (1684), and the integral appeared in (1686).† Three aspects of Leibniz's representation of the curve C in terms of differentials should be noted.

(1) The symbols d and $\int$, applied to a finite quantity x, generate an infinitely little and an infinitely great quantity respectively. So, if x is a finite angle or a finite line, dx and $\int x$ are, respectively, an infinitely little and an infinitely great angle or line. So the two symbols d and $\int$ change the order of infinity but preserve the geometrical dimensions. Notice that Newton's dot symbol does not do that. If x is a finite flowing line $\dot{x}$ is a finite velocity.

(2) Since geometrical dimension is preserved, the symbols d and $\int$ can be iterated so that higher-order infinitesimals and higher-order infinites are obtained. So ddx is infinitely little compared with dx, and $\int\int x$ is infinitely great compared with $\int x$. A hierarchy of infinitesimals and infinites is thus obtained. Higher-order differentials were denoted by repeating the symbol d. It became usual, from the mid-1690s, to abbreviate $dd \cdots d$ (n times) as d^n, so that the nth differential of x is $d^n x$.

(3) The representation of the curve C in terms of differentials can be achieved in a variety of ways. One can choose the progressions of x_n, y_n and s_n such that dx is constant, or dy is constant, or ds is constant. For instance, the choice dx constant

† *LMS*, **5**: 220–6 and 226–33.

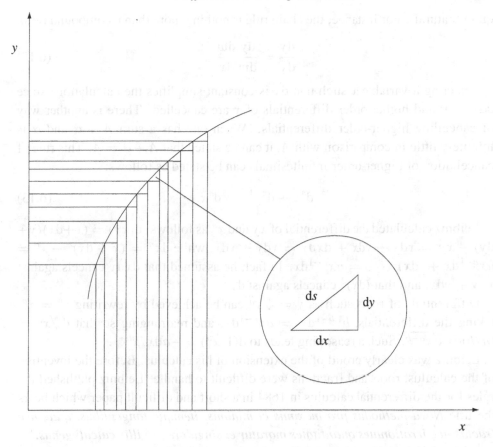

Fig. 6.3. Differential representation of a curve

(i.e. the x_n are equidistant) generates successions of y_n and s_n where ds and dy are not (generally) constant. As Bos has shown (1974) the choice of dx constant is equivalent to selecting x as the independent variable and s and y as dependent variables. (The Newtonian equivalent is to choose $\dot{x}$ constant, i.e. x flows with uniform velocity.)†

Bos has furthermore stressed that the Leibnizian calculus is not concerned with 'functions' and 'derivatives', but with progressions of variable quantities and their differences. Therefore we should not read, for instance, dy/dx as the derivative d/dx of $y(x)$, but as a ratio between two differential quantities, dy and dx. The conception of dy/dx as a ratio renders the algebraical manipulation of differentials

† Bos (1974) is still the best study on the Leibnizian calculus. In §6.2 I owe much to this paper.

quite 'natural'. For instance, the chain rule is nothing more than a compound ratio:

$$\frac{dy}{dx} = \frac{dy}{dw}\frac{dw}{dx}. \tag{6.12}$$

Selecting a variable x such that dx is constant simplifies the calculations since $ddx = 0$ and higher-order differentials of x are cancelled. There is another way of cancelling higher-order differentials. When one has a sum $A + \alpha$ and α is infinitely little in comparison with A, it can be stated that $A + \alpha = A$. This rule of cancellation of higher-order infinitesimals can be stated as follows:

$$d^n x + d^{n+1} x = d^n x. \tag{6.13}$$

Leibniz calculated the differential of xy and x^n as follows: $d(xy) = (x+dx)(y+ dy) - xy = xdy + ydx + dxdy = xdy + ydx$, while $dx^n = (x + dx)^n - x^n = nx^{n-1}dx + (dx)^2(\cdots) = nx^{n-1}dx$. In fact, he assumed that $dxdy$ cancels against $xdy + ydx$, and that $(dx)^2$ cancels against dx.

Differentials of roots such as $y = \sqrt[b]{x^a}$ can be achieved by rewriting $y^b = x^a$, taking the differentials, $by^{b-1}dy = ax^{a-1}dx$, and rearranging so that $d\sqrt[b]{x^a} = (a/b)dx\sqrt[b]{x^{a-b}}$. Such a reasoning leads to $d(1/x^a) = -adx/x^{a+1}$.

Leibniz was clearly proud of the extension of his calculus. Before the invention of the calculus, roots and fractions were difficult to handle. Leibniz published the rules for the differential calculus in 1684 in a short and difficult paper which bears the title *Nova methodus pro maximis et minimis, itemque tangentibus, quae nec fractas nec irrationales quantitates moratur et singulare pro illis calculi genus.*[†]

In the 1690s and during the first decades of the eighteenth century one of the main research areas of Leibnizian calculus consisted in the development of rules of differentiation and integration for transcendental quantities: trigonometric quantities, the logarithm, and the exponential (see §5.2 and §8.6.3.5).[‡] Johann Bernoulli was particularly active in this field. For instance, he expressed the following rules of what he called the 'exponential calculus': $d\ln u = du/u$ and $d(u^v) = vu^{v-1}du + u^v \ln u dv$.[§]

Integration was performed by Leibniz generally by reductions of $\int ydx$ through methods of variable substitution or integration by parts. These methods could be worked out in a purely analytical way. Instead of building complex geometrical constructions of auxiliary curves (as in the method of transmutation), the new notation allowed algebraical manipulations.

The most powerful means for performing integrations came from the understanding of the fundamental theorem of the calculus. Even at first glance the

---

[†] *LMS*, **5**: 220–6.
[‡] Breger (1986).
[§] Bos (1996).

notation d and $\int$, for difference and sum, suggests the inverse relationship of differentiation and integration. As Leibniz wrote in 1679 to von Tschirnhaus:

Finding tangents to curves is reduced to the following problem: to find the differences of series; whereas, finding the areas of figures is reduced to this: given a series to find the sums, or (to explain this better) given a series to find another one, whose differences coincide with the terms of the given series.†

Leibniz conceives $\int y\mathrm{d}x$ as the 'sum' of an infinite sequence of strips $y\mathrm{d}x$. From his research on infinite series he knew that a sum of an infinite sequence can be obtained from the difference sequence. In order to reduce $\int y\mathrm{d}x$ to a sum of differences one must find a z such that $\mathrm{d}z = y\mathrm{d}x$. Thus, at once,

$$\int y\mathrm{d}x = \int \mathrm{d}z = z. \tag{6.14}$$

Once the inverse relation of differentiation and integration is understood several techniques of integration follow. For instance the rule of transmutation (integration by parts) comes by inverting $\mathrm{d}(xy) = x\mathrm{d}y + y\mathrm{d}x$. In fact $xy = \int \mathrm{d}(xy) = \int x\mathrm{d}y + \int y\mathrm{d}x$.‡

As an example of Leibniz's inverse algorithm we can consider the application of the transmutation theorem to the quadrature of the cycloid generated by a circle of radius a rolling along the vertical line $x = 2a$ (see fig. 6.4). The ordinate BC is equal to $BE + EC = BE + \stackrel{\frown}{AE}$, where $\stackrel{\frown}{AE}$ is the length s of the circular arc. Since $\mathrm{d}s/a = \mathrm{d}x/\sqrt{2ax - x^2}$, it follows that $s = \int_0^x a\mathrm{d}u/\sqrt{2au - u^2}$. (Nowadays we have a notation for elementary transcendental functions and we would write $s = a \cdot \arccos((a - x)/a)$.) The equation of the cycloid is thus

$$y = \sqrt{2ax - x^2} + \int_0^x a\mathrm{d}u/\sqrt{2au - u^2}. \tag{6.15}$$

Since $\mathrm{d}y/\mathrm{d}x = (2a - x)/\sqrt{2ax - x^2}$, from (6.11):

$$\int_0^{x_0} y\mathrm{d}x = x_0 y_0 - \int_0^{x_0} \sqrt{2ax - x^2}\mathrm{d}x. \tag{6.16}$$

For $x_0 = 2a$ and $y_0 = \pi a$, formula (6.16) gives $3\pi a^2/2$ for the area subtended under the half-arch.

6.3 Dynamics

Leibniz had great interest in the applications of his calculus to geometry and dynamics. In this context he wrote and solved several differential equations.

---

† 'Invenire Tangentes curvae, reducitur ad hoc problema: inveniri seriei differentias; invenire autem aream figurae, reducitur ad hoc: datae seriei invenire summas, vel (quod magis instruit) data seriei invenire aliam, cujus differentiae coincidant terminis seriei datae.' *LMS*, **4**: 479.

‡ *LMS*, **5**: 408.

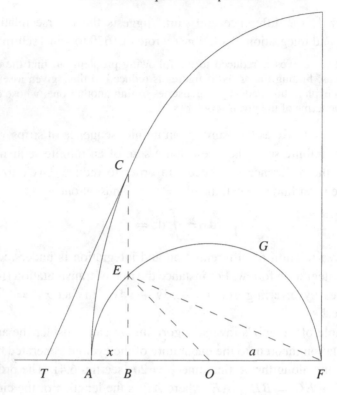

Fig. 6.4. Leibniz's quadrature of the cycloid

This very important subject entered into the world of Continental mathematics thanks to Leibniz's development of integration techniques. Since 1673 he had been interested in problems such as the determination of centres of gravity, while in 1687–8 he faced the problem of determining the isochrone curve. This last problem (which consists in determining the curve along which a body slides without friction in a constant gravitational field in such a way that the vertical component of velocity is constant) led him to formulate and integrate a differential equation.

The first systematic application of the differential calculus to natural philosophy is to be found in two papers by Leibniz published in January and February 1689 in the *Acta eruditorum*. They bear the titles *Schediasma de resistentia medii & motu projectorum gravium in medio resistente* and *Tentamen de motuum coelestium causis*. As Bertoloni Meli has shown, the latter was written in 1688 after a cursory study of Newton's *magnum opus*.† Leibniz maintained that he had seen the *Principia* for the first time only in April 1689 in Rome, during his diplomatic mission in Italy, when his papers had already been published. When he sent them

† *LMS*, **6**: 135–43, 144–61 and 161–87. See Bertoloni Meli (1988), (1991), (1993a).

for publication, so he explained to Otto Mencke, he had read only the review of Newton's work in the June 1688 issue of the *Acta*.† The evidence provided by Bertoloni Meli, however, disproves Leibniz's attempt to maintain the independence of his *Tentamen* from the *Principia*.

Nonetheless, Leibniz's papers are extremely original. Their physics, especially in the *Tentamen*, is different from Newton's. They are notable as the first published attempts to apply the new calculus to complex problems in natural philosophy. These papers should be studied in the broad context of Leibniz's effort to build a general theory of motion and live and dead forces. In 1690, Leibniz referred to this theory for the first time as 'dynamics', a term which he introduced in his technical vocabulary during his trip to Italy.‡

6.3.1 First applications of Leibniz's calculus to dynamics

The *Schediasma* is concerned with motion in resisting media. Both the case in which resistance is proportional to velocity and that in which it is proportional to the square of velocity are considered.§ All the problems faced in the *Schediasma* had already been considered by Huygens and by Newton. Furthermore, in the published *Schediasma* the calculus does not occur: only the solutions are given. However, Leibniz states that his calculus was essential in the process of discovery. In fact from a group of manuscripts it is possible to reconstruct Leibniz's mathematical demonstrations.¶

In the first article of the *Schediasma* Leibniz considers the rectilinear motion of a body retarded by a resistance proportional to velocity. We have seen Newton's solution of this problem in Proposition 2, Book 2 (§3.11). Leibniz proceeds as follows. He denotes the velocity by v, the time by t, the space covered by p (for convenience we will use s), the initial maximum velocity by a, the maximum distance traversed by b. Leibniz states a 'general principle' according to which the elements of space are in a compound ratio of the velocities and the elements of time, i.e. ds is proportional to vdt. Leibniz writes

$$\frac{ds}{dt} = \frac{v}{a}. \tag{6.17}$$

By hypothesis (when dt is chosen constant),

$$\frac{-dv}{a} = \frac{ds}{b}, \tag{6.18}$$

---

† *Correspondence*, **3**: 3–5.
‡ Fichant (1994): 9.
§ The best study on Leibniz's *Schediasma* is Aiton (1972b). I rely heavily on Aiton's work in what follows and I refer the reader to Aiton's paper for details.
¶ For details see Blay (1992): 129. In §6.3.1 I owe many insights to Blay (1992).

whose integral, taking into consideration initial conditions, is

$$\frac{a-v}{a} = \frac{s}{b}. \tag{6.19}$$

The above equations lead to the following first-order differential equation:

$$a\,dt = \frac{b\,ds}{b-s}. \tag{6.20}$$

From this Leibniz concludes that 'if the distances remaining are as numbers, the times taken will be as logarithms'.†

In the next articles Leibniz deals with rectilinear motion retarded by resistance proportional to the square of velocity. He further considers motion under the action of constant gravity and resistance (proportional to velocity and to the square of velocity). Let us briefly see how Leibniz deals with rectilinear vertical motion under the action of gravity and resistance.

In the case of resistance proportional to velocity he writes:

$$dv = dt - ds. \tag{6.21}$$

I again remind the reader that Leibniz's differential equations are not equivalent to present day differential equations. They have to be read as equations between differential quantities; in this case differentials of time, velocity and space. The above equation should be read to mean that, for equal infinitesimal intervals of time ($dt = $ constant), the infinitesimal change of velocity is proportional to the infinitesimal change of time minus the infinitesimal change of space.‡ Nowadays we would rather write in terms of functions and derivatives

$$m\frac{dv}{dt} = mg - k\frac{ds}{dt}. \tag{6.22}$$

Using (6.17) to eliminate dt from equation (6.21):

$$ds = \frac{v\,dv}{a-v}, \tag{6.23}$$

or

$$ds = \frac{a\,dv}{a-v} - dv. \tag{6.24}$$

Integrating (6.24) and taking into consideration that, because of (6.21), $s + v = t$, one has that:

If the complements of the acquired velocities to the maximum velocity are as numbers, then the times taken will be as logarithms.§

This result corresponds to Newton's Proposition 3, Book 2.

† Translation in Aiton (1972b): 262. We would write $at = b\ln(b/(b-s))$. Note that for Leibniz $\ln(0) = +\infty$.
‡ On this matter see Bos (1974): 50–1.
§ Translation by Aiton (1972b): 263. We would write $s = a\ln(a/(a-v)) - v$ and $t = a\ln(a/(a-v))$.

Employing the two results considered above, Leibniz is able to solve the problem of determining the trajectory of a body which is projected into a medium whose resistance is proportional to the velocity, and which is acted on by a constant gravitation. In fact he decomposes motion into a horizontal and a vertical component, obtains the equations of motion, and eliminates time in order to obtain the equation of the trajectory. These two examples from the *Schediasma* should suffice to give the reader an idea of Leibniz's algorithmic approach to the study of motion in resisting media.

The *Schediasma* is concluded by an article devoted to the determination of the trajectory of a body projected into a medium which resists as the square of the velocity and on which acts a constant gravitation. Newton did not face this general problem in the *Principia*. Leibniz, however, failed to reach a solution which could satisfy his more careful readers, since he assumed that the composition of motions – employed above for the case in which resistance is proportional to velocity – could be used in this case too to derive the equation of the trajectory of a projectile. But in modern terms, while velocities add vectorially, their squares do not.† Huygens criticized Leibniz on this point. The latter had to admit his mistake. The general problem faced in the last article of the *Schediasma* was considered a very important one by late-seventeenth-century natural philosophers. Huygens considered it too difficult to be solved. It was proposed as a challenge by John Keill and was solved in 1719 by Johann Bernoulli.

The second paper by Leibniz which deserves our attention is the *Tentamen de motuum coelestium causis*. It has been the object of extensive study carried on by Aiton and Bertoloni Meli. Since there is no need to replicate their historical analysis here, I will deal very briefly with Leibniz's essay on planetary motions. As Bertoloni Meli has shown, Leibniz wrote the *Tentamen* after reading the *Principia*. It is notable that Leibniz tried to reduce Newton's results achieved in the first three sections of Book 1 into the framework of his own cosmology. Furthermore, Leibniz did not hesitate to apply the differential calculus to planetary motions. I will briefly turn to Leibniz's cosmology and to his mathematization of celestial mechanics, referring the reader to Aiton and Bertoloni Meli for further details.‡

Leibniz rejected Newton's conception of planetary motion. In the *Principia* planets move in void space and interact through a gravitational force which acts at a distance and instantaneously. Like many of his contemporaries, Leibniz could not accept such a view. He opted rather for a Cartesian cosmology according to which planets are driven around the Sun by a fluid vortex. More specifically, Leibniz thought that three forces acted on the planets: a transradial force, directed

---

† Aiton (1986): 140. In modern symbols one can write $|\vec{v_x} + \vec{v_y}|^2(\vec{v_x} + \vec{v_y})/|\vec{v_x} + \vec{v_y}| \neq |\vec{v_x}|^2\vec{v_x}/|\vec{v_x}| + |\vec{v_y}|^2\vec{v_y}/|\vec{v_y}|$.

‡ Aiton (1960), (1962), (1972a), (1984); Bertoloni Meli (1988), (1991), (1993a).

at right angles to the radius vector joining the Sun and the planet, and two opposite radial forces. The transradial force is caused by a 'harmonic vortex'. The particles composing the harmonic vortex describe circles around the Sun in such a way that their velocity is inversely proportional to the distance from the Sun. Instead of deriving the area law from first principles, as Newton had done in Propositions 1 and 2, Book 1 (see §3.4.1), Leibniz imposed this law by the assumption of the harmonic vortex. In fact, since the planets have a transradial velocity inversely proportional to distance from the Sun, they obey Kepler's area law. When the planet is closer to the Sun, its transverse velocity, i.e. its 'harmonic motion', increases; when it is further away, it decreases, in such a way that equal areas are spanned by the radius vector in equal times.† The two radial forces, namely the outward 'centrifugal conatus' and the inward 'solicitation of gravity', are responsible for the 'paracentric motion'. In practice the resultant of the two radial forces pulls and pushes the planet along the radius vector.

Leibniz represents the trajectory of the planet as a polygonal (see fig. 6.5) with infinitesimal sides M_1M_2, M_2M_3, etc. The planet moves with uniform velocity along one side, until it reaches a vertex. Here the velocity changes abruptly, and the planet performs another uniform rectilinear motion along the next side. The abrupt changes occur at equal infinitesimal intervals of time. These uniform motions can be decomposed in an infinity of ways. One can, for instance, decompose uniform motion along M_2M_3 into two uniform rectilinear motions:

- M_2L (the inertial motion at M_2), and LM_3,
- D_2M_3 (strictly T_2M_3, the circulation) and M_2D_2 (the paracentric motion).

According to Leibniz, the first decomposition is hypothetical, the second is real since each component corresponds to real forces acting on the planet.‡ It should be noted that motion along M_2M_3 is decomposed into uniform motions.

Leibniz shows that the centrifugal conatus is proportional to the square of transverse velocity and inversely proportional to the radius vector r. Since the circulation is harmonic (i.e. since the transradial velocity is inversely proportional to the radius vector), the centrifugal conatus is inversely proportional to the cube of the radius. He expresses the centrifugal conatus in calculus terms as

$$\frac{aa\theta\theta}{2r^3} \qquad (6.25)$$

where θ is the differential of time and a a constant related to areal velocity.

Leibniz was able to show that if the planet orbits in an ellipse, then the 'solicitation of gravity' (as he named it) is inverse square. In fact, from the

---

† One should also add that the planet's motion is planar.
‡ Aiton (1989a): 212–13.

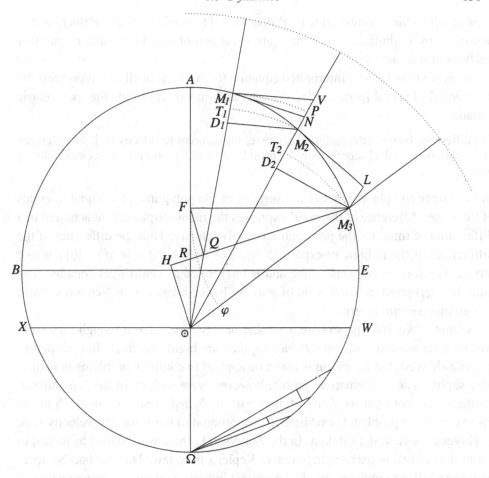

Fig. 6.5. Leibniz's model for planetary motion. After *The general history of astronomy,* **2A** (= *Planetary astronomy from the Renaissance to the rise of astrophysics, Part A: Tycho Brahe to Newton*), R. Taton & C. Wilson (eds.), 10. Cambridge: Cambridge University Press: 1995

geometry of the ellipse, Leibniz derived an equation that can be adapted for the reader's convenience as†

$$\mathrm{d}^2 r = \left(\frac{h^2}{r^3} - \frac{\mu}{r^2}\right)\mathrm{d}^2 t. \tag{6.26}$$

† Leibniz had a wrong factor 2, used the constants ambiguously, and denoted d*t* by *θ*. In fact, as Bertoloni Meli shows in (1993a): 121–2, from the geometry of the ellipse Leibniz could write the following proportion: $\theta a/r : \mathrm{d}r = b : \sqrt{(e^2 - p^2)}$, where *p* and *q* are the minor and major axis, *e* the eccentricity, *a* the principal *latus rectum*, and $p = 2r - q$. By differentiation, one obtains the equation of paracentric motion: $\mathrm{dd}r = \theta^2 a^2 : r^3 - \theta^2 a^2 : r^2$, which appears above in modernized form. For a more complete discussion of these problems see the works by Aiton and Bertoloni Meli already mentioned.

This result, which corresponds to Proposition 11, Book 1, of the *Principia* was described by Leibniz as a 'not inelegant specimen of our differential calculus or analysis of infinites'.[†]

Let us see how Leibniz interpreted equation (6.26). First of all, d^2r represents the differential of radial [paracentric] velocity, and thus it represents the 'paracentric conatus'.

The difference between the radii $[dr]$ expresses the paracentric velocity $[v_r]$, the difference of the differences $[d^2r]$ expresses the element [i.e. the differential] of paracentric velocity $[dv_r]$.[‡]

In fact, since the planet reaches the vertices of the polygonal after equal intervals of time, the 'difference of the radii' expresses the radial displacement acquired in a differential of time, i.e. the paracentric (radial) velocity. Thus the difference of the differences of the radii (d^2r) expresses dv_r. The right-hand side of (6.26) has two terms. The former is inverse cube and is interpreted as centrifugal conatus. The latter is interpreted as solicitation of gravity. In accordance with Newton's result, it is an inverse square term.

Leibniz's two 1689 papers are a notable achievement. Even though they were written after reading Newton's *Principia*, they are highly original. In both papers Leibniz showed that the calculus could be applied to complex problems in natural philosophy. The mathematical proofs, however, were lacking in the *Schediasma*. Furthermore, both papers were a failure. In the *Schediasma* the approach to the inverse ballistic problem for resistance proportional to the square of velocity was, as Huygens observed, mistaken. In the *Tentamen* Leibniz was unable, as he had to admit in the closing paragraph, to derive Kepler's third law. Thus Leibniz's papers very soon fell into oblivion and had no direct influence on the mathematization of resisted and planetary motion. However, they provided the first published example of application of the calculus to some of the problems that Newton had successfully addressed in the *Principia*. As a matter of fact, Leibniz's equation for paracentric motion is the first published differential equation applied to planetary motions. It is a beautiful equation which, independently from the *Tentamen*, reappeared in the mathematical literature in the middle of the eighteenth century in studies by Euler, Clairaut and d'Alembert devoted to motion in rotating frames.[§]

6.3.2 *Comparison of Leibniz's and Newton's mathematizations of trajectories*

A notable difference between Leibniz's and Newton's mathematizations of natural philosophy has been highlighted by Aiton and Bertoloni Meli. This difference

---

† Translation by Bertoloni Meli (1993a): 137.
‡ Translation by Bertoloni Meli (1993a): 135.
§ On this topic see Bertoloni Meli (1993b).

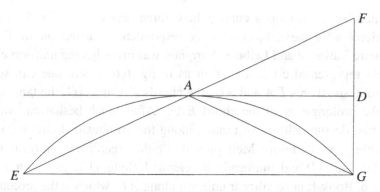

Fig. 6.6. Varignon on uniform circular motion

concerns the mathematical representation of trajectories. While in most of the *Principia* trajectories are continuous and are locally represented by parabolic or circular arcs, in the *Tentamen* and other writings by Leibniz trajectories are discontinuous and are locally represented by infinitesimal polygonal sides. Furthermore, Bertoloni Meli underlines a different approach related to the use of infinitesimals and limits. In the *Tentamen* motion and force are mathematized in terms of infinitesimals, in the *Principia* in terms of finite quantities and first and ultimate ratios.† Aiton and Bertoloni Meli made important new interpretations of Leibniz's dynamics and the reader is warmly advised to consult their works.

It is worth considering these topics in some detail. In this subsection I will show that the difference in the mathematization of trajectories studied by Aiton and Bertoloni Meli reveals a pragmatic difference in the programmes of Leibniz and Newton. Both men were aware that the choice between the polygonal and the continuous model was a matter of convention. However, Leibniz preferred the former model, since it led more straightforwardly to a formulation in terms of calculus, while Newton employed the latter extensively, since he accorded great importance to the geometrical and visual representation of the real world. I postpone to the next section (§6.4) a discussion of the vexed question of the foundation of calculus.

According to Aiton, Leibniz represents the orbit 'as a polygon with first-order infinitesimal sides'.‡ In the *Tentamen* accelerated motion is 'represented as a series of infinitesimal uniform rectilinear motions interrupted by impulses'. In the case of Newton, instead, 'accelerated motion, whether rectilinear or curvilinear,

† Bertoloni Meli writes: 'Variables employed by Newton [in the *Principia*], such as velocities and accelerations, are rigorously finite'. (1993a): 91.
‡ Aiton (1989a): 208.

is represented by a continuous curve where force acts continually'.† The two representations were the subject of a correspondence carried on in 1704–06 between Pierre Varignon and Leibniz. Varignon was investigating uniform circular motion. He represented circular motion as in fig. 6.6, where one can see two circular infinitesimal arcs EA and AG, the chords EA and AG, the tangent AD, and AF (the prolongation of the chord EA). After much hesitation, Varignon noted that two decompositions of motion along the infinitesimal circular arc AG were possible. As Bertoloni Meli puts it: 'either rectilinear uniform motion along the tangent AD and uniformly accelerated along DG, as Newton did in Proposition 6, Book 1; or rectilinear uniform along AF, which is the prolongation of the chord EA, and [rectilinear uniform] along FG'.‡ In the former case the orbit is represented by a sequence of first-order parabolic arcs traversed by uniformly accelerated motion, in the latter by a polygon with first-order infinitesimal sides traversed by uniform rectilinear motion.§ Varignon explained to Leibniz that only the first representation corresponds to reality, since force acts continuously: however, he agreed that both lead to correct results, provided that one does not mix them together and proceeds consistently within one of the two models. Leibniz, on the contrary, preferred the polygonal model. In 1706 he wrote to Varignon:

It is simpler not to put the acceleration in the [infinitesimal] elements, since this is not necessary.¶

Newton employed the parabolic approximation in the *Principia*, most notably in Proposition 6, Book 1, and in the numerous demonstrations that depend upon it (see §3.4.2 and §3.4.3). In Proposition 6, we need scarcely to be reminded, the central force is measured by the displacement QR from inertial motion. This displacement is seen as covered by uniformly accelerated motion: it is a small Galilean fall. In fact, Newton assumes that in a small interval of time, the central force can be regarded as constant (in magnitude and direction): thus the body's trajectory is a Galilean parabola. It is, however, also true, as Westfall and Cohen have taught us, that Newton in many sections of the *Principia* switched to the impulsive, polygonal, model.‖ There is no ambiguity in the *Principia*, since Newton uses either one model or the other. I can refer the reader to Propositions 1 and 2, Book 1, or to the first two Sections of Book 2, where accelerated motion

---

† Bertoloni Meli (1993a): 91.
‡ Bertoloni Meli (1993a): 80–1.
§ For a study of the correspondence between Leibniz, Varignon and Johann Bernoulli on the representation of trajectories see Aiton (1960), (1962) and (1989a). See also Fraser (1985) for the position taken by d'Alembert on this matter.
¶ 'la voye est plus simple qui ne met pas l'acceleration dans les elemens, lorsqu'on n'en a point besoin'. *LMS*, **4**: 151.
‖ Cohen (1971), Westfall (1971).

is modelled as a series of inertial uniform motions interrupted by instantaneous impulses which abruptly change the velocity (see §3.4.1, §3.11 and §3.13).

As far as Leibniz is concerned, it should be noted that he was fully aware of the fictitious character of the polygonal model, as Bertoloni Meli underlines.† And this is so for two reasons. First of all the shape of the polygonal trajectory depends on the arbitrary choice of the progression of variables. For instance, if the differentials of the arc are chosen to be constant one gets a different polygonal representation from that achieved when differentials of time are chosen as constant. But, more deeply, Leibniz did not think that abrupt changes could occur in Nature. According to the principle of continuity, which played such an eminent role in shaping his conception of motion and force, no such thing as an instantaneous change of velocity could actually occur. For that matter, the parabolic representation also is fictitious, since there is a discontinuity in the slopes of the parabolic arcs at the points where they join up.‡ Note that Leibniz did not reply to Varignon by saying that the continuous model does not correspond to reality or that it is mathematically unsound. He just stated that the polygonal model is *preferable*. It was a matter of convention. We have thus to ask ourselves what rendered the polygonal representation preferable for Leibniz.

The answer can be found by looking at the different argumentative strategies adopted in the *Principia* and in the *Tentamen*. While Newton aims at a geometrical representation of motion and force, Leibniz is interested in giving 'not inelegant specimens' of the applicability of his calculus. His aim is to write a differential equation of motion. Leibniz knew very well how useful it could be to represent curves as polygonals: for example, it was by using the characteristic triangle that he was able to advance integration techniques. The representation of geometrical objects in terms of infinitesimals was a preliminary to obtaining differential equations, as we have seen in the example of the quadrature of the cycloid (§6.2.4). Similarly, in dynamics the infinitesimal representation of the trajectory as a polygonal was a preliminary to obtaining differential equations of motion (§6.3.1).

If, on the other hand, the purpose is, as in the *Principia*, to offer a geometric representation of actual force and motion, the parabolic model is a convenient choice. According to Newton, in *rerum natura* trajectories are smooth and force continuous (§2.3.2 and §3.3). Smooth trajectories and continuous forces are approximated by Newton either via the polygonal model, as in Propositions 1 and 2, Book 1, and the first sections of Book 2 (see §3.4.1, §3.11 and §3.13), or via the parabolic model, as in Sections 2 and 3, Book 1. Newton's figures in Section 2 and 3, Book 1, represent, to second-order approximation, the real orbit traversed by

† Bertoloni Meli (1993a): 83.
‡ Erlichson (1992b).

the body under the action of a central force. In order to represent the central force geometrically, the actual displacement of the body can be approximated locally by a curve which has second-order contact with the real trajectory (e.g. the osculating circle (see §4.5) or the parabola of Proposition 6 (see §3.4.2)).

Leibniz's preference for the polygonal model and Newton's usage of the parabolic one do not reveal incommensurable approaches to the mathematization of natural philosophy. They emerge as choices motivated by different aims. Both Newton and Leibniz understood that these choices were a matter of convention. Newton was interested in the geometrical representability of motion and force: he thus drew figures which are approximations in terms of polygons, circles or parabolas of actual displacements of bodies in motion. Leibniz was interested in writing a differential equation of motion. He thus found it more useful to focus on the infinitesimal representation of trajectories.

In this section I have shown that both Leibniz and Newton considered the choice between the polygonal and the continuous model one of convention. As a matter of fact, they switched from one to the other, generally without confusion. Leibniz's preference for the polygonal model depended on the mathematical programme defended in the *Tentamen*, a programme which differed from that adopted in the *Principia*. In the next section, §6.4, I turn to the question of the foundations of calculus. Another point raised by many historians concerns the use of infinitesimals and limits. At this level too, profound conceptual differences between Leibniz and Newton should not be seen. As we shall see, the divide between Leibniz and Newton is not to be easily found at the level of conceptual schemes, but rather at the level of pragmatic choices.

6.4 Leibniz on the foundations of the calculus

6.4.1 *'Not equivalent in practice'*

During the famous priority dispute with Leibniz, the British were to maintain not only that Newton was the first inventor of the calculus, but also that he was the first to lay down its true foundations. They were to argue that the fact that the differential and integral calculus had no valid foundations was proof positive of Leibniz's plagiarism. He could copy and change the notation, but he was incapable of understanding the foundational aspects of the technique. As is well known, during his later years Newton spent a great deal of time and energy preparing various accounts of the dispute with Leibniz, and of the issues it raised with respect to the intrinsic nature of the calculus.

Notwithstanding the efforts of the Newtonians, establishing a comparison between Leibniz's and Newton's calculi is not a trivial task. The reason is that Leibniz and Newton presented several versions of their calculi. Leibniz never published a systematic treatise, but rather divulged the differential and integral calculus in a series of papers and letters. He changed his mind quite often, especially on foundational questions. Newton, as we know from §2.3, in the 1670s gave up his earlier version of the calculus (the analytical method of fluxions), based on infinitesimal moments, in order to develop a foundation in terms of first and ultimate ratios. Once he had achieved this mature version, he did not abandon it. But a further complication arises when we consider the fact that Newton produced a geometric, as well as an analytical, version of the method of first and ultimate ratios: the former appearing as the synthetic method of fluxions employed in the *Geometria curvilinea* and the *Principia*, the latter as the method of limits of *De quadratura*.

Modern historians of mathematics have attempted to contrast too sharply Leibniz's and Newton's calculi. For instance, it has been said that in the Newtonian version variable quantities are seen as varying continuously in time, while in the Leibnizian one they are conceived as ranging over a sequence of infinitely close values.† It has also been said that in the fluxional calculus 'time', and in general kinematical concepts such as 'fluent' and 'velocity', play a role which is not accorded to them in the differential calculus. The remark often occurs that geometrical quantities are seen in a different way by Leibniz and Newton. For instance, for Leibniz a curve is conceived as a polygonal with an infinite number of infinitesimal sides, while for Newton curves are smooth.‡

These sharp distinctions, which certainly help us to capture part of the truth, are possible, however, only after simplifying the two calculi. As a matter of fact, they are more applicable to a comparison between the simplified versions of the Leibnizian and the Newtonian calculi codified in textbooks such as l'Hospital's *Analyse des infiniment petits* (1696) and Simpson's *The Doctrine and Application of Fluxions* (1750), rather than to a comparison between Newton and Leibniz. As I will try to argue, important aspects of their mathematics are thus ignored in these historical interpretations.

The differences between the Leibnizian and the Newtonian calculi should not be overstressed. And especially, as I will maintain in this section, *they should be sought not at the syntactic or at the semantic level, but rather at the pragmatic level*. The two calculi (i.e. the Leibnizian calculus and the Newtonian *analytical* method) had much in common both at the syntactic level of the algorithm, and at the semantic level of the interpretation of the algorithm's symbols and justification of the algorithm's rules. There is in fact a possibility of a translation between the fluxional and the differential algorithms (through correspondences between $\dot{x}o$

---

† In several works Bos has masterfully underlined this aspect of the Leibnizian calculus. See, e.g., Bos (1980a): 92.

‡ Bertoloni Meli (1993a): 61–73.

and dx). These translations were conducted by the Leibnizian and the Newtonian mathematicians: they were aware that there is not a single theorem which can be proved in one of the two calculi and which cannot have a counterpart in the other. It is exactly this 'equivalence' which gave rise to the quarrel over priority.

In discussing the question of equivalence, quite appropriately Hall writes:

Did Newton and Leibniz discover the same thing? Obviously, in a straightforward mathematical sense they did: [Leibniz's] calculus and [Newton's] fluxions are not identical, but they are certainly equivalent. [...] Yet one wonders whether some more subtle element may not remain, concealed, for example, in that word 'equivalent'. I hazard the guess that unless we obliterate the distinction between 'identity' and 'equivalence', then if two sets of propositions are logically equivalent, but not identical, there must be some distinction between them of a more than trivial symbolic character.†

In order to explore this more subtle and concealed level, where a comparison between Newton's and Leibniz's calculi can be established, Skuli Sigurdsson has proposed using the category 'not equivalent in practice'. Despite the equivalence of the two calculi:

[this] equivalence breaks down once it is realized that competing formalisms suggest separate directions for research and therefore generate different kinds of knowledge.‡

Similarly Ivo Schneider has remarked that the starting point, the main emphasis and the expectations of the two pioneers were not at all identical.§ Bertoloni Meli has drawn a comparison between a Newtonian and a Leibnizian mathematician and two programmers who use different computer languages:

Even if the two programmes are designed to perform the same operations, the skills required to manipulate them may differ considerably. Thus subsequent modifications and developments may follow different routes, and this is precisely what happened in Britain and on the Continent in the eighteenth century: despite the initial 'equivalence' of fluxions and differentials.¶

I agree with the approach of the above-mentioned scholars. Rather than looking for sharp distinctions between the two calculi, we should look for subtler, less evident aspects. As I will try to argue in the following pages, Newton and Leibniz had two 'mathematically equivalent' symbolisms. At the syntactical level they could translate each other's results, while at the semantic level they agreed on important foundational questions. Nonetheless, at the pragmatic level, they oriented their researches along different directions. Thus, belonging to the Newtonian or to the Leibnizian school meant having different skills and different expectations. It meant stressing different lines of research and different values. After all, it often happens in the history of mathematics that the difference between

---

† Hall (1980): 257–8.
‡ Sigurdsson (1992): 110.
§ Schneider (1988): 142.
¶ Bertoloni Meli (1993a): 202.

two schools does not lie in logical or conceptual incommensurabilities, but rather in more pragmatic aspects: the teaching methods, the training of mathematicians, the expectations for future research, the system of values which supports the view that one method of proof is preferable to another, etc.

6.4.2 Infinitesimals and vanishing quantities

The problem of foundations did not exist in the seventeenth century in the form which it took in the early nineteenth century. Basically, one of the most important foundational questions faced by seventeenth- and eighteenth-century mathematicians was a question concerning the referential content of mathematical symbols (typically 'do infinitesimals exist?').This 'ontological' question was followed by a 'logical' question about the legitimacy of the rules of demonstration of the new analysis (typically 'is $x + dx = x$ legitimate?').[†] To these two questions Newton and Leibniz gave similar answers. They both stated that:

- actual infinitesimals do not exist, they are useful fictions employed to abbreviate proofs;
- infinitesimals should rather be defined as varying quantities in a state of approaching zero;
- infinitesimals can be completely avoided in favour of limit-based proofs, which constitute the rigorous formulation of calculus.

6.4.2.1 Actual infinitesimals do not exist

To the question 'do differentials exist?', many seventeenth-century 'geometers' answered in the affirmative. Leibniz did not. From his very early manuscripts to his mature works, it is possible to infer that for him actual differentials were just 'fictions', symbols without referential content. For instance, in a letter to Des Bosses of 1706 he stated:

Philosophically speaking, I no more believe in infinitely small quantities than in infinitely great ones, that is, in infinitesimals rather than infinituples. I consider both as fictions of the mind for succinct ways of speaking, appropriate to the calculus, as also are the imaginary roots in algebra.[‡]

The question of the existence of infinitesimals is faced in a particularly interesting exchange of letters with Johann Bernoulli dating 1698. The Swiss mathematician interpreted differentials as actually existing components of continuous quantity. Much to Bernoulli's dismay, the master revealed that he conceived differentials as

---

[†] Bos (1980a).

[‡] 'Ego philosophice loquendo non magis statuo magnitudines infinite parvas quam infinite magnas, seu non magis infinitesimas quam infinutuplas. Utrasque enim per modum loquendi compendiosum pro mentis fictionibus habeo, ad calculum aptis, quales etiam sunt radices imaginariae in Algebra.' *LPS*, 2: 305.

useful symbolic devices, but that he was not prepared to commit himself on their existence.†

Leibniz repeated many times that for him the question of the existence of infinitesimals had to be distinguished from that of their usefulness as algorithmic devices. While he was leaning, for philosophical reasons, towards a denial of the existence of infinitesimals, he wanted also to stress that this ontological question was somewhat extraneous to mathematics. A typical statement, written in the early years of the eighteenth century, is the following:

We have to make an effort in order to keep pure mathematics chaste from metaphysical controversies. This we will achieve if, without worrying whether the infinites and infinitely smalls in quantities, numbers and lines are real, we use infinites and infinitely smalls as an appropriate expression for abbreviating reasonings.‡

Leibniz was thus leaving to his disciples the choice of maintaining, *philosophice loquendo*, different approaches to the ontological question on the existence of infinitesimals. What he wished to defend was their utility as symbols in mathematical calculation.

As we have seen in §2.3.2, Newton held similar views: he denied the existence in *rerum natura* of infinitesimals. However, he admitted their usefulness as abbreviations. For instance, in 1715 – after having defended the safer ontological content of first and ultimate ratios, when compared with infinitesimals – he noted:

But when he [Newton] is not demonstrating but only investigating a Proposition, for making dispatch he supposes the moment o to be infinitely little, & forbears to write it down & uses all manner of approximations wch he conceives will produce no error in the conclusion.§

6.4.2.2 *Infinitesimals as variables*

The use of infinitesimals was justified, according to Leibniz, since correct results could be derived by employing the algorithm of differentials. As Leibniz said, differentials are 'fictions', but 'well-founded fictions'. Why 'well-founded'? Leibniz seems to have had the following answer. He denied the actual infinite and actual infinitesimal and conceived the differentials as 'incomparable quantities': varying quantities which tend to zero. In his writings of the 1690s Leibniz described these 'incomparables' as magnitudes in a fluid state which is different from zero, but which is not finite. These quantities would give a meaning to dy/dx as a ratio of two quantities. In fact, if dy and dx are zeros, one ends with the

† Leibniz wrote: 'In Calculo haec utiliter assumimus; sed hinc non sequitur extare posse in natura.' *LMS*, **3**(2): 516.

‡ 'Dandam operam est, ut Mathesis pura a controversiis Methaphysicis illibata conservetur. Hoc nos facturos si parum curantes an realia sint infinita et infinite parva in quantitatibus, numeris, lineis: utamur infinitis et infinite parvis tanquam expressione apta ad cogitationes contrahendas'. Niedersächsische Landesbibliothek (Hanover) Lh 35 VIII 21, f. 1r quoted in Pasini (1993): 149n.

§ *Mathematical Papers*, **8**: 572.

difficulty of attributing a value to $0/0$, but if they are finite, they cannot be neglected (thus $x + dx = x$ would be invalid).

The idea of interpreting differentials as 'incomparable quantities' was to play a prominent role in the *Tentamen de motuum coelestium causis* (1689). Leibniz wrote:

In the demonstrations I have employed *incomparably small quantities*, such as the difference between two finite quantities, incomparable with the quantities themselves. Such matters, if I am not mistaken, can be exposed most lucidly as follows. Thus if someone does not want to employ *infinitely small* quantities, one can take them to be as small as one judges sufficient as to be incomparable, so that they produce an error of no importance and even smaller than allowed.†

Since it is highly probable that Leibniz wrote the *Tentamen* after reading the *Principia*, it is tempting to conceive the 'incomparable quantities' as an adaptation of Leibniz's infinitesimals to Newton's vanishing quantities.

After 1689, in a period in which he certainly had access to and was carefully reading the *Principia*, Leibniz defined his position more precisely, getting closer and closer to Newton. In fact, he often spoke of differentials in a way which is very close to Newton's 'vanishing quantities' or to 'moments'. That is, he had recourse to the concept of continuous time variation and defined differentials as 'vanishing' quantities or as 'momentary' increments. For instance in the first volume of *Miscellanea* of the Berlin Academy published in 1710 he wrote:

Here dx means the element, that is the (momentaneous) increment or decrement of the quantity x (continuously) growing. It is also called *difference*, clearly between two successive elementary (or inassignably) different x, as one is obtained from the other which (momentaneously) increases or decreases.‡

In a manuscript written in about 1702, bearing the title *Défense du calcul des différences*, Leibniz employed expressions which even more closely resemble the Newtonian method of first and ultimate ratios. We read, for instance: 'I would like therefore to add that the foundation of all this can be explained by taking z and x, i.e. bA, hA, *in the very act of* vanishing and falling on A'.§ Here bA and hA are defined by Leibniz as two vanishing magnitudes ('grandeurs evanouissantes') which vanish simultaneously when b and h fall on A ('tombent en A'). The parallelism with Newton's first and ultimate ratios is quite evident. Leibniz is here trying to justify the use of infinitesimals by recourse to the concept of continuous variation and limiting ratios proper to Newton's theory.

† Translation by Bertoloni Meli (1993a): 130–1.
‡ 'Hic *dx* significat elementum, id est incrementum vel decrementum (momentaneum) ipsius quantitatis x (continue) crescentis. Vocatur et *differentia*, nempe inter duas proximas x elementariter (seu inassignabiliter) differentes, dum una fit ex altera (momentanee) crescente vel decrescente'. *LMS*, 7: 222.
§ 'Je veux pourtant adjouter que le fondement de tout cecy se peut expliquer en prenant z et x, c'est à dire bA, hA, *dans l'acte même* d'evanouir et de tomber en A'. Niedersächsische Landesbibliothek (Hanover) Lh 35 VI 22, f. 2v. Quoted in Pasini (1993): 145n and edited in Pasini (1988): 708.

6.4.2.3 From infinitesimals to limits

In several mature writings Leibniz stated that differentials are well founded, since they are symbolic abbreviations for limit procedures. The calculus of differentials would than be a shorthand for a calculus of finite quantities and limits, equivalent to Archimedean exhaustion. For instance, in 1701 he wrote:

In fact, instead of the infinite or the infinitely small, one can take magnitudes that are so large or so small that the error will be less than the given error, so that one differs from the style of Archimedes only in the expressions, which are, in our method, more direct and more apt to the art of discovery.†

Jacob Bernoulli in a reply, planned with Leibniz, to the mathematician Clüver – an early opponent of the differential calculus – endorsed a limit-based approach which would have satisfied Newton:

I believe [...] that your illusion arises from the fact that you think of dy as something determined in itself while it is not but a mere fiction of the mind and consists merely in a perpetual fluxion ['fluxio'] towards nothingness. This is why the ratio $(2y^2 + dy^2)/(4y^2 - dy^2)$ is always variable and does not become fixed but when it is perfectly nothing and does not differ in any way from one half.‡

The use of the term 'fluxio' as well as the content of this statement bears a strong Newtonian flavour.

Once the calculus is reduced to limit-based proofs, the logical question takes the form 'are limit-based proofs legitimate?' In order to answer this question, both Newton and Leibniz employed the concept of continuity. However, the former legitimated limits in terms of intuition of continuous flow, while the latter referred to a philosophical 'principle of continuity'. In the *Principia*, in order to demonstrate the existence of limits, Newton referred to the intuition of continuous motion (§3.3). Leibniz, on the contrary, in order to justify the limiting procedures, referred to a metaphysical principle of continuity which he expressed in several forms and contexts.§ The 'law of continuity' pervades Leibniz's thought. He made use of it in cosmology, in physics and in logic. Thus, invoking the law of continuity, he affirmed that rest can be conceived as an infinitely little velocity or that equality can be conceived as an infinitely little inequality. In 1687 Leibniz stated this principle in his difficult philosophical prose as follows:

When the difference between two instances in a given series or that which is presupposed [*in datis*] can be diminished until it becomes smaller than any given quantity whatever,

---

† 'Car au lieu de l'infini ou de l'infiniment petit, on prend des quantités aussi grandes, et aussi petites qu'il faut pour que l'erreur soit moindre que l'erreur donnée, de sorte qu'on ne diffère du stile d'Archimède que dans les expressions, qui sont plus directes dans nôtre méthode et plus conformes à l'art d'inventer'. *LMS*, **5**: 350.

‡ *LMS*, **3**(1): 55. Translation in Mancosu & Vailati (1990).

§ Breger (1990).

the corresponding difference in what is sought [*in quaesitis*] or in their results must of necessity also be diminished or become less than any given quantity whatever.†

In order to explain the meaning of this general principle, Leibniz refers to the geometry of conic sections. An ellipse, he says, may approach a parabola as closely as one pleases, so that the difference between the ellipse and the parabola (the difference between what 'results') may become less than any given difference ('moindre qu'aucune difference donnée'), provided that one of the foci (what is 'presupposed') is removed far enough away from the other. Consequently, the theorems valid for the ellipse can be extrapolated to the parabola, 'considering the parabola as an ellipse when one of the foci is infinitely distant, or (in order to avoid this expression) as a figure which differs from a certain ellipse less than any given difference'.‡ It is the continuous dependence between what is 'presupposed' and what 'results' that justifies limit-based reasonings – reasonings in which, for instance, one extrapolates to the parabola what has been proved for the ellipses: 'in continuous magnitudes the exclusive *extremum* can be treated as inclusive'.§

The principle of continuity could thus be used to justify the infinitesimal concepts employed in the calculus, once these infinitesimals were understood as vanishing quantities and the procedures of the calculus were read in terms of limits. For instance, the identification between a curve and a polygon with infinitely many infinitesimal sides was justified by Leibniz in an exchange of letters with Varignon as follows:

However, even though it is not rigorously true that rest is a kind of movement, or that equality is a kind of inequality, as it not true at all that the circle is a kind of regular polygon: nonetheless, one can say that rest, equality and circle terminate the movements, [in]equalities and regular polygons, which by a continuous change there arrive vanishing. And despite the fact that these terminations are exclusive, i.e. not included rigorously in the varieties which they limit, nonetheless they have the properties, as if they were included, according to the language of the infinites or infinitesimals, which considers the circle, for instance, as a regular polygon with an infinite number of sides.¶

As we know, this is the polygonal conception of a curve which is employed in the *Tentamen* in the study of planetary trajectories.

† The original French is in *LPS*, **3**: 52. A Latin version is in *LMS*, **6**: 129. Translation by Leroy E. Loemker in Leibniz (1976): 352.
‡ 'en considerant cellecy comme une Ellipse dont un des foyers est infiniment eloigne ou (pour eviter cette expression) comme une figure qui differe de quelque Ellipse moins que d'aucune difference donnée.' *ibid.*
§ 'in continuis extremum exclusivum tractari possit ut inclusivum.' *LMS*, **5**: 385.
¶ 'Cependant quoyqu'il ne soit point vray à la rigueur que le repos est une espece de mouvement, ou que l'égalité est une espece d'inégalité, comme il n'est point vray non plus que le Cercle est une espece de polygone regulier: neantmoins on peut dire, que le repos l'égalité, et le cercle terminent les mouvemens, les égalités, et les polygones reguliers, qui par un changement continuel y arrivent en evanouissant. Et quoyque ces terminaisons soyent exclusives, c'est à dire, non-comprises à la rigueur dans les varietés qu'elles bornent, neantmoins elles en ont les proprietés, comme si elles y estoient comprises, suivant le langage des infinies ou infinitesimales, qui prende le cercle, par example, pour un polygone regulier dont le nombre des costés est infini'. *LMS*, **4**: 106.

6.5 Conclusion: different policies

The similarities between Newton's and Leibniz's approach to the foundation of the calculus are striking. Newton's approach to the question of the existence of infinitesimals is similar to Leibniz's. For Newton too, infinitesimals ('moments' or 'indefinitely little quantities') can be used as a shorthand for longer and more rigorous proofs given in terms of limits (§6.4.2.1). Both Newton and Leibniz speak of infinitesimals as 'vanishing' quantities in such a way that these quantities seem to be defined as something in between zero and finite, as quantities in the state of disappearing, or coming into existence, in a fuzzy realm between nothing and finite (§6.4.2.2). More often they make it clear that infinitesimals can be replaced in terms of limits (§6.4.2.3).

I do not find a strong conceptual opposition between Leibniz and Newton, but rather different 'policies'. Both agreed that limits provide a rigorous foundation for the calculus. However, for Leibniz this was more a rhetorical move to defend the legitimacy of the differential algorithm, while for Newton this was a programme that should be implemented. Newton developed explicitly a theory of limits (see §2.3 and §3.3), publishing it in analytical and synthetic forms. Leibniz simply alluded to the possibility of building the calculus on the basis of such a theory. As we have now learned from Knobloch's studies, Leibniz in an unpublished long treatise on quadratures completed in 1676 gave what he conceived as a rigorous geometric foundation of the calculus in terms of *ad absurdum* Archimedean procedures. The method proposed in the manuscript can be defined as 'Leibniz's synthetic method': a geometric version of his calculus.† After 1676, however, Leibniz could live with the infinitesimal quantities, and did not publish his treatise. Newton made a serious effort in published works, such as the *Principia* and *De quadratura*, to eliminate infinitesimals.‡ It is relevant that the mature Leibniz felt the need to elaborate publicly the theory of limits mainly as a retrospective justification: typically when occupied in defending the calculus from critics such as Nieuwentijt, Clüver, Jenisch and some members of the Paris Academy of Sciences. His idea was that such 'metaphysical' questions should not interfere with the successful development of the algorithm of differentials.

We thus begin to grasp why it is fruitful to conceive the Newtonian and the Leibnizian calculi as 'not equivalent in practice'. Notwithstanding the similarities regarding the justification of the algorithm, *in practice* the approach of the two men was different. While Newton spent much effort in developing limits as his 'rigorous' language, Leibniz preferred to promote the use of infinitesimals as a means of discovery of new truths.

---

† Knobloch (1989), Leibniz (1993).
‡ See §2.3 and §3.3. Cf. also Lai (1975), Kitcher (1973) and Guicciardini (1993).

Also, the argument of continuity with the 'geometry of the Ancients' played a different role in Newton's and in Leibniz's conceptions of calculus. For Newton, showing a continuity between his method and the methods of the Ancients was a crucial step in order to guarantee the acceptability of his mathematical discoveries. Leibniz stressed this continuity in ways which are significantly different. The German philosopher also referred in several places to the Renaissance Neoplatonic tradition of the *prisca philosophia*. Several studies have established how deeply Leibniz was impressed by the myth of a *philosophia perennis*. Other scholars have researched the influence on Leibniz's thought of a mystical Platonic–Pythagorean philosophy, of the alchemical tradition or of the Kabbala.† However, one can compare the opening of Newton's *De mundi systemate* with the opening lines of Leibniz's *Tentatem de motuum coelestium causis*. While Newton was squarely placing himself in the Neoplatonic tradition, citing as precursors the Chaldeans, the Pythagoreans, Numa Pompilius, and so forth (see §4.2), Leibniz referred to much more contemporary sources. After having stated that the Pythagoreans 'seem perhaps to have proffered tentatively rather than correctly determined' the heliocentric system, he moved on to chart at length the contributions of Copernicus, Tycho Brahe, Kepler, Gian Domenico Cassini, Descartes, Galileo and Evangelista Torricelli.‡ In a few words: there were important cultural traditions that were common ground for Newton and Leibniz. But the *uses* of these traditions were different.

As far as mathematics is concerned, the rhetoric on the novelty of the calculus pervades Leibniz's writings. Reference to the ancient mathematicians generally took the rather abused form of a tribute to Archimedes' 'method of exhaustion'. In several circumstances Leibniz stated that the limiting procedures characteristic of the calculus are reducible to *ad absurdum* Archimedean demonstrations: i.e. to the so-called 'method of exhaustion'. He generally also affirmed that such a reduction, even if possible in principle, could be avoided in the *ars inveniendi*.§ As we mentioned above, in an important manuscript written at the end of his Paris stay and edited recently by Knobloch, Leibniz showed at length how the quadrature procedures of the integral calculus could be rewritten in the style of the *ad absurdum* Archimedean proofs, but did not feel the need to publish these results.¶

Leibniz in most of his declarations concerning the calculus wished to highlight the novelty and the revolutionary character of his algorithm, rather than continuity

---

† Politella (1938), Schmitt (1966), Couzin (1970), Belaval (1975), MacDonald Ross (1978) and (1983).
‡ Bertoloni Meli (1993a): 126–8.
§ For instance in a reply to Nieuwentijt published in 1695 Leibniz wrote: 'Et quae tali quantitate [incomparabiliter parva] non differunt, aequalia esse statuo, quod etiam Archimedes sumsit, aliique post ipsum omnes. Et hoc ipsum est, quod dicitur differentiam esse data quavis minorem. Et Archimedeo quidem processu res semper deductione ad absurdum confirmari potest'. *LMS*, **5**: 322. See also *LMS*, **7**: 391.
¶ Knobloch (1989), Leibniz (1993).

with ancient exemplars. Significantly his first publication on the calculus is entitled *Nova methodus*. This approach is quite at odds with Newton's 'classicism'. Furthermore, Leibniz often refers to the heuristic character of the calculus understood as an algorithm independent from geometrical interpretation. It is exactly this independence that would render the calculus so efficacious in the process of discovery. The calculus, according to Leibniz, should be seen also as an *ars inveniendi*: as such it should be valued by its fruitfulness, rather than by its referential content. We can calculate, according to Leibniz, with symbols devoid of referential content (for instance, with $\sqrt{-1}$) provided the calculus is structured in such a way as to lead to correct results. Newton could not agree: for him mathematics devoid of referential content could not be acceptable.

Writing to Huygens in September 1691, Leibniz affirmed with pride:

It is true, Sir, as you correctly believe, that what is better and more useful in my new calculus is that it yields truths by means of a kind of analysis, and without any effort of the imagination, which often works as by chance, and it gives us the same advantages over Archimedes, which Viète and Descartes gave us over Apollonius.[†]

Similarly, three months later he remembered his successful study of the cycloid:

I remember that when I considered the cycloid, my calculus gave me, almost without meditation, the great part of the discoveries which have been made concerning this subject. For what I love most about my calculus is that it gives us the same advantanges over the Ancients in the geometry of Archimedes, that Viète and Descartes have given us in the geometry of Euclid or Apollonius, in freeing us from having to work with the imagination.[‡]

Leibniz was thus praising the calculus as a *cogitatio caeca* and promoted the 'blind use of reasoning' among his disciples. Nobody, according to Leibniz, could follow a long reasoning without freeing the mind from the 'effort of imagination'.[§]

Newton required different standards of rigour: according to him interpretability could not be abandoned in 'good geometry'. Newton did not value mechanical algorithmic reasoning. He always spoke in the highest terms of the geometrical demonstrations of Huygens and contrasted the elegant geometrical methods of the 'Ancients' with the mechanical algebraic methods of Descartes. Furthermore, he made it clear that the symbols of the 'analytical method of fluxions' had to be interpreted in terms of the 'synthetic method' (see §2.3.2). Newton understood his

---

[†] 'Il est vray, Mons., comme vous jugés fort bien, que, ce qu'il ij a de meilleur et de plus commode dans mon nouveau calcul c'est qu'il offre des verités par une espece d'analyse, et sans aucun effort d'imagination, qui souvent ne reussit que par hazard, et il nous donne sur Archimede tous les avantages que Viete et Des Cartes nous avoient donnés sur Apollonius'. *LMS*, **2**: 104 (also Huygens *Oeuvres*, **10**: 157).

[‡] 'Je me souviens qu'autres fois lors que je consideray la cycloide, mon calcul me presenta presque sans meditation la pluspart des decouvertes qu'on a faites la dessus. Car ce que j'aime le plus dans ce calcul, c'est qu'il nous donne le même avantage sur les anciens dans la Geometrie d'Archimede, que Viete et des Cartes nous ont donné dans la Geometrie d'Euclide ou d'Apollonius; en nous dispensant de travailler avec l'imagination.' *LMS*, **2**: 123 (also Huygens *Oeuvres*, **10**: 227).

[§] Pasini (1993): 205.

method as more than a mere heuristic tool: geometrical interpretability guaranteed ontological content.†

It should be stated that, notwithstanding his declarations in favour of a calculus as 'blind reasoning', Leibniz always embedded the algorithm into a geometrical interpretation. Leibniz's differentials and integrals, as much as Newton's fluents and fluxions, were referred to geometrical objects. It is revealing that Leibniz always paid attention to the geometrical dimensions of the symbols occurring in a differential equation. In fact, it is by studying the geometry of differentials (e.g. the characteristic triangle) that Leibniz and his immediate followers could extract differential equations. But once a differential equation was obtained, it was handled as far as possible as an algebraic object. From time to time, recourse to geometrical thinking on the model under study was necessary. Leibniz was compelled to do so (as much as Newton), since the rules of the calculus alone did not yet allow one to solve in a completely algorithmic way the problems in geometry and dynamics that he faced (especially when transcendental quantities occurred). A complete algebraization of calculus was achieved only in the late eighteenth century. In Leibniz's times the calculus as 'blind reasoning' was thus more a *desideratum* than a reality. Reinterpretation of the symbolism into the geometric model was possible, and in many cases necessary. But, contrary to Newton's opinion, this reinterpretation was not seen by Leibniz as a benefit, as a strategy to be pursued.

Leibniz's attention to symbolism led him to develop an algebra of differentials (see §6.2.4). His main target was the construction of a set of algorithmic rules: a *calculus*. The rules of the calculus are thus instructions on how to manipulate the ds and the $\int$s, and they allow algorithmic procedures which are as much as possible independent of the geometrical context (Leibniz even considered $d^\alpha x$, for a fractional α). For instance, the chain rule in Leibnizian terms takes a form (see formula (6.12)) which is suggested by the notation itself. Everything can be done, of course, also in the notation of the analytical method of fluxions. Newton, however, preferred to give examples which show the rule, rather than the rule itself. For instance, he would introduce the chain rule with an example, as a set of instructions applied to the solution of a particular problem.

Leibniz valued highly the possibility of applying his algorithm to natural philosophy. The *Schediasma* and the *Tentamen* were written in response to the *Principia*, but also in order to show the power and extension of the calculus. Whereas Newton's *magnum opus* had been published in a language which seemed at first sight not revolutionary, Leibniz, in his first and unique attempt to mathematize natural philosophy, was eager to contrast his new 'blind' algorithm with Newtonian geometrical methods. He did not know that Newton was already in possession of

---

† In 1715 Newton wrote with disdain: 'Mr Leibniz's [method] is only for finding it out'. *Mathematical Papers*, **8**: 598.

an equivalent algorithm. He did not know that Newton had already employed his sophisticated techniques of the analytical method of fluxions in handling advanced topics such as central force motion, attraction of extended bodies and the solid of least resistance.

Part three

Two schools

7

Britain: in the wake of the *Principia*

7.1 Purpose of this chapter

In this chapter I will try to chart the failures and successes that the British Newtonians encountered in reading Newton's *magnum opus*. I will abandon technical matters and I will adopt a descriptive level of presentation, while in the next chapter the reader will find some information on the mathematical approach to natural philosophy developed in the British school. Even though my analysis is far from complete, I hope that the material collected allows some generalizations on the Newtonian school. I am particularly interested in identifying the validation criteria that were shared among Newton's closest associates. After the publication of the *Principia*, and particularly during the priority dispute with Leibniz, Newton surrounded himself with a small group of mathematicians with whom he shared his discoveries. They adhered to many of Newton's methodological values. As a bonus they were often helped by Newton in their academic careers: e.g. David Gregory was appointed to the Savilian Chair of Astronomy in Oxford through Newton's recommendation. The same holds true for the appointment of Edmond Halley to the Savilian Chair of Geometry, Colin Maclaurin to the Chair of Mathematics in Edinburgh, or William Whiston to the Lucasian Chair in Cambridge. We will have to pay some attention to the unpublished, private, side of their scientific production. As will become clearer in the next chapter, the values and expectations that the Newtonians shared oriented their research along lines different from those pursued by the Leibnizians. These values can sound new and unexpected to the modern reader. This happens frequently in the history of science. For instance, as Kuhn has argued in (1957), the refusal to use the equant circle was an important aspect of the Copernican astronomical technique, an aspect linked to a constellation of values that the historian must unearth. Similarly, in the Newtonian school, the refusal of infinitesimals and of uninterpreted symbols, and the preference accorded to geometrical proofs in the published writings, are linked to their conception of mathematics, to the cosmology that they accepted, to the competence of the audience that they had in mind.

7.2 Philosophers without Mathematicks

The publication of the *Principia* was preceded by some well-orchestrated propaganda by Halley at the Royal Society.† The book won immediate fame for its author and was adopted by most of the fellows of the Royal Society as containing the right answer, as well as the right method of obtaining it, to the problems concerning cosmology and planetary motions which had occupied people such as Hooke, Halley and Wren in the early 1680s. However, praising the *Principia* was easy, reading it was another matter. A well-known legend says that a student in Cambridge, while Newton was passing by, was heard to utter, 'There goes the man who has writt a book that neither he nor any one else understands.'‡ Who understood the *Principia*'s mathematics? Who were its readers? In this chapter I will deal with Newton's fellow countrymen. In the next we will move to the Continent, in order to consider the reactions in two important centres of mathematical research: Basel and Paris.

The historian who is looking for mathematical readers of the *Principia* and their reactions faces two problems. The first is that the number of people competent enough to follow Newton's demonstrations was, and still is, very limited. The second is lack of sufficient information. Of course, absence of evidence is not to be confused with evidence of absence. It is particularly sad that we know so little, or more accurately that we know nothing, about the reactions to the *Principia* of two of the most eminent British natural philosophers of Newton's times: Wren and Wallis.

Wren was a competent mathematician, who, for instance, in the 1660s had rectified the cycloid. He was interested in planetary motions and had embraced a magnetic explanation of the action of the Sun on its planets. He speculated about a model in which the planet's inertial motion is deflected by the Sun's attractive force. There are reasons to think that he even thought that this force is inverse square. As a matter of fact, Newton himself credited Wren, together with Halley

---

† Cook (1991).
‡ King's College, Cambridge, MS Keynes 130.5, no. 2. Quoted in Axtell (1969): 166.

and Hooke, with the inverse square law.† It is, in fact, very simple to obtain the inverse square law, by combining Huygens' law for circular uniform motion [$F = v^2/\rho$] and Kepler's third law. Newton certainly wished to quote Halley and Wren in order to refute Hooke's claims of having being plagiarized by Newton on this crucial matter. As Bennett says, 'this little remark was the sole concrete result of much passion and ill-feeling.'‡ It is probable that Wren formulated the idea, taken up by Hooke, of employing conical pendula as a model of planetary motion. Wren also investigated cometary paths and had a role in the edition of Flamsteed's *Historia coelestis*. In a few words, he was well equipped to read the *Principia* with a competent eye. Unfortunately, I have been unable to locate any trace of Wren's reactions to Newton's masterpiece.§

A similar situation holds for Wallis. One of the inspirers of Newton's mathematical activity, Wallis was also keenly interested in the applications of mathematics to natural philosophy. Since the 1660s, not only had he produced works on topics close to Newton's *Principia*, such as mechanics and gravitation, but he had even formulated theories on the flow and ebb of tides and on the air's gravity. One of his last mathematical essays devoted to natural philosophy, *A Discourse concerning the Measure of the Air's Resistance to Bodies moved in it* (1687), was concerned with projectile motion in a resisting medium. Here Wallis dealt with the case in which resistance is proportional to the velocity, reaching results similar to Newton's. Wallis sent his paper to the Royal Society in January 1687. Halley consulted Newton in order to avoid a further priority dispute. He reassured Newton that Wallis's 'result is much the same with yours', adding that 'he had the hint from an account I gave him of what you have demonstrated.'¶ Halley had written to Wallis in December 1686, asking his opinion about the results obtained by Newton:

You were pleased to mention some thoughts you had of communicating your conclusions concerning the opposition of the Medium to projects moving through it; the Society hopes you continue still inclined so to do, not doubting but that your extraordinary talent in matters of this nature, will be able to clear up this subject which hitherto seems to have been only mentioned among Mathematicians, never yet fully discussed. Mr. Isaac Newton about 2 years since gave me the inclosed propositions, touching the opposition of the Medium to a direct impressed Motion, and to falling bodies, upon supposition that the opposition is as the Velocity: which tis possible is not true; however I thought any thing of his might not be unacceptable to you, and I begg your opinion thereupon, if it might be (especially the 7th problem) somewhat better illustrated.‖

The 'inclosed propositions' are Problems 6 and 7 of the *de Motu* – the tract that Newton sent to the Royal Society in 1684 – which, with some alterations, were

---

† *Variorum*: 100. *Principles*: 46.
‡ Bennett (1982): 55. See the exchange of letters between Halley and Newton in *Correspondence*, **2**: 431–43.
§ On Wren see Bennett (1982).
¶ *Correspondence*, **2**: 469.
‖ *Correspondence*, **2**: 456.

included in Section 1, Book 2 of the *Principia*. Wallis's results on projectile motion were equivalent to Newton's and Huygens'. However, Newton in Book 2 of the *Principia* was going to deal with the more difficult case in which resistance is proportional to the square of velocity. As Halley announced to Wallis on the 9th of April 1687:

Sr. I give you many thanks for your thoughts upon ye opposition of the medium to projectiles; it will serve to adorn and recommend the Transactions: Mr. Newton in his second book has handled this subject copiously and considered the opposition in a direct and duplicate proportion of the Velocity, and shown what are the curves that will result therefrom; and as truth is alwaies [sic] the same, I find your conclusions (as farr as you goe) aequivalent [sic] to his.†

Certainly Wallis must have realized that the younger Cambridge scholar had reached much more general results than his. After the encounter with the *Principia* Wallis never returned to applied mathematics. He did, however, play a relevant role in rendering public the mathematical works of the British, especially of Newton, with his highly chauvinist *Opera*, which he edited from 1693 to 1699.

Much more is known about Wren's two companion planetologists, Hooke and Halley. Much has been said about the former's role in awakening, in 1679–80, Newton's interest in planetary motion and about his attempt to establish his priority for the inverse square law and its relation to elliptic Keplerian motion. Whatever might be Hooke's contribution in motivating Newton to consider orbital motion as due to rectilinear inertia plus a centripetal force, it is certain that he was not mathematically suited to read and comment on the *Principia*.‡ From the point of view of our research, a more interesting case is that of Edmond Halley.

Halley was the man who, after having discussed with Wren and Hooke the problem of determining the trajectories described by a planet under the influence of an inverse square force directed towards the Sun, paid the famous visit to Newton in 1684. It was through his intervention and financial support that the *Principia* was eventually published. During the years 1685–87 he revised the proofs and corresponded with Newton, trying to mediate between the Lucasian Professor and the Royal Society. He is thus often, and with some reason, depicted as the hero who unearthed from the closet of a secretive Newton the truth of universal gravitation.§ The ways in which he promoted and employed the *Principia* are not, however, limited to proof editing and financing. As is well known, Halley made momentous contributions to Newtonian cometary and stellar astronomy.

Halley also played a relevant role in rendering the *Principia*'s cosmology and physics accessible to a wide audience and in promoting the idea that Newton's

---

† Bodleian MS Lister 3, f. 119ᵛ. MacPike (1932): 81.
‡ For further details see, e.g., Westfall (1971): 427 *passim*; Erlichson (1992a).
§ Ronan (1970): 69ff.

highly speculative work responded to the ideal of useful knowledge that motivated many late-seventeenth-century fellows of the Royal Society.

In 1686, Halley read a paper at the Royal Society in which some features of Newton's *Principia* were presented. It was entitled *A Discourse concerning Gravity, and its Properties, wherein the Descent of Heavy Bodies, and the Motion of Projects is briefly, but fully handled: Together with the Solution of a Problem of great Use in Gunnery*. The first part is devoted to a simple account of the properties of gravitational attraction. After having dismissed Descartes', Huygens' and Vossius's explanations of gravity, Halley enumerates some properties of Newtonian gravitation. Halley presents what he views as the correct 'affections or properties of gravity' as revealed in the works of Galileo, Torricelli, Huygens 'and now lately by our worthy Country-man, Mr. Isaac Newton (who has an incomparable Treatise of Motion almost ready for the Press)'.† These properties are that gravity tends to the centre of the Earth, is equal at all distances from the Earth's centre, and varies inversely as the square of distance. From this last 'Mr. Newton has made out all the Phaenomena of the Celestial Motions, so easily and naturally that its truth is past dispute'.‡ The second part is devoted to projectile motion and it is presented as being related to Newton's *Principia*. However, the mathematical content of Halley's paper reveals how far he was willing to present advanced mathematical techniques.

It is surprising, for instance, to see that the 'Problem of great Use in Gunnery' depended upon a treatment of parabolic motion which was well within Galileo's and Torricelli's conceptual framework. Halley's problem (how to calculate the elevation of a mortar, given the velocity of projection, in order to hit a given target) had already been treated by Anderson and Blondel. Halley ignored air resistance and assumed gravity to be constant.§ There is no way to maintain that such a simple ballistic problem could depend upon Newton's *Principia*. However, Halley's paper served the purpose of announcing in print universal gravitation and in conveying the idea that Newton's work had a practical and technological dividend.

A similar approach was taken in Halley's review of the *Principia* which appeared in 1687. This was the text of a letter accompanying an advanced copy of the *Principia* sent to James II. After a brief and clear presentation of the basic elements

---

† Halley (1686): 6.

‡ *ibid.*: 8.

§ 'Mr. Isaac Newton has shewed how to define the Spaces of the descent of a Body, let fall from any given heighth [sic], down to the Center. Supposing the Gravitation to increase, as in the fifth property; but considering the smallness of height, to which any Project can be made ascend, and over how little an Arch of the Globe it can be cast by any of our Engines, we may well enough suppose the Gravity equal throughout, and the descents of Projects in parallel lines, which in truth are towards the Center, the difference being so small as by no means to be discovered in Practice. The Opposition of the Air, 'tis true, is considerable against all light bodies moving through it, as likewise against small ones (of which more hereafter) but in great and ponderous Shot, this Impediment is found by Experience but very small, and may safely be neglected'. Halley (1686): 8. See also Halley (1695).

of the cosmology contained in Book 3, Halley underlined the practical use of Newton's discoveries. He mentioned lunar theory as a possible way to determine the longitude:

And tho' by reason of the great Complication of the Problem, he [i.e. Newton] has not yet been able to make it purely Geometrical, 'tis to be hoped, that in some farther Essay he may surmount the difficulty: and having perfected the Theory of the Moon, the long desir'd discovery of the Longitude (which at Sea is only practicable this way) may at length be brought to light, to the great Honour of your Majesty and Advantage of your Subjects.†

Halley knew what he was talking about, having great experience as an astronomer as well as a seaman (e.g. he had been to St. Helena and later, in 1701, he was to survey the English Channel).

The next topic, i.e. the theory of tides, chosen by Halley was again well within his competence. As Ronan observes, this was also a diplomatic choice since the King, as Duke of York, had been Lord High Admiral during the reign of his brother, Charles II, and had at one time commanded the British fleet in the war with the United Provinces.‡ Halley was thus addressing the King as an expert on this matter, as a person able to understand the usefulness of Newton's work. So he could offer himself, if admitted to the Royal presence, for further clarifications. Halley had studied in 1684 the anomalous behaviour of the tides in Tonkin. Newton himself in the *Principia* had taken into consideration Halley's numerical results. The letter to King James ended with an exposition of Newton's explanation of this anomaly, 'so that the whole appearance of these strange Tides, is without any forcing naturally deduced from these Principles, and is a great Argument of the certainty of the whole Theory.'§

In subsequent years Halley made use of the *Principia* in some papers he read at the Royal Society. Referring to Proposition 37, Book 2, he dealt with fountains and efflux of water from reservoirs, thus continuing research interests that had been cultivated at the Royal Society.¶ In a way his effort was to render these researches Newtonian. In January 1692 he read a paper about 'an instrument for measuring the way of a ship'. One ingredient of Halley's instrument was a 'ball let down into the water'. In order to measure the force of the flow of water against the ball, Halley referred to Proposition 35/4, Book 2.‖

Halley was the first to try to render available to other natural philosophers the contents of the *Principia*, a work which many found very difficult to read. He considered the third book as particularly suited to the propagation of Newtonian

† Halley (1697): 459. Halley urged Newton to improve his lunar theory in order to determine longitude. See *Correspondence*, **2**: 482.
‡ Rigaud (1838): 88; Ronan (1970): 87.
§ Halley (1697): 457.
¶ MacPike (1932): 135–70.
‖ MacPike (1932): 160.

theory of gravitation. When Newton threatened to withdraw Book 3 from publication, Halley, who also had an economic interest in the matter, implored Newton:

Sr I must now again beg you, not to let your resentments run so high, as to deprive us of your third book, wherein the application of your Mathematicall doctrine to the Theory of Comets, and severall curious Experiments, which, as I guess by what you write, ought to compose it, will undoubtedly render it acceptable to those that will call themselves Philosophers without Mathematicks, which are by much the greater number.†

In Britain, in particular around the Royal Society, there were many who were desirous to accept this book as the basis of a new natural philosophy. The great majority, as Halley says, consisted of 'Philosophers without Mathematicks'. The early attempts to come to terms with Newton's highly esoteric work have been masterfully described by Axtell:

When the Earl of Halifax asked Newton 'if there was no method to make him master of his discoveries without learning Mathematicks, Sir Isaac said No, it was impossible', but someone 'recommended [John] Machin to his Lordship for that purpose, who gave him 50 Guineas by way of encouragement'.‡

Some, such as Gilbert Clerke, approached Newton directly. Clerke had problems in understanding Newton's terminology related to proportion theory.§ Others, such as Colin Campbell, asked the advice of mathematicians experts in the new methods. Campbell asked John Craig, the author of a book on quadratures and a reputed mathematician, for an account of the *Principia*. Craig found it 'a task of no small trouble' and simply sent the book to his perplexed correspondent. The theologian and classicist Richard Bentley, through the intermediary William Wotton, also asked Craig for some help. Bentley had indeed realized that the *Principia* could be employed for apologetic reasons. As is well known, he was to employ Newton's work in his Boyle lectures. Craig, however, did not prove very helpful: he replied with a list of more than 45 preliminary readings. The theologian turned to Newton himself, who replied with interesting suggestions. As we already know (see §5.3), Newton stressed the usefulness of Huygens's *Horologium*. He then added:

When you have read the first 60 pages, pass on to ye 3d Book & when you see the design of that you may turn back to such Propositions as you shall have a desire to know, or peruse the whole in order if you think fit.¶

This is what Newton suggested to interested literate, but innumerate, readers. The first 60 pages, i.e. the first three sections of Book 1, were, according to him, within the capability of a reader with little mathematical background, who could

---

† Halley to Newton (29 June 1686) in *Correspondence*, **2**: 443.
‡ King's College, Cambridge, MS Keynes 130.5, no. 2, p. 4. Quoted in Axtell (1969): 171.
§ Sylla (1984).
¶ *Correspondence*, **3**, 156.

afterwards approach the cosmological propositions of Book 3. He was close to the truth. As a matter of fact 'the first three sections of Newton's *Principia*' became standard basic reading for British university students in the eighteenth century.

It seems, however, that Bentley never even tried to penetrate the demonstrations in the first three sections. Instead, he studied carefully the cosmology of gravitation in Book 3, reaching a good qualitative understanding of the Newtonian 'System of the World'. Even though a 'Philosopher without Mathematicks', Bentley was able to raise interesting questions. In a series of letters dating from 1692, he posed to Newton the problem of reconciling universal gravitation with the stability and order of the Universe. He suggested that since the 'fixed' stars exert a gravitational pull on each other, they would eventually collapse into one spherical body. Newton's replies, analysed carefully by Hoskin, are fascinating, since he touched on themes ranging from concepts of equilibrium to the infinity of the Universe and God's providence. Here it will suffice to say that one of his answers was that the infinity, distance and uniform distribution of the stars accounted for their near stability. Furthermore many Newtonians accepted that God's providential intervention prevented the collapse of the system of the stars, and hence stability was not guaranteed by natural causes since without divine 'reformation' the stars would eventually collapse.†

The case of Bentley is similar to Locke's, the first 'Newtonian Philosopher without the Help of Geometry'.‡ Locke began to read the *Principia* two months after publication. He took excerpts and notes. The influential review published in the *Bibliothèque Universelle* in 1688 is now generally attributed to him.§ It does not seem, however, that he could go very far with the mathematics. Desaguliers reports that Locke, after discovering that he was unable to follow Newton's demonstrations, asked Huygens whether they were true. Since Huygens assured him that they were so, he 'took them for granted [. . .] carefully examined the Reasonings and Corollaries drawn from them, [and] became Master of all the Physics'.¶ Locke must have had some interest in the mathematics, since in March 1689/90 he received from Newton an alternative demonstration of the basic Proposition 11 of Book 1. Furthermore, in 1691 Newton gave him a copy of the *Principia* with corrections in the margin. In the same year Locke wrote a notebook called *Adversaria Physica* in which he commented on several propositions.‖

---

† Hoskin (1977).
‡ This definition is in Desaguliers (1734): *Preface*.
§ Axtell (1965) and Cohen (1971): 145–7.
¶ Desaguliers (1734): *Preface*. Quoted in Axtell (1969): 177.
‖ Bodleian, MS Locke, d.9: folios 168, 83. See Axtell (1969): 176–7.

7.3 The best geometers and naturalists

The understanding of the mathematics contained in the *Principia* created problems not only for 'Philosophers without Mathematicks', but also for the 'best geometers and naturalists'.† The need for a second amended edition began to be felt both in Britain and on the Continent. Newton himself, in the early 1690s, began to think about a new edition. The Swiss mathematician Nicolas Fatio de Duillier was the first to make these projects known to the literary world. Since Rigaud's *Historical Essay*, Fatio's correspondence with Huygens on this would-be second edition is well known. During the 1690s rumours about a new and, hopefully, more understandable edition of the *Principia* are echoed in letters by, amongst others, Huygens, l'Hospital, Leibniz, Johann Bernoulli, David Gregory and Nicolas Malebranche.‡ Rigaud also documented Fatio's intention of being the editor.§ However, in December 1691 Fatio was already despairing of Newton's willingness to proceed with the new edition.¶

We do not know how far Fatio actually went in his editorial enterprise. In a letter to Huygens dated April 1692 he confessed that he had mastered only the first five sections of Book 1 and the part on comets in Book 3.‖ His extant manuscripts related to the *Principia* are rather meagre. In the Bodleian there are his annotated copies of the first and third editions of the *Principia*. In the Basel University Library there is a memorandum concerning some corrections that Newton communicated to him. Furthermore, in April 1690, Fatio went to Holland where he left in Huygens' hands a list of Newton's *errata* and *corrigenda*. These, when found in Huygens' *Nachlaß*, were interpreted by Groning as being Huygens' criticisms of Newton.††

Through his correspondence with Huygens, Fatio was selling himself as the man who could render the *Principia* more intelligible. He could boast of his close friendship with Newton, whom he could consult personally for clarification when he wished to do so: a privilege not accorded to many. As a matter of fact, it is well known that Newton had a very special sympathy for the young Swiss, to whom he showed, in 1689, some of his unpublished manuscripts. Newton also gave him a detailed idea of his project of a complete restructuring of the *Principia*.

---

† David Gregory, in a letter to Newton in early Autumn 1687, wrote that the *Principia* would have aroused the admiration of the 'best geometers and naturalists'. *Correspondence*, **2**: 311.

‡ Edleston (1850): xii–xv.

§ Rigaud (1838): 89–96.

¶ Fatio wrote to Huygens: 'Il est assez inutile de prier Monsieur Newton de faire une nouvelle édition de son livre. Je l'ai importuné plusieurs fois sur ce sujet sans l'avoir jamais pu fléchir. Mais il n'est pas impossible que j'entreprenne cette édition; à quoi je me sens d'autant plus porté que je ne croi pas qu'il y ait persone qui entende à fonds une si grande partie de ce livre que moi, graces aux peines que j'ai prises et au temps que j'a emploié pour en surmonter l'obscurité. D'ailleurs je pourrois facilement aller faire un tour à Cambridge et recevoir de Mr. Newton même l'explication de ce que je n'ai point entendu.' *Correspondence*, **3**: 186.

‖ Cohen (1971): 180 and Huygens *Oeuvres*, **22**: 158–9.

†† On Fatio's *marginalia* see Rigaud (1838): 91–5.

Fatio had met Newton in June 1689 at a meeting of the Royal Society in which Huygens, then on a tour to England, spoke about his forthcoming *Traité de la lumière*. The 25-year-old Swiss already had a respectable scientific career. Educated in Geneva, he had been travelling Europe introducing himself in the foremost centres of the time. In Paris he had participated in the observations of the zodiacal light carried out by Cassini. He had met Jacob Bernoulli in Basel and Huygens in Holland. To the latter he communicated a mathematical mistake he had found in Tschirnhaus's *Medicina mentis*. Fatio maintained contacts with both Huygens and Bernoulli, corresponding with them about the new calculus. After 1689 he became one of Newton's closest *protégés*. He tried to capitalize on Newton's friendship as much as he could. Having access to Newton's private papers, he could indeed propose himself as editor of the second edition of the *Principia*, and thus play a prominent role in the scientific European *milieu*, establishing himself as the intermediary between Newton and the Continentals. However, his intentions soon became less diplomatic and more ambitious.

Since the 1680s, Fatio had maintained a mechanical theory of gravitation based on the movement of ether particles.† It seems that Newton, at first, considered such a theory with some sympathy. Indeed, Fatio was able to copy a manuscript where Newton defined Fatio's theory as the only possible mechanical explanation of gravity.‡ Despite his close connection with Newton he did not give up this theory and there are reasons to suspect that he was planning to use the second edition of the *Principia* as a vehicle for promoting his ideas.§

Fatio was not a quiet disciple. In his *Lineae brevissimi descensus investigatio geometrica*, published with the imprimatur of the Royal Society in 1699, he made it clear that he was certain that Leibniz had plagiarized, with his differential and integral calculus, Newton's algorithm of fluxions. This is the rude statement that initiated the famous priority dispute which was to divide the Newtonian and the Leibnizian mathematicians.¶ In this tract Fatio dealt with the brachistochrone and with the solid of least resistance. Newton in Proposition 35/4, Book 2, had simply stated the geometrical properties for such a solid. Fatio provided a proof in terms of fluxional calculus.

Fatio's plan for a second edition of the *Principia* was not implemented. His relationship with Newton deteriorated for reasons that are difficult to detect. One

---

† Zehe (1983).

‡ Newton (1962): 313.

§ In a copy of the third edition of the *Principia* he wrote the following verses: 'Optima fundamenta jacis, Newtone; sed exis/ Infelix operis summa. Systemata mundi/ Corrumpis; nec digna Deo, nec consona fingis./ Tu nebulas remove: facies tua clara nitebit./ Newtoni Responsum./ Si bona fundamenta; tamen structura bene illis/ Non quadret, sed obliqua labet minitata ruinam:/ Ergo cadat: templumque novum coeleste resurgat./ Sed maneat jactum solide fundamen in aevum./ Insculptoque basi NEWTONI nomine; in ipso/ Culmine scribatur, FACIUS multum addidit aedi:/ Aedi, quae immensi typus est templi Omnipotentis.' Quoted in Rigaud (1838): 95.

¶ For further information see Hall (1980) and *Mathematical Papers*, **8**: 469–697.

possibility is that Newton felt that Fatio would have edited a work too committed to his own ether theory of gravitation. Newton, in fact, soon lost any esteem for Fatio's theory. Later, the Swiss scholar adhered to the religious sect of the so-called 'Cévennes prophets' and his life became rather troubled.

The Scottish mathematician David Gregory was the next to approach Newton with a proposal for a second edition. Gregory had begun to read the *Principia* carefully in the early autumn of 1687. As he had probably already done with Huygens' *Horologium*† he wrote a detailed commentary on the whole work. The result was a 213-page manuscript entitled *Notae in Newtoni Principia Mathematica Philosophiae Naturalis*. Gregory's *Notae* are extremely important, since they allow us to see how a well-trained mathematician could understand Newton's *Principia*. Despite this, they have been analysed only superficially by historians of science.‡

David Gregory was the nephew of James Gregory, a mathematician who had obtained outstanding results. After having been Professor of Mathematics in Edinburgh, David moved, in 1691, to Oxford where he was appointed to the Savilian Chair of Astronomy. He soon became, in connection with his compatriots Archibald Pitcairne and John Craig, a convinced Newtonian. The *Notae* provide a thorough commentary on the *Principia*, from beginning to end. They were written in three stages, as it is evident from the dating, as well as from the paper and ink used. The original manuscript in Gregory's hand is kept in the Royal Society (London), shelved as MS 210. The first 30 pages, a commentary on the first nine sections of Book 1, are dated from September 1687 to April 1688. The remaining pages are dated from December 1692 to January 1694. There are also later additions written on slips of paper affixed by paste or wax. The last addition was made in 1708. There are also three transcripts: in Christ Church (Oxford), in the University Library (Edinburgh), and in the Gregory Collection of the University of Aberdeen.§ The Christ Church exemplar seems to be prepared for the press, since Gregory's 'notes' are numbered as footnotes. In fact Gregory wished to publish the *Notae* as a running commentary to the *Principia*. He may have communicated this project to Newton at the end of a visit to Cambridge in May 1694. The May visit was very important for Gregory, since it occurred after a period of tension between him and Newton. The five-day meeting, which probably occurred thanks to Fatio's intercession,¶ indeed changed Gregory's scientific life. During these days Newton showed him his projects for a revision of the *Principia*, and let him study his tracts on fluxions (including a version of *De quadratura*).

---

† Eagles (1977): 28.

‡ Wightman (1953) and (1957) give some description of the manuscript and its transcriptions. Cohen (1971) provides a much more detailed treatment. The only serious work on the *Notae* is Eagles's Ph.D. thesis (1977).

§ Royal Society Library, MS 210. Christ Church Library, MS B.13.cxxxi. University Library (Edinburgh), MS DC.4.35. University of Aberdeen Library, MS 465.

¶ This is suggested by Eagles (1977): 398–9.

The two men also discussed the Ancients' *prisca sapientia* and the mathematics of the ancient geometers. Gregory took memoranda of this visit which have been widely used by Newtonian scholars.† When he met Newton in May, Gregory had already completed his commentary on the *Principia*. As he was told that Newton was planning a second edition, and was allowed to see the details of the reworkings, corrections and restructurings necessary, Gregory thought of his *Notes* as a possible addition to the projected second edition. There is no trace that Newton ever took into consideration such a proposal by Gregory. However, some publicity of Gregory's project must have existed since both Huygens and Malebranche refer to Gregory's notes.‡ As late as 1713 the blind Lucasian Professor, Nicholas Saunderson, was writing to William Jones about a proposal for publishing the *Notae*.§

Any reader of Gregory's *Notae* will not be surprised that, as they stand, they would not have been welcomed by Newton. As a matter of fact, in many places Gregory tends to be rather critical, while in others he attributes to his uncle or to himself the mathematical discoveries necessary to develop certain demonstrations of the *Principia*. What is written in the slips added after the May 1694 visit is completely different in character: here Gregory shows a great deference to Newton and often adds sentences such as 'ut Candidissimus Newtonus mihi narravit'. In a memorandum written in July 1694, Gregory spelled out his project about the *Notae*. It appears that he was not yet willing to defer too much to Newton, since he still wanted to insert parts related to his and his uncle's discoveries, as well as to Leibniz's differential calculus:

If my Notes on the Newtonian Philosophy are published (as indeed I heartily wish and expect, and made the proposal to the author himself on 8 May 1694) a great deal that serves to detect slips or even mistakes of Newton and is contained in my Notes is to be omitted if a new edition of the actual work is made by the author: otherwise the notes are to be inserted in their proper places.

I should also insert: 1. the problem of curvatures of curves at Schol. Lem. 11, prop. 24 [read page 24]. 2. the problem of the locus of the image in a spherical mirror, when the eye is placed off the optical axis (if my Optics are not published) at prop. 97 or 98, Book I, page 233, 234. 3. The theorem on the indefinite quadrature of the lunulae of Hyppocrates of Chios. 4. Numerous astronomical problems among my uncle's manuscripts, and in general whatever I still keep to myself such as these notes contain; and whatever things hitherto as yet unpublished are kept close by geometers or sons of the Art, these are to be inserted in their proper places, as the Differential Calculus of Leibniz and other matters.¶

No doubt Newton would have disliked seeing his *Principia* published with notes spelling out slips and mistakes. He would have also refused to accept Gregory's

---

† See esp. *Correspondence*, **3**: 311–22, 327–48, 384–9.
‡ Huygens *Oeuvres*, **10**: 614. Royal Society *Orig. Lett. Book*, M II.10 quoted in Edleston (1850): xiii.
§ Rigaud (1841), **1**: 264–5.
¶ Translation from the original Latin in the Edinburgh University Library in *Correspondence*, **3**: 386.

willingness to insert his and his uncle's scientific results on curvatures, optics, Hyppocrates' lunulae and astronomy. One should remember that Newton had had fears that the two Gregories could deprive him of his mathematical discoveries. It goes without saying that the proposal of inserting Leibniz's differential calculus into the body of the *Principia* was absolutely out of question. It seems that Newton had mixed feelings towards the author of the *Notae*. On one hand Newton played a prominent role in Gregory's scientific career. Most notably, Gregory was appointed to the Savilian Chair thanks to Newton's recommendation, and he was entrusted by Newton with the task of publishing the *Nova et accurata motuum lunarium theoria*, i.e. Newton's mature attempt to tackle lunar inequalities. Gregory was in the committee which eventually published Flamsteed's *Historia Coelestis*. On the other hand there was a period of tension between Gregory and Newton. Gregory, in a work published in 1688, attributed to himself one of Newton's most important methods of quadrature, i.e. the quadrature of $y = x^\theta (e + f x^\eta)^\lambda$ contained in the *epistola posterior* to Leibniz (see §2.2.7), a method that he came to know through John Craig. Furthermore, he often defended the priority of his uncle James as far as infinite series were concerned. So tensions between Gregory and Newton occurred, especially in the years 1691 to 1694. According to a memorandum by Thomas Hearne in 1705, Newton complained that Gregory in his *Astronomiae physicae et geometricae elementa* (1702) had borrowed many results from him without due recognition.†

We will see below that Gregory's *Notae* are an active reading of the *Principia*: the reading of a critic as well as disciple. Basically we will find:

- criticisms about Newton's obscurity or inconclusiveness;
- attributions of methods or results to other authors (among them, James Gregory and David Gregory himself);
- demonstrations worked with symbolic, rather than with geometrical tools.

The comment on Lemma 1, Book 3, where Newton explained, with studied brevity, how the precession of the equinoxes could be calculated, is typical of Gregory's frustration over Newton's concise style:

I labour to be brief, I become obscure. He could at least have indicated the method by which the ratios of the sums of the forces could be investigated.‡

In other instances, Gregory reworks some of the demonstrations trying a style which appears to him more rigorous or clearer. The basic Proposition 6, Book 1, is an example.§ Here Gregory follows the same path as Huygens. That is, in order to establish proportions, he evaluates the centripetal force at two different points of

---

† Rigaud (1838): 102.
‡ Gregory *Notae*: 170. Translation by Eagles (1977): 30.
§ Gregory *Notae*: 5.

the orbits. He then states a proportion among forces and displacements acquired in equal intervals of time.† There is a possibility that Huygens and Gregory discussed this matter during the latter's visit to Holland. This reworking of Proposition 6 appears also in Gregory *Astronomiae* (1702).‡ However, on a pasted slip (i.e. an addition later than May 1694, and probably later than 1702, the year of publication of his book on astronomy), Gregory retracts this emendation and gives a version equivalent to the one to be found in the first edition of the *Principia*. Since the slips added to the *Notae* generally reflect the result of discussions with Newton, it might well be that here Gregory corrected his emendation of Proposition 6 after receiving explanations from the author of the *Principia*. As we have explained in §5.6, Newton's innovative use of proportion theory allows him to restate calculus techniques in the geometrical limit procedures of the method of first and ultimate ratios, an innovative characteristic of Newton's procedures that both Huygens and Gregory missed.

Another interesting criticism concerns Corollary 1 to Propositions 11–13, Book 1. We remind the reader that this Corollary states that, since conic section trajectories (when one focus coincides with the force centre) imply an inverse square central force, then the converse is true (see §3.4.4 and §8.6.2.1). Gregory writes:

This is the converse of the three preceding [propositions]. But I wished that the author had demonstrated that a body could not, given this [inverse square] law of centripetal force, be carried in any curve other than conic sections. This is likely to be true, since there are no curves of the same degree, other than conic sections.§

Here Gregory detects the presence of a logical problem: the solution of the direct problem of central forces does not imply the solution of the inverse problem! His tentative explanation concerning the degree of the trajectory's equation is wrong, even though it is an interesting attempt to reach a solution of the inverse problem via algebraical means.

Gregory reacted critically also to the famous Lemma 2 of Book 2 concerning the moment of the product of two fluent quantities (§3.12). Newton had wrongly stated that the moment of AB could be calculated without having recourse to limiting procedures or cancellation of infinitesimals. Gregory rightly denies this.¶

In the *Principia* Newton had been extremely sparing in recognizing his debt towards other natural philosophers. In the *Notae* Gregory provides such missing

---

† See §5.5 for details.

‡ Gregory (1702): 45, Propositio 37. This procedure was also followed by Hayes (1704): 299.

§ 'Est conversa trium propositionum praecedentium. Sed velim demonstrasset auctor non posse corpus hac lege vis centripetae, in alia curva praeter conisectionem deferri. Hoc verisimile est, quoniam nulla, praeter conisectiones, datur curva ejusdem gradus'. Gregory *Notae*: 8.

¶ Gregory wrote: 'Hoc caute factum ut res exacte procedat, & si aliter fiat, et latera A et B ipsorum momentis a, b, aucta ponantur, rectanguli AB momentum $Ab + aB + ab$ minime differt a momento $Ab + aB$; nam productus ex infinite parvis, scilicet ab, negligendus est'. Gregory *Notae*: 60.

references. In many cases he simply gives the reader the appropriate information on Euclid or Apollonius. But Gregory also gives references which, to a certain extent, deprive Newton of complete originality. For instance, he observes that Newton should have mentioned Kepler in connection with the third law.† Lemma 12 of Book 1 concerning the parallelograms circumscribed to a conic is attributed to Philippe de La Hire and to Grégoire de Saint-Vincent (§3.4.3).‡ Gregory attributes to de Witt Newton's important results on projective geometry expounded in Section 5, Book 1.§

One of the most interesting characteristics of Gregory's commentary is the presence of several reworkings of Newton's geometrical demonstrations in terms of the analytical method of fluxions. Often Gregory translates Newton's connected prose into algebraic terms, putting, e.g., T for period or v for 'velocitas'.¶ Such a translation has the effect that Newton's geometrical reductions to quadratures (e.g. in Sections 8, 12 and 13, Book 1) lead clearly to the expression of a fluxional equation. In the *Principia* many propositions are solved by reduction to quadratures. In many cases, Newton does not perform the quadratures but simply says that the problem is solved 'granting the quadrature of curvilinear figures'. In the *Notae* Gregory provides such quadratures, but often attributes to himself the method for performing such 'integrations'. As we have already remarked, Gregory got this method through Craig who had copied it from Newton's manuscripts. One hopes that Newton did not see the *Notae*.

In several cases, however, Gregory was unable to 'square the curves'. For instance, he observes that in the case of Proposition 39 'the author's reasoning is analytical'.‖ This and its subsequent propositions lead to a fluxional equation in polar coordinates which, when solved, gives the general solution of the inverse problem of central forces. As we will discuss in §8.6.2.2, Gregory, during the May meeting, had to consult Newton on this problem (in the case of inverse cube forces) and received as an answer the solution of the pertinent fluxional equation.†† This solution is inserted on a pasted slip in the *Notae*. However, Gregory did not give up his intention of claiming some originality. In fact, he states that the problem is solved by 'squaring the curve [...] by our method edited for the first time by Archibald Pitcairne in 1688 and then by Wallis in his *Algebra* on pag. 378'.‡‡

---

† 'Mirum est Auctorem Kepleri non meminisse.' Gregory *Notae*: 5 [Christ Church].
‡ 'Probatur hoc Lemma prop: 21: Lib. 5, et prop. 43: lib: 4 Conicae de la Hire prop: 72 Lib: 4ti, et prop: Lib: Gregorii a Sto Vincentio'. Gregory *Notae*: 6. In the *Astronomiae* (1702): 47 the reference is only to Grégoire de Saint-Vincent.
§ Gregory *Notae*: 29 [Christ Church].
¶ See, e.g., Gregory's *Notae* to Propositions 4, 16, 30, 41, 50, 81 and 91, Book 1.
‖ 'Ratiocinatio auctoris hoc loco est analytica'. Gregory *Notae*: 25.
†† *Mathematical Papers*, **6**: 435–8 and *Correspondence*, **3**: 348–9.
‡‡ 'quadrando curvam cujus haec est ordinata per Methodum nostram ab Arch: Pitcarnio anno 1688 primo editam deinde a D:s: Wallisio in sua Algebra pag: 378 &c'. Gregory *Notae*: 28. Cf. Wallis (1693–99), **2**: 377–80.

Another demonstration based on the analytical method of fluxions sent by Newton to Gregory concerns Proposition 35/4, Book 2, dealing with the solid of least resistance.† In a letter dated July 1694 Gregory received Newton's highly inventive solution. As Eagles has shown, he did not understand it properly.‡ Gregory inserted Newton's solution in a tract on fluxions he wrote in 1694. It appears also on page 134 of the Christ Church transcript of the *Notae*. Again he refers to his method of quadratures.

Another reference to Gregory's method occurs on page 52 of the *Notae* in the discussion of Corollary 1 to Proposition 91 (concerning the attraction of a homogeneous cylinder on a point situated on the prolongation of the axis) (§3.8). Again, Gregory translates Newton's prose into symbolic terms and performs the necessary quadrature invoking 'his' method. It is, however, interesting to observe that he was unable to provide a demonstration of the subsequent Corollary 2, dealing with the attraction exerted by an homogeneous ellipsoid on a point situated on the prolongation of the axis of revolution. As Eagles underlines, Gregory's attempts were blocked by his inability to 'integrate' $ax/\sqrt{(bx^2 + cx + d)}$.§

Summing up, we can say that Gregory was interested in analytical fluxional versions of the *Principia*'s demonstrations. He was aware that some of these are translatable in terms of fluxions, but lacked the mathematical means necessary to perform some integrations. He had to consult Newton on Corollary 3 to Proposition 41, Book 1 and on Proposition 35/4, Book 2, and never solved Corollary 2 to Proposition 91, Book 1.¶

It is interesting to note that the first extant letter (presumably the first) of Roger Cotes to Newton concerns Corollary 2 to Proposition 91, Book 1. The young Plumian Professor had been just chosen, thanks to Bentley's recommendation, as editor of the second edition of the *Principia*. In presenting himself to Newton, he chose this difficult Corollary as proof of his skill in mathematics. On the 18th of August 1709 he wrote:

Some days ago I was examining the 2d Cor: of Prop. 91 Lib. I and found it to be true by ye Quadratures of ye 1st and 2d Curves of the 8th Form of ye second Table in Yr Treatise *De Quadrat.*‖

Cotes continued his letter by correcting some of Newton's 'Forms' of the *De quadratura*, which ended with some 'tables' (in Leibnizian terms 'integral tables') which could be used in some Propositions of the *Principia*. Cotes in his editorial

---

† *Mathematical Papers*, **6**: 466–80 and *Correspondence*, **3**, 380–83.
‡ Eagles (1977): 419–22.
§ Eagles (1977): 423.
¶ An annotated copy of Newton's *Principia*, first edition, has been found in Moscow University Library by Vladimir S. Kirsanov. In Kirsanov (1992) the annotations are attributed to Gregory.
‖ Edleston (1850): 3–4.

work made ample use of the analytical method of fluxions.† In this case (see fig. 3.18), applying the procedure developed in Proposition 91, Book 1 (see §3.8), it can be shown that the attraction at point P situated externally on the axis of revolution of the ellipsoid $AGBC$ is proportional to the area subtended to the curve with ordinate

$$1 - \frac{PE}{PD},\tag{7.1}$$

calculated from PA to PB. On taking $PE = x$, it is possible to show that formula (7.1) reduces to a fluent of the following form:

$$1 - \frac{x}{\sqrt{e + fx + gx^2}},\tag{7.2}$$

where e, f and g are constants.‡

That Cotes was using analytical quadrature techniques in order to check the correctness of several Propositions of the *Principia* appears also from the exchange of letters with Newton, which took place in May 1710, concerning the corrections to Propositions 29 and 30, Book 2.§ Proposition 29 studies the motion of a body which oscillates along an inverted cycloid under the action of a constant gravitational force and a resistance proportional to the square of velocity. Cotes translates Newton's solution into calculus terms. This problem had already been tackled by means of the integral calculus by Johann Bernoulli in 1695 and by Leibniz in 1696.¶ Cotes observed:

When I first look'd over this passage upon account of it I thought the whole construction erroneous. I therefore set my self, after the following manner, to examine how it ought to be, which I have put down for a further use I have of it.‖

Cotes takes 'x, z and v for quantitys analogous to the Force [the component tangent to the cycloid] arising from ye gravity of ye Pendulous body [which is, as Huygens had demonstrated, proportional to the displacement x along the cycloid's arc from the equilibrium position], the force of resistance, & ye velocity'.†† He then denotes with $\dot{t}$ the 'moment of time' and affirms that

$$v \propto \frac{-\dot{x}}{\dot{t}},\tag{7.3}$$

i.e. in Leibnizian terms, $v \propto dx/dt$. Furthermore:

$$\dot{v} \propto (x - z)\dot{t},\tag{7.4}$$

---

† On Cotes's editorial work see Edleston (1850), Hall (1958) and Cohen (1971).

‡ In Leibnizian terms, one has to find the integral: $\int \left(1 - x/\sqrt{e + fx + gx^2}\right) dx$, with suitable integration limits. For a discussion see *Mathematical Papers*, **6**: 225–6.

§ Edleston (1850): 21–3 and Gowing (1983): 55.

¶ *Acta Eruditorum* (1695): 533 and (1696): 145.

‖ Edleston (1850): 21.

†† Note that Cotes is writing proportions, so that, e.g., v is proportional ('analogous') to velocity.

i.e. in Leibnizian terms $dv \propto (x - z)dt$. Equation (7.4) means that the acceleration is proportional to the tangential component of gravity minus the resistance. Since, by hypothesis, the resistance z is proportional to the square of velocity v^2, one has

$$\dot{z} \propto v\dot{v}. \tag{7.5}$$

Combining the above results, the following fluxional equation is obtained:

$$\dot{z} \propto v\dot{v} \propto -\dot{x}(x - z) \propto z\dot{x} - x\dot{x}. \tag{7.6}$$

Another source to be taken into consideration in order to appreciate Cotes's awareness of the translatability into calculus terms of the *Principia* is his only published paper *Logometria* (1714), together with the posthumous extension of it *Harmonia Mensurarum* (1722). These two works have been analysed in detail by Gowing.[†] Several problems solved in *Logometria* and *Harmonia mensurarum* have their origin in the editorial work on the *Principia*. Developing earlier results obtained by Halley and De Moivre, Cotes was able to establish the existence of a 'harmony' between 'measures of angles' (i.e. trigonometric quantities) and 'measures of ratios' (i.e. logarithms). In his peculiar notation he expressed a result equivalent to

$$i \ln(\cos\theta + i\sin\theta) = \theta, \tag{7.7}$$

which has a sign error.[‡] Cotes was able to improve Newton's methods of 'integration' of rational and irrational functions. Newton reduced these quadratures to the quadrature of a conic area. Cotes, instead, obtained these quadratures in terms of logarithmic, trigonometric and hyperbolic functions. He applied his integral tables (which he called 'continuation formulae') to the solution of several problems. Some of them have to do with the *Principia*. He dealt with vertical ascent and descent in resisting media, with the gravitational attraction of spheroids, with cycloidal motion in resisting media, with the density of the atmosphere, with spiral orbits in an inverse cube central force field, and with the attraction due to a sphere composed of particles that exert an inverse cube attraction.[§] Both in his correspondence with Newton and in his research work, Cotes made it clear that the hidden analysis that lay behind the *Principia* was the analytical method of fluxions.

We have already described in §4.5 how the analytical method of fluxions was applied to central forces by Newton, Keill, Halley and De Moivre. We will return to Keill and Cotes in §8.6.2.5.

† Gowing (1983). On Cotes and Varignon see Gowing (1992).
‡ Schneider (1968): 236.
§ Gowing (1983): 51–5, 64–6, 195–203.

7.4 Guides to the *Principia*

In Britain the Newtonian mathematical philosophy very soon became dominant. The need to have access to the intricacies of the *Principia* began to be felt by the numerous neophytes who did not have enough mathematics to understand Newton's demonstrations. Keill's famous introductions to the 'true physics' and the 'true astronomy' proved to be useless from the mathematical point of view. The *Introductio ad veram physicam*, published in 1702, based on Keill's lectures in Oxford, was translated into English and served as a model for many other Newtonian treatises. Even though Keill was one of the most vehement defenders of Newton in the priority dispute with Leibniz on the invention of the calculus, it seems that he did little to promote knowledge of the applications of mathematics to natural philosophy. Mathematical methods were employed in some of his lectures. In the Lectio XV of his *Physicam* on motion on inclined planes and motion of pendulums we find geometrical theorems on the isochronism of the cycloid, while we find some use of fluxional calculus in the treatment of the brachistochrone. In the Lectiones XXIV and XXV of the *Introductio ad veram astronomiam* (1718) one finds a treatment of Kepler's problem. But generally Keill preferred to acquaint his readers with Newtonian natural philosophy while avoiding mathematical demonstrations. It should be noted that the *Physicam* could be defined an introduction to Newtonian as well as to Huygenian natural philosophy. In fact, few of the topics of the *Principia* are covered by Keill (the method of philosophizing, the laws of motion and absolute time and space). Most of the book is devoted to Huygenian topics such as pendular motion and Huygens' force law for uniformly circular motion. As we will see in §8.6.2.5 and §8.6.2.6, Keill was able to participate with some competence in the mathematization of central force.

The first thorough mathematical introduction to the *Principia* came from the pen of Keill's colleague at Oxford, David Gregory, whose ponderous *Astronomiae physicae & geometricae elementa* appeared in 1702. Gregory began writing the *Astronomiae* in 1697, probably when he lost any hope of contributing to the second edition of the *Principia*. Gregory dealt only with those parts of Newton's *magnum opus* which had directly to do with astronomical and cosmological matters. He further gave detailed astronomical data in support of Newtonian gravitation. He thus omitted any treatment of the more abstract or theoretical parts. More precisely, in Gregory's treatise one cannot find any discussion of the laws of motion, the definitions of force, mass and quantity of motion, the first nine lemmas on first and ultimate ratios (only Lemmas 10 and 11 are dealt with), Propositions 7, 8, 9 and 10 on the direct problem of central forces (other than inverse square), the geometrical Sections 4 and 5, Sections 7 and 8 on the general inverse problem of central forces, Section 10 on pendular motion, Sections 12 and 13 on the

gravitation of extended bodies, or any of the second book (with the exception of the final section on the refutation of Cartesian vortex theories). Considering how detailed are the unpublished *Notae* that Gregory had composed before his 1694 visit to Cambridge, one has the measure of the amount of previous analysis of the *Principia* that he did not employ in the *Astronomiae*. Gregory's astronomical treatise can be better defined as an introduction to the third book of the *Principia*, enriched by a presentation of some relevant propositions from the first book. Gregory thought it necessary to acquaint his readers with Propositions 1 and 2, on the equivalence between Kepler's area law and central force motion, and with Proposition 4, Book 1. He further dealt with Propositions 11–13, which state that Keplerian motion along a conic section implies an inverse square force. Here, as in the *Notae*, he reframed Newton's demonstration in a style which would have suited Huygens. The treatment of the inverse problem of central forces was at variance with the text of the *Principia*. Gregory also dealt with the motion of apsides as presented in Section 9 and gave a rather detailed discussion on the theory of perturbations developed by Newton in Section 11. The *Astronomiae* was opened by a transcription of Newton's 'classical scholia' on the *prisca sapientia*. Another Newtonian work was included in the body of the treatise: on pages 332–6 Gregory reproduced some attempts by Newton to perfect the theory of the Moon.†
The reader of Gregory's treatise on astronomy was thus given a professional and detailed account of the Newtonian cosmology and planetary theory. Some relevant mathematical propositions from Book 1 were presented in a paraphrase basically respectful of Newton's text.

A much more personal, and actually mathematically incorrect, paraphrase was proposed in William Whiston's Cambridge lectures. The *Praelectiones Physico-Mathematicae Cantabrigiae in Scholis Publicis Habitae: Quibus Philosophia Illustrissimi Newtoni Mathematica Explicatius traditur, & facilius demonstratur* were published in 1710 as a sequel to the *Praelectiones Astronomicae* (1707). They were based on Whiston's Lucasian lectures given from 1704 to 1708, and very soon became popular as introductions to the *Principia*. Whiston was chosen by Newton as his successor to the Lucasian Chair. He was expelled from the University in 1710 because of his religious views. Newton not only supported Whiston's career, but also concurred with many of his religious views.‡
The *Praelectiones Physico-Mathematicae* were devoted, according to the author, to the 'Mathematicians of the lower Form, and who only understand the first Elements of Geometry and Astronomy'. It was thus Whiston's purpose to find an easier demonstration, as the title goes, of the basic propositions of the *Principia*. Whiston's proposal is to avoid Proposition 6, Book 1, which is the foundation of

† For a detailed analysis of Gregory's *Astronomiae* see Eagles (1977): 477–582. See also Newton (1975).
‡ Farrell (1981), Force (1985).

Newton's treatment of central force! His alternative theorem states that centripetal force is proportional to curvature times velocity: 'Let it therefore be laid down for a certain Truth, That the Proportions of centripetal Forces are every where to be estimated from the Proportions of the Curvatures and Velocities conjunctly'.† Furthermore, we learn that velocity is inversely proportional to distance from the force centre and that in an ellipse 'in Arches described in the same time, the Curvature is reciprocally as the distance from the Focus'. Whiston concludes that for Keplerian motion the force (proportional to velocity times curvature) is inversely proportional to the square of distance.‡ When George Cheyne published, in 1703, his decent treatise on fluxions *Fluxionum Methodus Inversa*, Newton reacted angrily, assigning to De Moivre the task of raising objections against the young friend of Archibald Pitcairne.§ Newton did not leave others, without his consent, the space to present Newtonian mathematics or natural philosophy. Despite their mathematical incorrectness, Whiston's *Praelectiones* were accepted without *animadversiones* in the Newtonian circle. The impression that Whiston was a protégé of Newton is reinforced by the historian of mathematics.

The British market of introductions to the *Principia* was clearly interesting, given the number of potential readers. The fact that there was some danger in proposing a book on the subject without previous negotiations with Newton is evident, not only from the Cheyne's affair, but also from Ditton's *The General Laws of Nature and Motion; with their Application to Mechanics. Also the Doctrine of Centripetal Forces, and Velocities of Bodies, describing any of the Conick sections. Being a Part of the Great Mr. Newton's Principles*, a small treatise where the English-speaking reader was given some instructions on the first three sections of Book 1. In the *Preface* Ditton was cautious in stating Newton's absolute right of property on the mathematization of central forces and appealed to freedom of interpretation only as far as the language was concerned:

The materials that this Book is composed of, are so absolutely Mr Newton's Property, that I dare hardly pretend to call any thing mine. The Principles most certainly are all his own: and if I have attempted any where to make any Use of them, or to draw any consequences from them; yet the indisputable Right that he has to the Former, gives him a title to the Latter also, where they are just and good. This is certain, that his Inventions are new and compleat; and equally exclude all the Additions and Claims of those that come after. [...] Further, To render what I have done more universally serviceable here at Home, I chose to make it appear in English rather than Latin. For if it be granted that Mr. Newton's Discoveries are but barely useful, there's no Reason why a Multitude of very capable Minds

† Whiston (1716): 133.
‡ Whiston (1716): 157.
§ An attempt by Cheyne to apply Newton's theory of tides to the atmosphere was also severely criticized. 'Mr. Halley & M. De Moivre told me that Dr. Cheyne gave a paper to Dr. Mead concerning the Tides in the Air (analogous to those in the Sea) made by the Moon, which they see; and that there was not one right or sensible word in it. Dr. Mead did not print it, being putt off it by Mr. Halley & Mr. De Moivre'. Memorandum by David Gregory in Hiscock (1937): 19.

shou'd be debarr'd from them meerly [sic] for want of a Language. [...] Thus indeed I confess, do some people argue for keeping the Sacred Books in an unknown Tongue: But we pretend to a Protestant Liberty, at least with respect to our Philosophy.†

An attempt to provide some introduction to the *Principia* can be found in Charles Hayes's *Treatise of fluxions* (1704). An expert in geography and cartography, Hayes must have had some competence in mathematics. In writing his treatise he referred both to Continental and to British work. In Hayes's *Fluxions* we find a treatment of conic sections based on James Milnes's *Sectionum Conicarum Elementa* (1702). Milnes's book was important since several properties of conic sections that Newton had used in Sections 2, 3, 4 and 5, Book 1, were demonstrated and made available to the most scrupulous readers of the *Principia*.‡ Hayes further dealt with the direct and inverse algorithm of fluxions, devoting attention to topics such as the tautochrone (p. 145) and the solid of least resistance (p. 313). Section 14, devoted to *the use of fluxions in astronomy*, is an incredible mixture of Newton and Leibniz. Here Hayes paraphrased Lemmas 10 and 11 and Propositions 1, 4, 6 and 11 from Book 1 of the *Principia*. However, he inserted into this Newtonian corpus, without any scruple, bits from Leibniz's *Tentamen de motuum coelestium causis*, which was of course based on completely different principles. The reader is led to think that Newton's and Leibniz's works on planetary motion are quite compatible.

A much more reliable commentary, published in 1730, is John Clarke's *Demonstration of some of the principal sections of Sir Isaac Newton's Principles of natural philosophy: in which his peculiar method of treating useful subjects, is explained, and applied to some of the chief phaenomena of the system of the world.* Clarke presented almost verbatim a large part of Book 1 (the first three sections and 'some of the most useful Propositions out of some of the other sections, as those of the direct ascent of Bodies, of the motion of bodies in Pendulums, and of the motion of bodies mutually attracting each other' §), interposing some of the applications to astronomy taken from Book 3. Clarke's commentary is useful since it gives the references to Euclid, Apollonius, Milnes and other texts necessary to understand Newton's geometrical arguments.¶ For instance, in explaining Section 1 on the method of first and ultimate ratios, Clarke made use of the *de Quadratura*. In the comments of Proposition 9, Book 1, on motion along an equiangular spiral, one finds reference to l'Hospital's *Analyse* (1696). He also commented on and explained the most difficult passages. For instance, in his commentary on Corollary 1, Proposition 13, Book 1, Clarke discussed correctly the

† Ditton (1705), *Preface*: pages unnumbered.
‡ Most notably, Milnes on pages 66–8 proves that the parallelograms circumscribed to an ellipse have equal areas – a property that Newton had used in the demonstrations of Propositions 10 and 11 (§3.4.3).
§ i.e. Props. 32–4, 37–9, 48–52, 57–60, 70–8, Book 1. Clarke (1730): v–vi.
¶ Milnes is referred to, for instance, in the commentary on Lemma 12 to Proposition 10, Book 1.

difficult question of the inverse problem of central forces and referred the reader to Keill's paper (1708) devoted to this topic.† Clarke made clear his intentions, which he successfully fulfilled, in the Preface:

In the Demonstrations I thought it sufficient barely to refer the Reader to the Elements of Euclid (except where the Proposition is an uncommon one, and then it is particularly specified,) lest they should be too much interrupted by perpetual References, and because I suppose every Person to be so much acquainted with the Elements of Geometry as is necessary to qualify him for the study of Natural Philosophy, otherwise it is like a Man's attempting to compose a Tune before he is able to distinguish the Notes. But all other Propositions with respect to Conick Sections, or any other Curves, are always particularly referred to, that the Reader may have Recourse to them as he sees fit; and I have made use of Mr. Milnes's Book principally, because his Demonstrations are geometrical and very clear, and because in composing it, he had the same View of explaining Sir Isaac Newton's Philosophy. And as the several Propositions were intended to solve some Phaenomenon in Nature, I have accordingly applied them to particular Instances taken out of the Third Book, that the Reader may at once see the Excellency and Usefulness of the Author's Method of treating Philosophy.‡

In 1730 another 400-page guide to the *Principia* was published: the *Philosophiae Mathematicae Newtonianae* by George Peter Domcke, an obscure scientific lecturer. In the Preface Domcke cited Whiston's and Machin's positive evaluations of his work. He further affirmed to have turned, in the study of curves, to the 'logarithmic canons' of Halley and La Hire, as well as to the methods of Cotes: 'Where, however, the geometric method of demonstration seems too long and to exceed the limits of abbreviation, I have called Algebra to my assistance'.§ He actually opened his treatise with a long treatment of mathematics, apparently to be used in dealing with the *Principia*. Domcke's *Philosophiae* is divided into two *Tomi*. The former is devoted to 'arithmetica universalis' (i.e. the four operations, fractions, extraction of roots, logarithms, the direct and inverse method of fluxions) and to geometry (i.e. conic sections, the cycloid, spirals and trigonometry). The latter is a commentary on the first and third Books of the *Principia*. Despite the above declaration in favour of the algebraical method, Domcke simply paraphrased Newton's geometrical demonstrations: he employed the analytical method of fluxions only in the study of the attraction of a point placed on the prolongation of the axis of an homogeneous cylinder (which is Corollary 1 to Proposition 91, Book 1, discussed in §3.8).¶

An interesting case is Willem Jacob 'sGravesande's introduction to New-tonian natural philosophy, which was translated into English in 1720–21 by

---

† On Keill (1708) see §8.6.2.5 and Guicciardini (1995).

‡ Clarke (1730): vi.

§ 'Ubi vero Geometrica demonstrandi methodus nimis longa limitesque compendii excedere videbatur, Alge-braicam in subsidium vocavi'. Domcke (1730): ix.

¶ We mention here some later commentaries on the *Principia* which were published in later years: the notes of Thorp (1777) on his English translation of Book 1; the commentary on the first three sections of Book 1 by Carr (1821); Wright (1830); Evans (1837); (1850); Frost (1854).

John Theophilus Desaguliers with the title *Mathematical elements of Natural Philosophy confirmed by Experiments, or an Introduction to Sir Isaac Newton's Philosophy.* As Desaguliers wrote in the dedication to Newton:

> as there are more Admirers of your wonderful Discoveries, than there are Mathematicians able to understand the first two Books of your *Principia*: so I hope You will not be displeased, that both my Author and myself have, by Experiments, endeavoured to explain some of those Propositions, which were implicitly believed by many Readers.†

In this treatise we do not find mathematical arguments, but rather experimental 'demonstrations' of some of the propositions of Newton's *magnum opus*. For instance, Proposition 1, Book 1, is 'verified' by means of rotating tables on which the experimenter places balls subject to various constraints. As the ball is pulled towards the table's centre, its angular velocity increases in accordance with Proposition 1. Similarly, there are experiments with rotating balls devoted to Huygens' law ($F \propto v^2/\rho$), others on resistance of fluids and projectile motion, or on the flattening of a rotating Earth (demonstrated with a little mechanical model). The truths of the *Principia* thus become tangible: it becomes possible to set up public experimental lectures where the innumerate public could be convinced of the veracity of Newton's mathematical discoveries.

Only after 1740 do we find elementary texts with some instructions on the applications of the analytical method of fluxions to the *Principia*: i.e. William Emerson's and Thomas Simpson's textbooks on fluxions. In Section 3 of Emerson's *Doctrine of Fluxions* (1743) one can find analytical fluxional solutions of problems taken from Newton's *Principia*: e.g. 'to find the velocity of a projectile moving in a given curve'; 'to find the force wherewith an infinite solid attracts a corpuscle, when the law of attraction is inversely as some power of the distance greater than 1'; 'to find the force wherewith a sphere attracts a corpuscle, when the force of every particle is reciprocally as the square of the distance'; 'to find the force wherewith a spheroid attracts a corpuscle placed at the pole'.‡ Simpson in his *Doctrine and Application of Fluxions* (1750) not only translated some of the basic propositions of the *Principia* (in Section 7 he dealt with motion in resisting media, in Section 8 with attraction of extended bodies) into analytical fluxional terms, but he also presented some innovative results on the attraction of ellipsoids.§

7.5 Conclusion: private exchanges, public expositions

Throughout this chapter we have followed the diffusion of the *Principia* in Britain. A characteristic of the British Newtonians catches the attention. While many

---

† 'sGravesande (1720–21): ii.
‡ Emerson also published in 1770 a *Short comment on Sir Isaac Newton's Principia*.
§ For further details on Emerson and Simpson see Guicciardini (1989).

of them (Keill, De Moivre, Halley, Fatio, Gregory, Cotes) were aware of the possibility of dealing with some propositions of the *Principia* in terms of the analytical method of fluxions, almost no effort was made, until the 1740s, to render this awareness public.† Most notably, David Gregory and Cotes discussed the *Principia* in calculus terms in their correspondence with Newton, who used to show his protégés from time to time how to apply the calculus to natural philosophy. However, in Newton's circle the application of calculus (the 'analytical method of fluxions') to natural philosophy was not the language in which results had to be published. The second and third editions of the *Principia*, despite some relevant variants, remained unchanged as far as the mathematical style was concerned. Cotes's *Logometria* (1714) presented the geometrical solution of several problems in natural philosophy, but the integrals necessary to achieve them were published posthumously in *Harmonia mensurarum* (1722). Gregory, who in the unpublished *Notae* had applied quadrature methods to central forces and attraction of extended bodies, in his *Astronomiae* (1702) paraphrased the geometrical demonstrations of the *Principia*. Rather than rendering his competence in the analytical method public, Gregory spent his last years in cooperating with Halley on the Oxford editions of Apollonius – an effort that Newton praised highly.

Geometry still dominated several works written in the middle of the century. Colin Maclaurin in *Treatise of fluxions* (1742) treated topics such as central forces, the attraction of ellipsoids, the Moon's motion and the tides following the style employed by Newton in the *Principia*. The same is true in broad of Matthew Stewart's studies of planetary motions in *Tracts, physical and mathematical* (1761) and Charles Walmesley's papers on the motion of the Moon's apse, on nutation and on planetary perturbations which appeared in the 1750s and 1760s.‡

Only after the middle of the century do we find British authors convinced that slavish adherence to the *Principia*'s geometrical methods would ultimately lead to total sterility. In 1771 John Landen, a proponent of an algebraical foundation of the calculus and famous for his contributions to the study of elliptic integrals, attacked Matthew Stewart's strictly geometrical study of the perturbing force of the Sun on the Moon's orbit.§

As late as 1757, Thomas Simpson, one of the main promoters in Britain of the analytical way of the Continentals, in the Preface of his *Miscellaneous tracts*

---

† An exception is the appendix to the first edition (1729) of Motte's English translation of the *Principia* (3rd ed.). Here one can find the fluxional equations, 'given by a Friend', relative to Corollary 2, Proposition 91, Book 1, and to the Scholium to Proposition 35/4, Book 2. Notice, however, how, in this case also the analytical method of fluxions was seen as something necessary to digest only some well localized demonstrations of the *Principia*.

‡ For a complete list of Walmesley's works see the bibliography in Guicciardini (1989).

§ Landen's *Animadversions on Dr. Stewart's computation of the Sun's distance from the Earth* (1771) and Dawson's *Four Propositions* (1769) were written as a criticism of Stewart's *The distance of the Sun from the Earth determined by the theory of gravity* (1763).

devoted to astronomy, had to justify his decision to depart from the geometrical style of the *Principia* in a rather apologetic manner. It is significant that he referred to Proposition 28, Book 3, as an example of analytical style in the *Principia*:

I have chiefly adhered to the analytic method of Investigation, as being the most direct and extensive, and best adapted to these abstruse kinds of speculations. Where a geometrical demonstration could be introduced, and seemed preferable, I have given one: but tho' a problem, sometimes, by this last method, acquires a degree of perspicuity and elegance, not easy to be arrived at any other way, yet I cannot be of the opinion of Those who affect to shew a dislike to every thing performed by means of symbols and an algebraical Process; since, so far is the synthetic method from having the advantage in all cases, that there are innumerable enquiries into nature, as well as in abstracted science, where it cannot be at all applied, to any purpose. Sir Isaac Newton himself (who perhaps extended it as far as any man could) has even in the most simple case of the lunar orbit (Princip. B.3. prop. 28) been obliged to call in the assistance of algebra; which he has also done, in treating of the motion of bodies in resisting media, and in various other places. And it appears clear to me, that, it is by a diligent cultivation of the Modern Analysis, that Foreign Mathematicians have, of late, been able to push their Researches farther, in many particulars, than Sir Isaac Newton and his Followers here, have done: tho' it must be allowed, on the other hand, that the same Neatness, and Accuracy of Demonstration, is not every-where to be found in those Authors, owing in some measure, perhaps, to too great a disregard for the Geometry of the Ancients.†

In the first decades of the eighteenth century, Newtonian analytical natural philosophy remained almost wholly hidden in manuscripts. After the inception of the priority dispute with Leibniz, Newton did publish his *De quadratura* (1704) and a collection of mathematical tracts which included the *De analysi* (1711). However, the version of *De quadratura* that Newton chose for publication did not include the analytical treatments of central force which we have discussed in §4.5. These formulas for central force based on the determination of curvature were communicated privately to De Moivre and Keill (§8.6.2.5). This is in sharp conflict with what was going on in Leibniz's circle, where the *Acta eruditorum* and the *Mémoires* of the Paris Academy were the vehicles for rendering the calculus public. It might be argued that the Newtonians were still following a publication policy which was not infrequent in the first half of the seventeenth century. Before the inception of scientific journals mathematicians revealed their truths in correspondence and often kept the methods hidden. Leibniz and his associates can thus be seen as innovators in publication practice, since they systematically published the analytical methods employed to achieve results and solve problems. On the Continent these analytical methods, discovered and extended by mathematicians such as Varignon, Hermann and Johann Bernoulli, aroused great excitement. It is to the Continental school that we turn in the next chapter.

† Simpson (1757): Preface. On Simpson and the 'analytical' fluxionists of the mid-eighteenth-century see Guicciardini (1989), Chapter 6.

8

Basel: challenging the *Principia*

8.1 Purpose of this chapter

In this chapter we will be concerned with the small group of enthusiasts who extended and promoted Leibniz's calculus at the turn of the century. A handful of mathematicians, mainly based in Basel and Paris, invested their intellectual strengths in the new and controversial algorithm first published in the *Acta eruditorum* for 1684. Their reaction to the *Principia* resulted in a clearly stated programme: Newton's demonstrations had to be translated into Leibnizian language. I am interested in understanding not only what these mathematicians achieved, but especially how they justified the relevance of their research programme.

In §8.2 and §8.3 the reader will find some introductory material concerning the Leibnizian school. Then we will move on to consider some results related to the *Principia* achieved by Varignon (§8.4), Hermann (§8.5) and Johann Bernoulli (§8.6). This rather selective choice of material will allow us to show the following.

- Varignon did not prove new results. Nevertheless, his work is important since he insisted publicly on the advantages of adopting Leibniz's calculus in dealing with some of the main themes of the *Principia*. Varignon stressed the generality of calculus methods.

- Hermann is a transitional figure. He was trained in the Leibnizian calculus. However, he was also respectful towards the *Principia*'s geometrical methods. We will focus on his proof of the law of areas. Newton had given a demonstration (Proposition 1, Book 1) based on the synthetic method of fluxions (method (ii) in §3.16). Hermann provided a demonstration in terms of calculus; a demonstration that, as he said, was 'based on a completely different foundation'. We have often underlined the fact that some of the synthetic demonstrations of the *Principia* are independent of analytical techniques. The divide between Hermann's and Newton's proofs of the law of areas is an example of this independence.

- Johann Bernoulli stated that Newton had not solved the inverse problem of central forces. He published a solution consisting of the integration of a differential equation. In Proposition 41, Book 1, Newton, through a geometrical procedure based on infinitesimals, reduced the inverse problem to a quadrature (method (iii*b*) in §3.16). I will show that Newton knew how to perform such a quadrature by the help of his analytical method of fluxions. However, he kept this solution hidden and insisted in publishing an *a posteriori* geometric solution in Corollary 1 to Propositions 11–13, Book 1 (method (i) in §3.16). This is an example of what we have termed 'quadrature avoidance': Newton knew the calculus solution, but preferred to publish a geometric one.

- Johann and Niklaus Bernoulli claimed that there was an error in Proposition 10, Book 2, dealing with resisted motion. Johann provided a calculus solution based on the integration of a differential equation. Newton relied on a representation of the trajectory in terms of geometrical infinitesimal elements. He also applied Taylor series expansions (method (iii*a*) in §3.16). Niklaus Bernoulli stated that Newton's mistake was in the use of the Taylor coefficients. On the contrary, Newton's mistake was caused by the difficulty of evaluating the order of approximation in geometrical terms.

In §8.7 we look ahead to the period in which Euler's *Mechanica* put an end to the competition between the analytical and the synthetic approaches to the science of motion. After Euler the *Principia* were read – as is evident in the annotated editions by Le Seur and Jacquier, or by Chastelet and Clairaut – in terms of the new symbolic Eulerian calculus.

8.2 The Bernoullian school

Leibniz's *Nova methodus* (1684), the first publication of the basic rules of the differential calculus, remained for several years almost unnoticed.† It was a difficult paper. Its innovations, *per se* a hurdle, were obscured by typographical errors. As Fontenelle says, the first to realize its importance were two gifted brothers domiciled in Basel, Jacob and Johann Bernoulli:

The Messieurs Bernoulli were the first who perceived the Beauty of the Method; and have carried it such a length, as by its means to surmount Difficulties that were before thought insuperable.‡

Actually Basel soon became the indisputable centre of diffusion of the Leibnizian calculus. The two Bernoullis became the promoters of the 'new method' on the Continent. They taught it first of all to their sons and nephews, so that a dynasty of Bernoullis was to dominate the European scene for half a century, but also to bright local students. Jacob Hermann and Leonhard Euler are the two most famous. Interested natural philosophers used to visit Basel, often as paying guests of the Bernoullis, in order to be instructed through private lessons on the mysteries of the calculus. For instance, the Italian Giuseppe S. Verzaglia decided in 1708 to move

† *LMS*, **5**: 220–26.
‡ L'Hospital (1696): ix. Translation by Stone (1730): viii.

from Modena in order to stay for a year and a half in Johann Bernoulli's house.†
In 1729 Maupertuis, already a reputed Parisian academician, humiliated himself
by matriculating as a student in the University of Basel in order to have access to
Johann Bernoulli.‡

The Basel group was efficient and powerful. However, it is somewhat misleading
to define it as a 'school'. Tensions between the members of the Bernoulli's circle
were the rule. Jacob and Johann, after their period of common mathematical
apprenticeship, quarrelled over priority issues concerned with the calculus of
variations. Daniel and Johann did the same over hydrodynamics. Euler very soon
proved to be too original a mathematician to be confined in the role of a disciple.
Hermann, after his stay in Italy, distanced himself from the Bernoullis, so that
Brook Taylor could write to William Jones in 1716:

He [Hermann] gives greater testimonies to Sir Isaac Newton than I have seen from any of
the foreigners; though in one place he seems to wish that Leibniz might have a share in his
discovery of universal gravitation. I would transcribe the passage, if I had not lent out the
book [the *Phoronomia*]. By my correspondence with M. Monmort I find Herman [sic] is
not of Bernoulli's party, but is rather under his displeasure.§

Despite tensions and disputes, the Basel group had a strong homogeneity. Mathematical methods, research aims, textbooks adopted (e.g. l'Hospital's *Analyse*
(1696)), systems of recruitment and education were shared by all its quarrelsome
members. In the next section §8.3 I will attempt a brief description of these
common elements.

8.3 'Une science cabalistique'

In Basel, at the turn of the century, the formation of a professional group of
mathematicians, who distinguished themselves by their ability in handling the new
Leibnizian calculus, takes place. Until the publication of l'Hospital's *Analyse* in
1696 only a privileged handful could penetrate the intricacies of the new analysis.
As Fontenelle remembers in his eulogy of l'Hospital read in 1704:

The geometry of the infinitely smalls was still a kind of mystery, and, so to speak, a
Cabbalistic Science whose knowledge was confined just to five or six persons.¶

This small initial group shared the persuasion that Leibniz had invented a
radically new mathematical method, a *Nova methodus*. The rhetorical accent was
clearly on its novelty, rather than on the continuity with ancient, perhaps lost,
exemplars. Johann Bernoulli, in a colourful description of his encounter with the

---

† This unfortunate stay has been described masterfully in Nagel (1991).
‡ Greenberg (1986): 67.
§ Rigaud (1841), 1: 280.
¶ 'la Géométrie des Infiniment petits n'etoit encore qu'une espece de Mystere &., pour ainsi-dire, une Science
   Cabalistique renfermée entre cinq, ou six personnes'. Quoted in Mancosu (1989): 225.

Leibnizian calculus, said that thanks to it he could abandon the familiar harbour for new unexplored territories.† Leibniz was encouraging such explorations:

All my purpose in producing this new calculus [...] was to open a new way, by which those who are more penetrating than me could find something of importance.‡

In the Preface to l'Hospital's *Analyse* we find a clear statement on the inadequacies of ancient geometry when compared with the Leibnizian mathematical methods:

What is extant in the works of the ancients, particularly of Archimedes, on these matters is surely worthy of admiration. But apart from the fact that they treat very few curves, and these only quite superficially, the propositions are almost entirely particular and without order, which fail to communicate any regular and applied method.§

The Bernoullian mathematicians felt that a new field of research was opened up by the differential and integral calculus, a field where many new results could easily be obtained by following as guidelines the analogies suggested by the calculus's notation. The algorithm allowed the discovery of new generalizations, new relations and new formulas. The mechanization and standardization of mathematical research rendered possible by the stress on the algorithm made the Basel school active and open to innovation. Johann Bernoulli, in his first encounter in 1691 with the French Oratorians headed by Malebranche, sold exactly this aspect of the Leibnizian calculus. He fascinated them by showing that with calculus techniques he could determine the catenary, a problem that had baffled Galileo, and that the calculation of the radius of curvature was 'mere child's play'.¶ Knowledge of this 'marvellous'‖ algorithm could be actually sold. One of the French mathematicians that Johann Bernoulli had met during his stay in Paris in 1691–92 was Guillaume François de l'Hospital who offered him 'one of the most surprising arrangements in the history of mathematics'.†† The Marquis wrote to the young Swiss mathematician:

I will be happy to give you a retainer of 300 pounds [...] I am not so unreasonable as to demand in return all of your time, but I will ask you to give me at intervals some hours of your time to work on what I request and also to communicate to me your discoveries, at the same time asking you not to disclose any of them to others. I ask you even not to send

---

† Dupont & Roero (1991): 139–41.

‡ 'Tout ce que je m'estois proposé en produisant le nouveau calcul [...] a esté d'ouvrir un chemin, où des personnes plus penetrantes que moy pourroient trouver quelque chose d'importance'. Leibniz to Huygens (October 1693) in Leibniz *Mathematikern*: 720. Quoted in Heinekamp (1982): 103.

§ 'Ce que nous avons des anciens sur ces matiéres, principalement d'*Archimede*, est assurément digne d'admiration. Mais outre qu'ils n'ont touché qu' à fort peu de courbes, qu'ils n'y ont même touché que légérement; ce ne sont presque par tout que propositions particulieres & sans ordre, qui ne font apercevoir aucune méthode réguliere & suivie'. L'Hospital (1696): iv–v.

¶ Johann Bernoulli to Montmort (May 1718), quoted in Feigenbaum (1992): 388.

‖ L'Hospital to Leibniz (December 1692) in *LMS*, **2**: 216.

†† Feigenbaum (1992): 388.

here to Mr. Varignon or to others any copies of the writings you have left with me; If they are published, I will not be at all pleased.†

This well-known contract shows clearly how knowledge of the calculus could be turned into profit. Usually, the profit was in terms of power in the Academies and the Universities. The Bernoullis and their students actually colonized chairs of mathematics all over Europe, from Groningen to St Petersburg, from Padua to Frankfurt-an-der-Oder. Their expansionist policy has been described by Robinet in his book devoted to the conquest of the Chair of Mathematics in Padua.‡ Hermann was elected there in 1707 after complicated negotiations with the Italians. The aim was that of establishing in Northern Italy a centre of diffusion of the new Leibnizian method. As Leibniz wrote to Hermann:

I also thought it would be useful for the public, and a source of honour for you, if our new analysis were introduced into Italy adorned with your genius.§

Which kind of calculus was promoted by the Bernoullian circle? In a sense this question is too wide to receive a schematic answer. However, in broad outline, one can say that three main research areas were developed: (i) methods of solution of ordinary differential equations; (ii) the first steps towards a formulation of the calculus of variations; (iii) the summation of infinite series. The calculus was a set of algorithmic rules which allowed the development of these three major research fields. The calculus was always interpreted in geometric terms, i.e. the differentials and the integrals were referred to geometric entities. Nonetheless, most of the techniques consisted in formal manipulations with symbols. The members of the Bernoullian circle were masters of such symbolic manipulations, which often seemed to others groundless and illogical. In order to establish their respectability, the Bernoullians had to defend themselves from critics. Actually one of their concerns was that of refuting critics, such as Detleff Clüver, Bernhardt Nieuwentijt and Michel Rolle, and obtaining victory against a much more fearful enemy: Isaac Newton.

Mancosu and Vailati have described the strategies with which Jacob Bernoulli and Leibniz tried, unsuccessfully, to convince Clüver, a minor but seemingly menacing mathematician, about the correctness of the infinitesimal calculus. Many reacted to Leibniz's infinitesimals with bewilderment or scepticism. At the end of the seventeenth century the Dutch theologian Nieuwentijt carried on a campaign, which anticipates aspects of Berkeley's *Analyst* (1734), against the calculus. Hermann was given responsibility for a reply which took book-length form in (1700). In Paris a number of anti-infinitesimalist academicians, including Rolle, La

---

† L'Hospital to Johann Bernoulli (March 1694). Translation by Truesdell in (1958): 61.
‡ Robinet (1991); Mazzone & Roero (1997).
§ 'Judicavi etiam in publicum utile, et tibi honorificum fore, si nova Analysis nostra tuo ingenio ornata in Italiam introduceretur'. Leibniz to Hermann (November 1704) *LMS*, **4**: 263.

Hire and Jean Galloys, carried on a campaign against the Malebranchists who, after Johann Bernoulli's instructions, were promoting the calculus. Varignon, assisted by Leibniz, was successful in establishing the legitimacy of the calculus at the Académie.†

Beyond doubt the greatest challenge to the reputation of the Leibnizians came from Britain. In 1699 Fatio accused Leibniz of having plagiarized Newton's algorithm. The famous priority dispute between Leibniz and Newton was thus caused by a Swiss *émigré*. There is no need to rehash here the details of this quarrel.‡ Newtonians and Leibnizians fought each other with all means, often hiding themselves behind anonymity. A climax was reached when John Keill reformulated Fatio's accusation in a paper devoted to central forces published in the *Philosophical Transactions of the Royal Society* for 1708. Leibniz, who was a fellow of the Royal Society, asked for a formal retraction. The result was the *Commercium Epistolicum* (1713), a report prepared by a committee secretly guided by Newton. The war fragmented into several minor battles. Typically, a mathematician belonging to one camp opposed himself to another belonging to the other party. Several issues were touched on: cosmology, physics, the foundations of calculus, and the roles of symbolism and geometry, for example. The atmosphere was poisoned and the contest was far from honest intellectual confrontation. Taking part in the fight was a duty, but sometimes also a chance to gain some space in the arena.

The *Principia* was often referred to in the debate. It became crucial for the Leibnizians to use the *magnum opus* as proof of the inadequacy and old fashioned character of Newton's mathematical methods. Johann Bernoulli was particularly active in this field. On the other hand, Newtonians wished to use the *Principia* as proof either of Newton's knowledge of calculus, or of the superiority of his geometrical methods over the algorithmic Leibnizian ones. From the point of view taken in our book this is extremely interesting. We have to take into due consideration the fact that the waters were muddied. Many of the arguments were designed *ad hoc* in order to bring an attack against the enemy. I believe that this is the reason why the priority dispute has been almost exclusively studied as a sociological fact. Little weight has been given to the content of the debate. Just to take an example, Keill's paper (1708) on central forces has been referred to by historians only for the three lines concerning Newton's right of priority over Leibniz. An attempt to read what Keill was doing, and what reasons he could offer in order to maintain the superiority of Newton's mathematization of central force over Leibniz's, has been obviously considered a useless task. Since Keill had polemical motivation, it has been tacitly assumed by historians of mathematics

† Mancosu & Vailati (1990); Mancosu (1989); Blay (1986); Peiffer (1988) and (1990).
‡ See Hall (1980) and Whiteside's commentary in *Mathematical Papers*, **8**: 469–697 for a detailed account.

that there is no valuable scientific content in his paper.† I am convinced that this approach to the priority debate is too narrow. In the remaining pages of this chapter we will devote some attention to the content of the confrontation between Leibnizians and Newtonians. We will focus on Varignon, Hermann and Johann Bernoulli.

8.4 Varignon on central forces and resisted motion

Pierre Varignon played a relevant role in the process of translation of the *Principia* into the Leibnizian calculus. During the first two decades of the eighteenth century he published on this topic a series of papers in the *Mémoires de l'Académie des Sciences*.‡ Varignon established his scientific reputation in 1687, when his work on statics, *Projet d'une nouvelle méchanique*, appeared. He had just moved to Paris from Caen. Because of this book, he was elected at the Académie and became Professor of Mathematics at the Collège Mazarin. In the capital he got in touch with the group of mathematicians gathered around Malebranche. As Peiffer convincingly argues, he was accepted as a member of the Malebranchiste équipe only after 1695. For instance, it is almost certain that he did not follow Johann Bernoulli's Parisian lectures in 1691–92.§

The Malebranchiste group included Varignon, l'Hospital, Carré, Reyneau, Montmort, Sauveur, Saurin, Privat de Molières, and Jaquemet.¶ These mathematicians were motivated by the desire to acquire a working knowledge of the new calculus. They shared a Leibnizian formation, as it appears from 'their' favoured textbook, l'Hospital's *Analyse*, which was based on Johann Bernoulli's lectures. However, many of them, such as Montmort and Varignon, functioned as mediators between the Leibnizian and the Newtonian camps.‖ Their activity fell into a period of cultural isolation of France caused by the War of the League of Augsburg (1689–97). They were interested in favouring as much as possible the difficult contacts with scholars abroad, Britain included.†† As a matter of fact, in certain respects, some of them leaned towards the British school. Varignon based his defence of calculus during the debate at the Académie on ideas which bear some resemblance to Newton's method of first and ultimate ratios. Traces of a Newtonian foundational approach can be found also in his posthumous *Éclaircissemens sur l'analyse des infiniment petits* (1725), a commentary on l'Hospital (1696). During the priority dispute, Montmort mediated between Brook Taylor and Johann

---

† On Keill's paper (1708) see Guicciardini (1995).
‡ These have been analysed in detail in Blay (1988) and (1992), to whom I owe a great deal in §8.4.
§ Peiffer (1990): 247–53.
¶ On the Malebranchists see Robinet (1960) and Fleckenstein (1948).
‖ Feigenbaum (1992).
†† Greenberg (1986).

Bernoulli. In general the French proved to be respectful towards both contending parties. As Greenberg has shown, interest in Newtonian mathematics increased in France after the 1720s.†

The *Principia* soon attracted the attention of the French. A review appeared in the *Journal des sçavants* for the 2nd of August 1688. There is evidence that it was discussed in the Malebranchist équipe. In the group's manifesto, the preface to l'Hospital's *Analyse* (1696), we find reference to Newton's work, which is described as 'almost entirely' based on the new calculus. In the correspondence of Jaquemet with Reyneau we find that the former studied the *Principia* from the early 1690s.‡ Malebranche too considered the *Principia*: he found difficulties in reading Newton's work. In July 1699 J. Monroe wrote that Malebranche 'mightily commends Mr. Newton, adding, at the same time that there were many things in his book that passed the bounds of his penetration, and that he would be very glad to see Dr. Gregory's critick upon it'.§ In November 1714 Charles-René Reyneau informed William Jones that Cotes's second edition was still very rare.¶ Notwithstanding their awareness of the importance of the *Principia* and their balanced position between Newton and Leibniz, the French, with the notable exception of Varignon, did not produce much work related to it.

Varignon began researching the mathematization of central force motion around 1695.‖ His programme was to deal with this topic in terms of the Leibnizian calculus. He was encouraged by Leibniz himself to pursue this research programme. The *Principia* was clearly his starting point. In a series of memoires published from 1700 to 1706 he actually reframed large parts of Sections 2, 3, 7, 8 and 9, Book 1, into the language of the differential calculus.

In 1700 Varignon published three papers in which he stated:

- the relation between force F, velocity v and displacement x expressed by Newton in Proposition 39, Book 1, of the *Principia*, i.e. $2 \int F \mathrm{d}x = v^2$;
- the expression of central force in terms of differentials:

$$F = \frac{\mathrm{d}s \mathrm{d}^2 s}{\mathrm{d}r \mathrm{d}t^2}. \tag{8.1}$$

Next he applied formula (8.1) to the solution of the direct problem of central forces.

---

† The Jesuit Castel openly promoted fluxional calculus in the *Journal de Trévoux* while Rondet translated into French in 1735 the second volume of Stone's *Method of fluxions* (1730), a sequel written in English to l'Hospital's *Analyse* (1696). Later in the 1730s a project of translating the volumes of the *Philosophical Transactions* into French was considered; in 1740 Buffon translated Newton's *Method of fluxions* (1736), and in 1749 Pézénas translated Maclaurin's *Treatise of fluxions* (1742).
‡ Malebranche (1967–78), **17**(2): 61–2.
§ Quoted in Ball (1893): 132.
¶ Rigaud (1841), **1**: 265–6.
‖ Varignon's first lost memoire on central forces was presented at the *Académie* in 1695.

As it was customary in the literature of this period the constant mass term was not made explicit, equations such as (8.1) being still read as proportionalities.

In 1701 Varignon published an alternative expression for central force, in which the radius of curvature occurs:

$$F = \frac{\mathrm{d}s^3}{\rho r \mathrm{d}\theta \mathrm{d}t^2}. \tag{8.2}$$

In equations (8.1) and (8.2) the polar coordinates are (r, θ), the radius of curvature is ρ, t is the time, and s is the trajectory's arc length. Leibniz and Newton had already achieved similar applications of calculus to central force motion.[†]

Varignon considered the trajectories dealt with by Newton in Propositions 7–13, Book 1 (§3.4.3). Most notably, he solved the direct problem for conic trajectories in the case in which the area law is valid for a focus, i.e. he provided a calculus equivalent of Propositions 11, 12 and 13, Book 1.[‡]

Immediately after the *Principia*'s publication, the status of the law of areas was not so definite as one might think. In fact, in (1700) Varignon dealt with central forces assuming different laws for the time of orbital motion, such as those proposed by Seth Ward and Cassini. He wrote:

My first purpose was to find the central forces of the planets according to the hypothesis of Kepler, Newton, and Leibniz, as it is more physical, making $\mathrm{d}t = r\mathrm{d}z$ [$\mathrm{d}t = rr\mathrm{d}\theta$]; but after having considered that this hypothesis [i.e. Kepler's area law] is not the unique one adopted in astronomy, here is how I can deal with all the hypotheses [...] whatever might be the hypothesis on the times, or the values of $\mathrm{d}t$.[§]

He then proceeded to consider cases in which the time is proportional to an angle measured from an equant point. David Gregory's comment was caustic:

In the History & Journales of the French Royal Academie for the year 1700, M. Varignon has a long treatise concerning the vis centripeta, filled with a great number of blunders [...] He investigates (&, as he thinks finds out) the law of gravity, whereby a body may revolve in an Ellipse, with a centripetal force tending to one focus, and describing angles proportional to the times at the other, not adverting that this is impossible, unless, the areae at one focus be proportional to the Angles at the other, which is not true.[¶]

As a matter of fact central force motion implies the law of areas, as Newton had demonstrated in Proposition 1, Book 1, a point that Varignon apparently missed.

From 1707 to 1711 Varignon published a series of memoirs devoted to resisted motion.[||] He explicitly referred to the first three sections of Book 2 of the *Principia*, as well as to the works by Leibniz, Huygens and Wallis.[††]

[†] Formula (8.2) occurs in a paper by Varignon published in the *Mémoires de l'Académie des Sciences* for 1701: 20–38. It can also be found in Varignon (1706): 189.

[‡] See Blay (1992): 197ff. for details on these works by Varignon.

[§] Varignon (1700): 225. Note that Varignon sets the constant areal velocity equal to 1.

[¶] Hiscock (1937): 18.

[||] For a detailed analysis of these papers see Blay (1992): 277ff.

[††] Varignon (1707): 382 and Blay (1992): 279.

Varignon aimed at deriving the various results on resisted motion from a 'general proposition'. Once again he had recourse to calculus. His general proposition is

$$\frac{\mathrm{d}v - \mathrm{d}u}{z} = \frac{\mathrm{d}t}{a} \tag{8.3}$$

where t is the time, z the resistance, u the velocity, v the 'primitive velocity' (i.e. the velocity that the body would have if there were no resistance), and a is a constant. This proposition can be rewritten in terms easier for a modern reader as

$$\frac{\mathrm{d}u}{\mathrm{d}t} = -\frac{z(u)}{a} + f, \tag{8.4}$$

where $f = \mathrm{d}v/\mathrm{d}t$ is a force, e.g. a gravitational force, and $z(u)$ is the resistance (a function of velocity).

In his subsequent papers Varignon applied the differential equation (8.3), to several particular cases. He considered rectilinear motion under different resistance laws and curvilinear motion under the action of constant gravitation and a resistance proportional to velocity [$z(u) = ku$]. He remarked that his results coincided with those obtained by Newton, Leibniz and Huygens. It is to be noted that Varignon did not deal with motion under the action of gravity and resistance proportional to the square of velocity, a problem which had baffled Huygens, Newton and Leibniz. It was Johann Bernoulli who was able to tackle this more advanced case (§8.6.3.4). We have seen from Leibniz's paper on resisted motion that the *homo universalis* in 1689 already knew and used Varignon's general principle (§6.3.1).

One of Varignon's main limitations was his imperfect knowledge of integration techniques. In general the French group, with the exception of Montmort, assimilated the direct differential calculus but showed little ability in handling the more difficult inverse integral calculus.† In the correspondence with Johann Bernoulli there are several letters in which Varignon implores the cantankerous Swiss geometer for some instructions on the integral calculus.‡ Varignon was probably convinced that, as there was a simple algorithmic method for the direct calculus, so there would have been an equally prodigious method for the inverse calculus, kept hidden by Bernoulli and l'Hospital.§ It is telling that Varignon was able to tackle the direct problem of central forces in 1700, but had to wait until 1710 for Hermann's and Bernoulli's solutions of the inverse problem. While the direct problem requires the differential calculus, the inverse problem of central forces requires the integral calculus. In 1710 Hermann's and Bernoulli's solutions

---

† See Greenberg (1986): 60–1 for an evaluation of the modest contributions to integration obtained by Carré, Bragelogne and Saulmon.
‡ Bernoulli, J. *Briefwechsel*, **2**: 29, 35, 61.
§ This idea was suggested to me by Jean Dhombres.

were communicated to the *Académie* (§8.5 and §8.6.2.3). Typically, a long and inconclusive paper by Varignon on the inverse problem followed. He was fair in acknowledging the fact that his contribution was merely a reformulation of the Swiss mathematicians' preceding 'letters'. It is notable that Varignon simply reformulated the differential equations of Bernoulli and Hermann, adding some other equivalent formulas, but did not proceed to integrate them!

As a matter of fact, Varignon, with his papers on central forces and resisted motion, was able to systematize results which were already known. He did not contribute any original theorem. However, his systematic work should not be underestimated. He carried out a programme of application of the calculus to dynamics that helped to reformulate most of Sections 2, 3, 7, 8 and 9, Book 1, and Sections 1 to 3, Book 2, of the *Principia* into the language and method of the Leibnizian calculus. This language and this method were to prevail among eighteenth-century Continentals. As a result the *Principia* on the Continent became a work written in an obsolete style, a work which could be read only through reformulations in calculus terms. Varignon played an important role in this process.†

8.5 Hermann's *Phoronomia*

The complexity of the process of mathematization of dynamics in the Continental school is particularly evident in transitional figures, such as Jacob Hermann (1678–1733). He belonged to the Basel school headed by the Bernoullis and, sharing their methodology, made important contributions to the analytical treatment of dynamics. However, he also leaned towards Newton and, on many occasions, preferred to deal with dynamical problems in terms of geometry. His methodology is thus quite eclectic. Hermann was a pupil of Jacob Bernoulli in Basel and was a remote relative of Leonhard Euler. Leibniz supported his career in various ways. Thanks to Leibniz's recommendation, Hermann held chairs of mathematics in Padua from 1707 to 1713 and in Frankfurt an der Oder until 1724, while from 1724 to 1731 he was connected with the Academy in St Petersburg. He was able to return home only in 1731 when he took up the professorship of ethics and natural law, the Chair of Mathematics still being occupied by Johann Bernoulli.‡

Hermann's main work is *Phoronomia*, which, written during his Italian period, was published in Amsterdam in 1716. This work is devoted to the dynamics of solid and fluid bodies and covers many problems dealt with in the first two books of the *Principia*. Leibniz complained in a letter to Bernoulli about

---

† On Varignon see also Gowing (1992).
‡ On Hermann see Mazzone & Roero (1997).

Hermann's excessive admiration for Newton.† In fact, Hermann's textbook was not entirely Leibnizian in character, since the calculus was not employed explicitly and extensively as Leibniz would have wished. As we know, in 1716 there were rumours that Hermann was not of 'Bernoulli's party' (§8.2). Newton, on the other hand, was extremely critical, not about the method, but rather about the content. In a note he observed with obsessive precision that the *Phoronomia* did not cover many of the subjects dealt with in the *Principia*.‡

As a matter of fact, the *Phoronomia* reflects rather different interests from those which motivated Newton to apply mathematics to natural philosophy. The Bernoullians shared a Cartesian cosmology in which action at a distance was not permitted. Most of the *Principia* (most of the pages that Newton enumerates in order to prove the incompleteness of Hermann's work) was devoted to the mathematization of a cosmology which had no physical meaning for the Bernoullis and their pupils. Furthermore, Hermann wrote the *Phoronomia* in Italy, where the interest of the Venetian and Bolognese mathematicians was focused on engineering problems related to hydraulics. Maffioli has documented the study of the *Principia*, especially Book 2, in Italy at the beginning of the eighteenth century. One of the first who could boast of being able to master Newton's *magnum opus* was Giuseppe S. Verzaglia, who made use of his stay in Basel at Johann Bernoulli's house. Eustachio Manfredi wrote to Guido Grandi in 1716 that Verzaglia was writing a commentary on the *Principia*. Manfredi added that for him Newton was 'speaking Arabic' and that he hoped for some explanations from Verzaglia.§ The Italians, as Maffioli has amply documented, were particularly interested in the 'science of waters': they were actually the specialists in the study of river flow. Consequently Eustachio Manfredi, Poleni, Michelotti and Jacopo Riccati discussed Proposition 37, Book 2, on the flow of water from a hole pierced in the bottom of a vessel. This proposition aroused many criticisms, e.g. from Huygens, Cotes and Varignon.¶ The fact that the *Phoronomia* is mainly devoted to the 'motion of fluid bodies' and the fact that it was written in Italy might be related.

Even though the cosmology of Newtonian central forces aroused scepticism on the Continent, the problems posed by the mathematization of central forces were deemed extremely interesting. The mathematical theory of central forces is touched on several times in Hermann's correspondence with the Italian

---

† Letter to Johann Bernoulli (November 1715): 'Multa sunt bona, sed [...] videtur nimium Anglis deferre.' *LMS*, **3**: 948.

‡ 'Hermannus omisit [in *Princip. Math.*] Sectionem totam primam [seu paginas] 10, quartam et quintam 46, decimam undecimam duodecimam, decimam teriam [et] decimam quartam 88 Libri primi [hoc est paginas] 144. ac tertiam & quartam 16, et Scholium generale sext[ae] et septimam 45 et nonam ll secundi [id est pag.] 72 & Librum totum tertium 129. Totum omissum [pag.] 345 [ubi] impressum [4]84'. *Mathematical Papers*, **8**: 443.

§ Maffioli (1994): 21.

¶ Maffioli (1994): 403ff.

mathematicians.† As we know, Newton in the first edition of the *Principia* (Corollary 1 to Propositions 11–13, Book 1) had simply stated, without offering proof, that conic sections are the only solutions to the inverse problem for inverse square central forces (§3.4.4). After an exchange of letters with several mathematicians, Hermann was able, in July 1710, to send a solution to Johann Bernoulli. In 1710 Hermann's letter was presented at the *Académie des Sciences* together with a 'Réponse' by Bernoulli.‡ Both Hermann and Bernoulli based their papers on Newton's *Principia*. However, Hermann focused on Proposition 6 (i.e. formula (3.1)), Bernoulli on Propositions 39–41 (see §8.6.2.3).

It can be proved that formula (3.1) in Cartesian coordinates is:

$$F \propto \frac{\mathrm{d}^2 x \sqrt{x^2 + y^2}}{x(y\mathrm{d}x - x\mathrm{d}y)^2}. \tag{8.5}$$

This is the starting point of Hermann's solution. For an inverse square law of force $F = -a/(x^2 + y^2)$, he obtained a second-order differential equation, which he could integrate. The general integral is the Cartesian equation of conic sections. As we will see in §8.6.2.3, Bernoulli criticized Hermann's solution in a rather unfair fashion. The latter was defended by the Italians, most notably by Jacopo Riccati. This affair was probably the reason why Hermann distanced himself from the Basel school.§

Actually, in the preface of *Phoronomia*, Hermann treats Newton with a reverence which is unusual for a pupil of the Bernoullis. He declares his intention of adhering to geometrical methods, since these seem to him more suitable for beginners. One should in fact remember that the *Phoronomia* was written in Italy with didactic purposes in mind, and the Italians were just learning calculus techniques. Hermann also praises Newton's adherence to the 'geometrical analysis' of the Ancients.¶ Despite Hermann's declaration in favour of geometry, his knowledge of calculus is evident by the way in which he deals with infinitesimals. In fact he manipulates geometrical infinitesimals according to the rules of the differential calculus. His geometry is immediately translatable into calculus terms.‖ He actually does so, from time to time, in the *Phoronomia*, especially in the Appendix. The *Phoronomia* is indeed representative of the process of transition that transformed dynamics in the first decades of the eighteenth century. One example can help

---

† Hermann's role in establishing the Leibnizian calculus in Italy is studied in Mazzone & Roero (1997), where several letters by and to Hermann are reproduced.

‡ Hermann (1710), Bernoulli, J. (1710).

§ Alain Albouy is currently working on Hermann's solution. My understanding of it improved considerably thanks to our discussions and correspondence.

¶ Hermann (1716): vii–viii.

‖ Christian Wolff noted that Hermann's decision to use the 'method of the Ancients' in the *Phoronomia* was unfortunate since Hermann was not sufficiently familiar with geometry. Wolff adds that Hermann would have done better service if he had instead used the calculus in which 'he is highly proficient'. Wolff (1732–41), **5**: 64.

us in understanding the nature of Hermann's mathematical style. I will discuss Hermann's demonstration of the law of areas, proved by Newton in the very first Proposition of the *Principia*.

Proposition 1, Book 1, plays a fundamental role in the treatment of central forces (§3.4.1). It says that if a body P is accelerated by a central force directed towards or away from a fixed centre S, then Kepler's area law holds, i.e. the radius vector from S to P sweeps equal areas in equal times. Proposition 1 is important since, for central force motion, it identifies a constant of motion: areal velocity (in modern terms, angular momentum). Furthermore, since the area swept out by SP is proportional to time, Newton could use this area as a geometric representation of time. In Proposition 2, Newton showed the inverse of Proposition 1. Proposition 2 says that if the area law holds (i.e. if there is a point S in the plane of the trajectory, and SP sweeps equal areas in equal times), then P is accelerated by a central force directed towards or away from S.

It is interesting to see how, in a modern textbook,† these two propositions are proved. The most natural choice is, under the assumption that motion is planar, to use polar coordinates (r, θ), so that the origin coincides with the centre of force. The radial and transverse acceleration are thus expressed by the following two formulas:

$$a_r = \mathrm{d}^2 r/\mathrm{d}t^2 - r(\mathrm{d}\theta/\mathrm{d}t)^2, \tag{8.6}$$

and

$$a_\theta = r\mathrm{d}^2\theta/\mathrm{d}t^2 + 2(\mathrm{d}r/\mathrm{d}t)(\mathrm{d}\theta/\mathrm{d}t). \tag{8.7}$$

Let $\mathcal{A}$ be the area swept out by the radius vector (therefore $2\mathrm{d}\mathcal{A}/\mathrm{d}t = r^2\mathrm{d}\theta/\mathrm{d}t$ and $2\mathrm{d}^2\mathcal{A}/\mathrm{d}t^2 = r^2\mathrm{d}^2\theta/\mathrm{d}t^2 + 2r(\mathrm{d}r/\mathrm{d}t)(\mathrm{d}\theta/\mathrm{d}t) = ra_\theta$). For a central force, a_θ is equal to zero. It follows that, by integration of (8.7), $\mathrm{d}\mathcal{A}/\mathrm{d}t = k$ (i.e. the areal velocity is equal to a constant k). Inversely, if $\mathrm{d}\mathcal{A}/\mathrm{d}t = k$, by differentiation it follows that a_θ is zero (i.e. the force is central). The double implication of Newton's Propositions 1 and 2 is thus embedded in (8.7).

The above demonstration is quite straightforward: mathematically speaking, it requires only elementary calculus and the use of polar coordinates. Equations (8.6) and (8.7) were only worked out in the 1740s, despite the fact that elementary calculus and polar coordinates were already in use in the late seventeenth century. Bertoloni Meli, in his essay (1993b), has given abundant evidence that the first expressions of (8.6) and (8.7) are to be found in the works of Daniel Bernoulli, Euler and Alexis-Claude Clairaut carried out in the 1740s. In their studies on constrained motion (typically a ball in a rotating tube) and planetary motions, these mathematicians arrived at expressions for radial and transverse acceleration.

† French (1971): 557–8.

Once this representation is achieved, the proof of Propositions 1 and 2 is simple. As we shall see, this representation was not available to Hermann, who resolved acceleration into a tangential and a normal component. The fact that the mathematical means were there, but their application to dynamics was lacking, teaches us something about the complexities of the history of mathematical dynamics. Progress comes not only from the discovery of new mathematics but also from the understanding of how mathematics can be applied to dynamical concepts.

Newton gave a proof of Proposition 1 based on the geometric limit procedures characteristic of the synthetic method of fluxions.† For the reader's convenience I will quote again the statement of Proposition 1 and I will present in some detail Newton's demonstration:

The areas which bodies made to move in orbits describe by radii drawn to an unmoving centre of forces lie in unmoving planes and are proportional to the times.‡

Newton's proof is as follows. Divide the time into equal and finite intervals. At the end of each interval the centripetal force acts 'with a single but great impulse',§ and the velocity of the body changes instantaneously. The resulting trajectory (see fig. 8.1) is a polygonal $ABCDEF$. The areas SAB, SBC, SCD, etc., are swept by the radius vector in equal times. Applying the first two laws of motion, it is possible to show that they are equal. In fact, if at the end of the first interval of time, when the body is at B, the centripetal force did not act, the body would continue in a straight line with uniform velocity (because of the first law of motion). This means that the body would reach c at the end of the second interval of time such that $AB = Bc$. Triangles SAB and SBc have equal areas. However, we know that, when the body is at B, the centripetal force does act. Where is the body at the end of the second interval? In order to answer this question, one has to consider how Newton, in Corollary 1 to the laws, defines the mode of action of two forces acting 'simultaneously': 'A body, acted on by two forces simultaneously, will describe the diagonal of a parallelogram in the same time as it would describe the sides by those forces separately.'¶ Invoking the above corollary, Newton deduces that the body will move along the diagonal of parallelogram $BcCV$, reaching C at the end of the second interval of time. The line Cc is parallel to VB, therefore triangles SBc and SBC have equal areas. Triangles SAB and SBC therefore have equal areas. One can iterate this reasoning and construct points C, D, E and F. They all lie on a plane, since the force is directed towards S, and the areas of triangles

---

† Aiton in (1989a) and Whiteside in *Mathematical Papers*, **6**: 37 maintain that a translation of Newton's geometrical demonstration of Proposition 1 into calculus language leads to a proof of Kepler's law not for finite, but only for infinitesimal arcs. I am aware that some Newtonian scholars are now casting doubts on the correctness of the Aiton–Whiteside criticism. See Erlichson (1992b) and Nauenberg (1998c): note 4.

‡ *Variorum*: 88. Cf. *Principles*: 40.

§ *ibid.*

¶ *Principles*: 14.

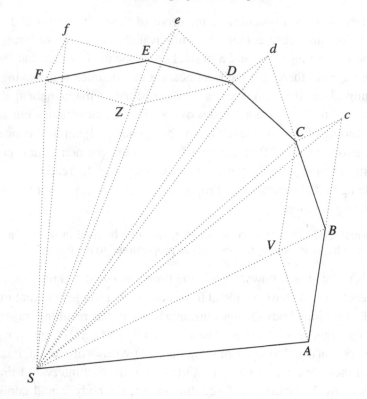

Fig. 8.1. Diagram for central force motion in Newton's *Principia*, Prop. 1, Book 1. After *Variorum*: 90

SCD, SDE, SEF, etc., are equal to the area of triangle SAB. The body therefore describes a polygonal trajectory which lies on a plane, and the radius vector SP sweeps equal areas SAB, SBC, SCD, etc., in equal times. Newton passes from the polygonal to the smooth trajectory with a limit argument based on the method of first and ultimate ratios. He writes:

Now let the number of triangles be increased and their width decreased *[a]in infinitum[a]*, and their ultimate perimeter ADF will [...] be a curved line; and thus the centripetal force by which the body is continually drawn back from the tangent of this curve will act continually, while any areas described, $SADS$ and $SAFS$, which are always proportional to the times of the description, will be proportional to those times in this case.†

That is to say, since Kepler's area law always holds for any discrete model (polygonal trajectory generated by an impulsive force), and since the continuous model (smooth trajectory generated by a continuous force) is the limit of the

---

† Notice that Cohen and Whitman translate *aa* as 'indefinitely'. *Variorum*: 90. Cf. *Principles*: 41.

discrete models for $\Delta t \to 0$, then the area law holds for the continuous model. The area swept by SP is proportional to time.

In Hermann's *Phoronomia*, central forces are treated in Chapter 2, Section 2 of Book 1, entitled 'On the curvilinear motions in void under any hypothesis on the variation of gravity'.† The motion of a point mass accelerated in void space by a central force is one of the main topics of the first book of the *Principia*. Hermann had achieved important results in this field during his Italian period. In dealing with central forces Continental mathematicians used to equate the infinitesimal element of time dt and the infinitesimal area dA swept by the radius vector so that:

$$d\!A = kdt, \tag{8.8}$$

where k is a constant (§8.4). The only proof then available that (8.8) holds for any central force was that given by Newton in the *Principia*. As Newton's proof was markedly geometrical in character, the analytical treatment of central forces ultimately relied on the synthetic methods of the *Principia*. Newton's geometric demonstration of the area law was assumed in the treatment of central forces until Hermann, in his *Phoronomia*, gave a proof of it in terms of differentials.‡ The same proof was restated, in a notation more accessible to a modern reader, in a 'letter' to John Keill that Hermann published in 1717 in the *Journal littéraire*.§ The difference basically consists in the fact that in the *Phoronomia*, Hermann expresses the differentials by referring to geometric points of a figure. For instance, if s is the arc length and p the perpendicular from force centre to tangent, the respective differentials are denoted in the *Phoronomia* with letters such as Nn and Cc, while in the *Journal littéraire* they appear as ds and dp (see fig. 8.2).

Hermann proves Newton's Proposition 1 as follows. First of all, he assumes that the trajectory of the body (whose mass is assumed, as customary in this period, as unitary) will be a plane curve ANB (see fig. 8.2). Hermann introduces the following geometrical constructions and symbols (the symbols occur only in the 1717 paper, and I will define them in parenthesis). The force centre is D, Nn ($=$ ds) is the infinitesimal element of arc, NC is the tangent at N, nc the tangent at n. The two tangents meet at e. The line DC ($= p$) is perpendicular to the tangent NC and meets it at C, while it cuts the tangent nc at c. Lines ON and On are perpendicular to the tangents NC and nc, so that O is the centre of the osculating circle (therefore $ON = \rho$). Finally, let nl ($=$ dα) and nm ($=$ dβ) be two infinitesimal segments meeting the tangent NC at l and m. There is an ambiguity about the inclination of nl and nm. In one representation they are the prolongations of Dn and On, respectively. However, in another representation, nl is parallel to

---

† 'De motibus curvilineis in vacuo, in quacunque gravitatis variabilis hypothesi.'
‡ Hermann (1716): 69–71. I have analysed Hermann's solution in Guicciardini (1996).
§ Hermann (1717).

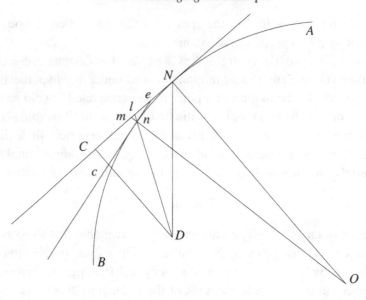

Fig. 8.2. Diagram for central force motion of a body of mass $m = 1$ adapted from Hermann's *Phoronomia*. After Guicciardini (1996): 175

DN and nm parallel to DC. This ambiguity is typical of infinitesimal techniques. The difference between the two representations can be ignored since higher-order infinitesimals are cancelled.† Other symbols will be defined when they occur.

Hermann resolves the central force F into two components, F_N normal to the trajectory and F_T tangent to the trajectory. He shows that

$$F_N\rho = v^2, \tag{8.9}$$

and

$$F_T ds = -v dv, \tag{8.10}$$

where a point of unit mass is considered, v is the velocity, s is the arclength and ρ is the radius of curvature. Equation (8.9) means that the resultant force normal to the trajectory is equal to the square of the speed divided by the radius of curvature. Equation (8.10) means that the force tangent to the trajectory is equal to the rate of change of the speed.

It is important to consider how Hermann deduces equations (8.9) and (8.10). Actually both were part of the standard repertoire of early eighteenth-century mathematicians. Hermann proceeds as follows. He begins with two 'general

---

† Notice also that the centre of the osculating circle is determined by the intersection of normals to two neighbouring points; ds is considered a straight segment, while Cc is assumed equal to dp, higher-order infinitesimals being neglected.

principles' valid for a uniform force G (a 'pesanteur uniforme') which accelerates from rest a body of unit mass in rectilinear accelerated motion. The 'first general principle already known to Newton and Varignon'† is

$$Gt = v, \tag{8.11}$$

where v is the velocity, and t the time. The second general principle is

$$(2l/G)^{1/2} = t, \tag{8.12}$$

which gives the time of fall from rest after the distance l is covered. These two principles are applied to the curvilinear motion caused by the centripetal force F. Hermann states that, during the infinitesimal interval of time ($= \mathrm{d}t$) required to traverse the infinitesimal arc Nn ($= \mathrm{d}s$), the force F can be assumed constant (both in magnitude and direction). He goes on: 'The tangential force is precisely that which causes the varied movement along the curve ANB'.‡ That is, the rate of change of speed is due to F_T. Therefore, the first principle (8.11) applied to the infinitesimal increment of speed $\mathrm{d}v$, acquired after $\mathrm{d}t$, yields:

$$F_T \mathrm{d}t = -\mathrm{d}v, \tag{8.13}$$

and (8.10) therefore follows (since $F_T v \mathrm{d}t = -v\mathrm{d}v$ and $v\mathrm{d}t = \mathrm{d}s$).

If during $\mathrm{d}t$ the body is accelerated by a constant force (which varies in neither magnitude nor direction), nl ($= \mathrm{d}\alpha$) may be conceived, according to Hermann, as a small Galilean fall, and from the second principle (8.12):

$$2\mathrm{d}\alpha/F = \mathrm{d}t^2 = \mathrm{d}s^2/v^2. \tag{8.14}$$

Therefore, from (8.14),

$$\mathrm{d}\alpha = (\mathrm{d}s^2 F)/(2v^2). \tag{8.15}$$

Furthermore,

$$nm/Nn = Nn/2ON, \tag{8.16}$$

or, in symbols,

$$\mathrm{d}\beta = \mathrm{d}s^2/2\rho. \tag{8.17}$$

Equation (8.16) (and its equivalent (8.17)) need some clarification. Since Hermann cancels third-order infinitesimals, he identifies the trajectory at N with the osculating circle at N. The versed sine nm is normal to the tangent NC, therefore triangles Nmn and QnN are similar (see fig. 8.3). And (8.16) follows.

† Hermann (1717): 412.
‡ ibid.: 413.

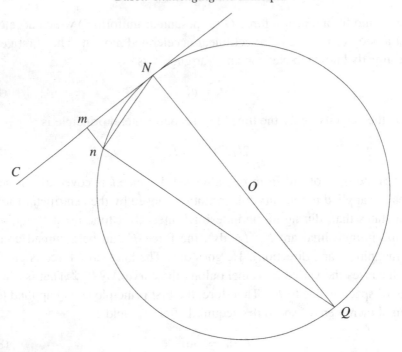

Fig. 8.3. Osculating circle at N of the trajectory ANB. Triangle Nmn is similar to triangle QnN. After Guicciardini (1996): 177

Now reconsider fig. 8.2. Since (by construction) $F/F_N = d\alpha/d\beta$,

$$\frac{F}{F_N} = \frac{d\alpha}{d\beta} = \frac{ds^2 F}{2v^2} \frac{2\rho}{ds^2} = \frac{F\rho}{v^2},\qquad(8.18)$$

and equation (8.9) follows.

Now consider $DN\ (= r)$, $NC\ (= q)$, and $DC\ (= p)$. Since, by hypothesis, the force is central with D as force centre:

$$F_T/F_N = q/p.\qquad(8.19)$$

Furthermore, from the similarity of triangles Cec and NOn (and cancelling higher-order infinitesimals),

$$dp/q = ds/\rho,\qquad(8.20)$$

where $Cc = dp$ and $Nn = ds$.†

Hermann divides (8.10) by (8.9) and gets

$$dv/v = -(F_T ds)/(F_N \rho).\qquad(8.21)$$

---

† The reader might want to prove (8.20) from $ds/dr = r/q$ and the expression for the radius of curvature $\rho = r\,dr/dp$.

The ratio $(F_T ds)/(F_N \rho)$, because of (8.19) and (8.20), is equal to dp/p. Therefore

$$dv/v = -dp/p. \tag{8.22}$$

At last, integrating (8.22), Hermann arrives at

$$pv = 2k, \tag{8.23}$$

which is equivalent to (8.8) since $pv = pds/dt = 2dA/dt$ (*pds* is twice the area $DNn = dA$ swept by the radius vector). Hermann in the *Phoronomia* observes that this is what the 'illustrious Newton for the first time proved in Prop. 1, Book 1, of *Principia Mathematica*, but from a completely different ground'.†

Newton's and Hermann's demonstrations that for any central force Kepler's area law holds are indeed quite different. While Newton relies on a geometrical limit procedure according to the synthetic method of fluxions, Hermann develops his proof in terms of differential equations of motion. The recourse to a limiting procedure gives a rigour to Newton's argument that is reminiscent of Archimedean demonstrations by 'exhaustion'. While Newton made rhetorical use of the similarities of his techniques with those of the 'Ancient Geometers' in order to defend the procedures employed in the *Principia*, it should be pointed out, as we have done repeatedly, that his geometrical methods are extremely innovative and cannot be identified with 'classical' techniques. In Proposition 1 geometry is applied to dynamical concepts in a way typical of seventeenth-century natural philosophy (Christiaan Huygens, rather than Archimedes, can be taken as a precursor of Newton from this point of view). Newton's limit procedure (which allows a transition from the discrete to the continuous model of the trajectory) could have been conceived only by a mathematician acquainted with seventeenth-century infinitesimal techniques and with the kinematical geometry introduced by, for example, Gilles Personne de Roberval and Isaac Barrow. However, it is true that Newton's geometrical demonstration of Proposition 1 is independent of the algorithm of the analytical method of fluxions. In fact, Newton's demonstration can be understood by a mathematician who knows nothing about the fluxional or the differential algorithms. Newton's demonstration of Proposition 1 is thus integrated into the scheme of seventeenth-century geometrical methods but does not belong entirely to the conceptual scheme of calculus.

Hermann's demonstration, on the contrary, was from the very beginning integrated into the conceptual scheme of Leibnizian calculus. The demonstrations in *Phoronomia* are 'geometrical' in the broad sense that the differential symbols are not employed. However, what Hermann actually does with his geometric symbols *nl*, *nm*, etc., is to write differential equations. His geometry is immediately

---

† 'Illustris Newtonus id primus demonstravit Prop. I Lib. I *Princ. Phil. Nat. Math.* sed ex diversissimo fundamento'. Hermann (1716): 71.

translatable into the language of the calculus, as the example considered above shows.

Hermann represents the trajectory locally in terms of differentials (ds, dp, $d\alpha$, $d\beta$, dv). Finite quantities, such as F_T, F_N and ρ, are then expressed in terms of ratios of differentials. The study of the geometrical and dynamical relationships of infinitesimals (see equations (8.13–8.20)) leads to differential equations (see equations (8.9), (8.10), (8.21), (8.22)) which can be manipulated algebraically until, thanks to an integration, the result sought is achieved (see equation (8.23)). The geometry of infinitesimals is thus the model from which one can work out differential equations. It is interesting to note that Hermann criticizes John Keill in the *Journal littéraire* because the Scottish mathematician in his treatment of central forces relies on 'the analogies of the common principle of proportionality that exists between times and areas' and stresses that he, Hermann, was the first to give a proof of this principle.† Notwithstanding his admiration for Newton, Hermann seems to imply that Proposition 1 was not really proved in the *Principia* and that reference to it could just be an 'analogy'. His approach, despite his declarations in the Preface of *Phoronomia* in favour of Newton and of the geometrical method, is typical of the Bernoullian school.

8.6 Johann Bernoulli's criticisms

8.6.1 *Johann Bernoulli and the* Principia

According to Fellmann the earliest proof of Bernoulli's reading of the *Principia* is to be found in a letter which l'Hospital sent him in January 1693.‡ This letter is about Lemma 25, Book 1, which concerns the nonquadrability in finite terms of 'ovals'. This Lemma aroused great interest in the 1690s and is referred to also in the correspondence between Leibniz and Huygens. It seems that Johann, during the last decade of the seventeenth century, did not devote much attention to the *Principia*, probably because Newton's natural philosophy was too distant from his Cartesian cosmology.§ When the priority dispute with Leibniz started, however, Johann began a campaign of demolition of Newton's *magnum opus*. He directed his criticisms towards the basic structure of Book 1: that is, he maintained that Newton had not solved the inverse problem of central forces. Propositions 10, 15, 16 and 26 of Book 2 were also attacked by Bernoulli. He maintained that there were mistakes with higher-order infinitesimals. According to Bernoulli, Newton

---

† Hermann (1717): 411.
‡ Fellmann (1988): 29. Fueter (1937).
§ It seems that Jacob Bernoulli devoted almost no attention to the *Principia*.

was unable to handle one of the basic constituents of the new calculus: higher-order differentiation.†

8.6.2 *Bernoulli's criticisms of Corollary 1 to Propositions 11–13, Book 1*

Johann Bernoulli in (1710) affirmed that Newton had not proved that conic sections, having a focus in the force centre, are necessary orbits for a body accelerated by an inverse square force. Section 8.6.2 is an attempt to assess what Bernoulli was criticizing and what were the immediate reactions of the Newtonians.

Keill, in a paper presented at the Royal Society in 1708, gave the first published application of the analytical method of fluxions to the solution of the inverse problem of central forces.‡ Furthermore, in a series of papers published from 1714 to 1719, he defended the correctness of Newton's geometrical solution (presented in Corollary 1 to Propositions 11–13, and in Proposition 17, Book 1, of the *Principia*) of this problem.

Keill's approach to central forces differed from that adopted by the Continentals. In 1710 Hermann, Johann Bernoulli and Varignon proposed their solutions of the inverse problem.§ They all started from Newton's *Principia*, to which now we turn. I will analyse in some detail the main Propositions to which the several contenders referred during the debate on the inverse problem of central forces.

8.6.2.1 *The inverse problem of central forces in the* Principia: *the public solution*

In Propositions 11, 12 and 13 of Book 1 Newton shows that if a 'body' orbits along a conic and the area law is valid for a focus S, then the body is accelerated by a force which is inversely proportional to the square of distance from S.

In Corollary 1 to Propositions 11–13 Newton states that the inverse is true. That is, if the force is inverse square, the trajectories are conic sections such that a focus coincides with the force centre. This Corollary is an example of inverse problem of central forces: the central force F and force centre S are given, and it is required, given initial position and velocity, to find the trajectory.¶ We recall that Corollary 1 (1687) reads as follows:

From the last three propositions [i.e. Propositions 11–13] it follows that if any body P departs from the place P along any straight line PR with any velocity whatever and is at the same time acted upon by a centripetal force that is inversely proportional to the square

---

† For a different view on Bernoulli's criticisms to Newton see Fleckenstein (1946).
‡ On Keill's solution (1708) see Guicciardini (1995).
§ Bernoulli, J. (1710), Hermann (1710), Varignon (1710).
¶ That is, one has to find a plane curve which is the orbit of the point mass for given initial conditions. The problem of determining the position in function of time is dealt with in Sections 6, 7 and 8 of Book 1 of the *Principia*.

of the distance[a] from the centre, this body will move in some one of the conics having a focus in the centre of forces; and conversely.[†]

In the following Proposition 17 Newton presents a constructive geometrical technique for determining the conic section which answers given initial conditions, when the absolute value at P and force centre S of the inverse square force, $F = -k/r^2$, are known. Proposition 17 is based on the fact that the initial position P, initial velocity at P, mass of the body and absolute value of central force at P determine the tangent and radius of curvature at P (the normal component of F at P is equal to mass times the square of initial velocity divided the radius of curvature). Since S, a focus, is given, the conditions of Corollary 1 determine uniquely a conic C. From Propositions 11–13 one knows that C is a possible orbit for $F = -k/r^2$. It should be noted that in order to invoke Propositions 11–13, Newton assumes implicitly that a Keplerian motion along C exists and is unique.

In 1709, after Keill's paper (1708) and before Johann Bernoulli's criticisms (1710), Newton realized that Corollary 1 needed some amplification. In a letter, dated October 1709, he gave instructions to Roger Cotes, who was editing the second edition of the *Principia*, to complete Corollary 1 with the following lines:

For if the focus and the point of contact and the position of the tangent are given, a conic can be described that will have a given curvature at that point. But the curvature is given from the given centripetal force [and velocity of the body]; and two different orbits touching each other cannot be described with the same centripetal force [and the same velocity].[‡]

This amplification was published in the second edition (1713) of the *Principia*. I have enclosed in square brackets further amplifications added in the third edition (1726). With these lines Corollary 1 acquires a clearer meaning. For a given inverse square central force, any initial condition determines a unique conic (identified constructively in Proposition 17). From Propositions 11–13 we know that Keplerian motion along this conic satisfies the equation of motion for the given force and given initial conditions. In the final lines a uniqueness theorem is invoked: two different orbits which satisfy the same initial conditions and which 'touch each another' (i.e. they have the same tangent and the same curvature) at P cannot exist. Therefore conic sections are the only possible orbits for an inverse square force.

[†] [a] 'of places' added in second edition. *Variorum*: 125. Cf. *Principles*: 61.

[‡] Newton's original Latin in the third edition of the *Principia* is: 'Nam datis umbilico, & puncto contactus, & positione tangentis, describi potest sectio conica, quae curvaturam datam ad punctum illud habebit. Datur autem curvatura ex data vi centripeta, [& velocitate corporis]: & orbes duo se mutuo tangentes eadem vi centripeta [eademque velocitate] describi non possunt'. *Variorum*: 125. Cf. *Correspondence*, **5**: 5–6 and *Principles*: 61. The logical correctness of this Newtonian procedure has been discussed in a number of papers. The literature devoted to this corollary is vast. See Aiton (1988) and (1989b); Cushing (1982); Arnol'd (1990); Erlichson (1990); Whiteside (1991a); Brackenridge (1995); Pourciau (1991) and (1992a). In what follows I am greatly indebted to Pourciau's reconstruction of the logic of Corollary 1.

The inverse problem of central forces was faced by Newton in a more general way in Propositions 39, 40 and 41 of Book 1.† These Propositions (which remained almost unaltered through the various editions) are based on the assumption that a method for 'squaring curilinear figures' is given. Newton had developed in his youth the 'analytical method of fluxions'. He was able to 'square' (in Leibnizian terms 'integrate') a large class of curves. However, as we have often repeated, in the *Principia* he chose not to make wholly explicit his mathematical discoveries in this field.

In Proposition 41 Newton reduces the general inverse problem of central forces to quadratures, but does not explain to the reader how the necessary integrations can be performed. He considers (see fig. 8.4) a 'body', fired at V along a given direction and with a given initial speed. The body is 'acted upon' by a centripetal force F directed towards force centre C. Proposition 41 reads as follows:

Supposing a centripetal force of any kind and granting the quadratures of curvilinear figures, it is required to find the trajectories in which bodies will move and also the times of their motions in the trajectories so found.‡

Let VIk be the curve sought and let IK be an infinitesimal arc traversed by the body in an infinitesimal interval of time ('Let the points I and K be very close indeed to each other'§). Since Kepler's area law holds, the infinitesimal interval of time is proportional to the sector $CIK = (CI \cdot KN)/2$. Newton denotes the constant areal velocity by $Q/2$; so that the infinitesimal arc IK is traversed in the infinitesimal interval of time $(CI \cdot KN)/Q$. The speed v along IK can be assumed constant. It is equal to the arc IK divided by the time: $v = (IK \cdot Q)/(CI \cdot KN)$.

Note that on the right of fig. 8.4 Newton adds the 'curves' that must be 'squared' in order to solve the inverse problem. The curve BF represents the magnitude of force as a function of radial displacement (DF is thus proportional to the force at I).

In Propositions 39 and 40 Newton had shown that the speed at I is proportional to the square root of the area $ABFD$ under the curve that represents force as a function of radial displacement CI:

$$v \propto \sqrt{ABFD}.\P \qquad (8.24)$$

One can thus write

$$(IK \cdot Q)/(CI \cdot KN) \propto \sqrt{ABFD}. \qquad (8.25)$$

† These propositions have been often analysed by historians. See De Gandt (1987) and (1995): 244ff. and Mahoney (1993).
‡ *Variorum*: 218. Cf. *Principles*: 130.
§ 'Sint autem puncta *I* & *K* sibi invicem vicinissima'. *Variorum*: 219. Cf. *Principles*: 130.
¶ Propositions 39 and 40 can be read in modern terms as the expression of the work–energy theorem. It should be stressed that Newton did not have the concepts of work and energy.

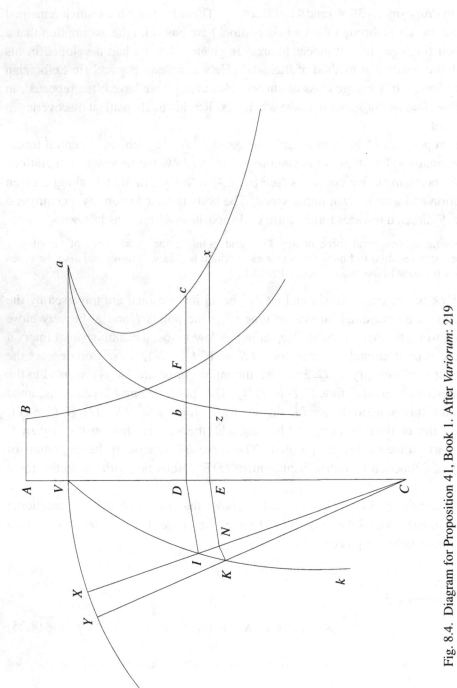

Fig. 8.4. Diagram for Proposition 41, Book 1. After *Variorum*: 219

Newton sets $CI = A$ and $Q/A = Z$, and then states:

$$IK/KN \propto \sqrt{ABFD}/Z. \qquad (8.26)$$

Equation (8.25) [= (8.26)] is the basic result that allows Newton to tackle the general inverse problem of central forces in Proposition 41. In order to facilitate the understanding of this geometrical formula, we will betray Newton and translate it into more familiar Leibnizian symbolic terms.

Substitute $Q = h$, $IK = ds$, $IN = -dr$, $KN = r d\theta$, $CI = r$, $CV = r_0$, and furthermore, following Propositions 39 and 40, put $ABFD = v_0^2 + 2 \int_r^{r_0} F dr = v^2$, where v is the speed at I, v_0 the initial speed at V. As always we assume the mass $m = 1$. We derive from (8.26)

$$\frac{ds^2 (= dr^2 + r^2 d\theta^2)}{r^2 d\theta^2} = \frac{r^2}{h^2}\left(v_0^2 + 2\int_r^{r_0} F dr\right). \qquad (8.27)$$

By means of simple algebraical manipulations, polar differential equations for the orbit are obtained. From (8.27):

$$d\theta = \frac{-h dr}{r^2 \sqrt{\left(v_0^2 + 2\int_r^{r_0} F dr - h^2/r^2\right)}}. \qquad (8.28)$$

Since $r^2 d\theta/dt = h$, from (8.28):

$$dt = \frac{-dr}{\sqrt{\left(v_0^2 + 2\int_r^{r_0} F dr - h^2/r^2\right)}}. \qquad (8.29)$$

It should be stressed that this translation, which is close to what Johann Bernoulli did in 1710, while it is helpful for the modern reader, does not yield correctly what can be read in the *Principia*. Newton always adheres to the need for geometrical representability and introduces an auxiliary circle $VXYR$. From (8.26), following steps similar to those which we have employed above, he deduces that

$$\text{area } CVX\left[= \frac{CX^2}{2}\theta\right] = \text{area under curve } acx, \qquad (8.30)$$

where the ordinate of the curve acx is

$$Dc = \frac{Q(CX)^2}{2A^2 \sqrt{\left(ABFD - Q^2/A^2\right)}}. \qquad (8.31)$$

Furthermore

$$\text{area } CVI\left[= \frac{Q}{2}t\right] = \text{area under curve } abz, \qquad (8.32)$$

where the ordinate of the curve *abz* is

$$Db = \frac{Q}{2\sqrt{(ABFD - Q^2/A^2)}}.^\dagger \tag{8.33}$$

If one 'squares' *acx*, the functional dependence of polar angle θ with distance r is given. If one 'squares' *abz*, the functional dependence of time t with distance r is given. Note that instead of working directly with polar coordinates, as in the above 'Leibnizian' translation, Newton relates the distance CI to geometrical quantities (CVI, CVX) proportional to time t and to polar angle θ, which are 'visualized' in fig. 8.4.‡

8.6.2.2 *Newton's unpublished analytical approach to the inverse problem of inverse cube forces*

Newton was aware that Propositions 39–41 could be translated into the language of the analytical method of fluxions. The techniques for the squaring of curves that Newton had devised in the late 1660s were ready for use. As Whiteside has indicated, in his 1671 tract on fluxions, generally known as *Tractatus de methodis serierum et fluxionum*, Newton had given the method for finding the 'value of the area' under the 'curve' $y(z) = c_1 z^{\eta-1}/\sqrt{c_2 + c_3 z^\eta + c_4 z^{2\eta}}$: for $\eta = -1$ one obtains the 'curve' which ensues from Proposition 41 applied to inverse square forces (cf. equations (8.28), (8.31) and (8.36)).§ Even though an analytical treatment of Proposition 41 applied to inverse square forces was within Newton's capabilities, there is no historical record that he performed it. However, a manuscript written in 1694 in which Newton tackles inverse cube forces is extant.

The inverse problem for inverse cube forces is faced in the *Principia* in Corollary 3 to Proposition 41. The problem is to determine the trajectory of a body accelerated by a central force which varies inversely as the cube of the distance. In the *Principia* Newton gives only the solution: he identifies some spiral trajectories which answer the problem. He was able to obtain this result only by application of his 'catalogues' (i.e. 'integral tables' in Leibnizian jargon). In fact equation (8.31) applied to an inverse cube force leads to the problem of squaring a curve contemplated by Newton in 1671. In the *Principia* he does not perform this quadrature explicitly, but simply states the result. He then adds:

All this follows from the foregoing proposition [41], by means of the quadrature of a certain curve, the finding of which, as being easy enough, I omit for the sake of brevity.¶

---

† In (8.30) and (8.32) I have added in square brackets symbolic expressions in terms of polar coordinate θ and time t.

‡ The role of visualization has been underlined by Erlichson (1994b).

§ *Mathematical Papers*, **6**: 348 and **3**: 252.

¶ *Variorum*: 223. Cf. *Principles*: 133.

The historian of mathematics cannot avoid a smile while reading Newton's conclusion to Corollary 3. The quadrature required was by no means 'easy'!

In 1694 David Gregory asked Newton to explain this Corollary. Newton replied as follows.† In the letter to Gregory he puts $CI = A = x$ (see fig. 8.4), and assumes the force to be a^4/x^3, where a is a constant. From Propositions 39 and 40 Newton derives that the square of velocity is proportional to the area $ABFD = 2a^4/x^2 - 2a^4/c^2$ (here the integral has a wrong factor 2). Note that c is chosen so that $-2a^4/c^2 = -2a^4/x_0^2 + v_0^2$. He then states that Corollary 3 is solved by means of the following fluxional equation, which can be compared to Newton's (8.30) and (8.31) and to Bernoulli's (8.35) for $F = -a^4/x^3$:

$$\square \frac{Qe^2}{2x^2\sqrt{(2a^4 - Q^2)/x^2 - 2a^4/c^2}}, \tag{8.34}$$

where the symbol $\square$ indicates 'fluent of' (in Leibnizian terminology 'integral of'), while e denotes CX. Newton performs the integration and finds the spiral trajectories constructed geometrically in the published Corollary 3.‡ Newton's solution of the inverse problem of inverse cube orbits is incomplete since he restricts the initial conditions so that initial radius vector and initial velocity are perpendicular. That Newton's treatment of the inverse problem of inverse cube forces is not exhaustive is evident from Proposition 9, Book 1. In this Proposition it is shown that a body orbiting in an equiangular (or 'logarithmic') spiral $[a\theta = \ln r]$ is accelerated by an inverse cube force, a case not included in Corollary 3 to Proposition 41. Notwithstanding the above-mentioned limitations, it is important to notice that the letter to Gregory on inverse cube orbits shows that Newton was able to translate Propositions 39–41 into the language of fluxional equations.

Gregory inserted this solution into his manuscript *Notae* in a space in his running commentary that he had left empty. Newton had clearly given him some much-needed assistance. It has often been repeated that Newton was unable to write the differential equations of motion. The case of Corollary 3 to Proposition 41 and the evidence that we have given in §4.5 clearly disprove this widely accepted judgment.

8.6.2.3 *Bernoulli's calculus approach to the inverse problem of central forces*

In 1710 Hermann's letter on the inverse problem was presented at the *Académie des Sciences* (see §8.5) together with a 'Réponse' by Bernoulli. Both Hermann and Bernoulli base their papers on Newton's *Principia*. However, Hermann focuses on Proposition 6, i.e. formula (3.1), Bernoulli on Propositions 39–41, i.e. formula (8.26).

---

† *Correspondence*, **6**: 435–7.
‡ Namely the following orbits: $r = \lambda \sec(\mu\theta)$ and $r = \lambda \operatorname{sech}(\mu\theta)$.

As we have noted above, Newton's treatment of the inverse problem of central forces developed in Propositions 39–41 is rather straightforwardly translatable into the language of calculus. Bernoulli obtains a first-order differential equation, equivalent to (8.28), applies it to inverse square forces and faces the mathematical problem of integrating it. Choosing Propositions 39–41, rather than Proposition 6, is more convenient since one gets a first-order differential equation instead of a second-order differential equation (in fact, in modern terms, not only conservation of angular momentum, but also conservation of energy is employed).

Bernoulli reduces Newton's Proposition 41 to the solution of a differential equation of the following form:

$$\mathrm{d}z = \frac{a^2 c\,\mathrm{d}x}{\sqrt{abx^4 - x^4 \int F\,\mathrm{d}x - a^2 c^2 x^2}}, \tag{8.35}$$

where a, b and c are constants, and the polar coordinates (r, θ) are $(x = r, z/a = -\theta)$.† For an inverse square force $F = a^2 g/x^2$ (g constant) Bernoulli gets the following differential equation:

$$\mathrm{d}z = \frac{a^2 c\,\mathrm{d}x}{x\sqrt{abx^2 + a^2 gx - a^2 c^2}}. \tag{8.36}$$

Integration leads to conic sections having the form

$$r = \frac{c^2/g}{1 + e\cos(\theta - \theta_0)}. \tag{8.37}$$

Bernoulli criticizes Hermann for having expressed his equation in such a way that the variables cannot be easily separated, at least in the general case when the force varies as any power of distance. He also underlines that Hermann had not given a general method for tackling the inverse problem. He writes: 'your solution seems conducted from design, arranged in function of what you are looking at, of what you know in advance'.‡ He stresses also that the generality of Hermann's solution is doubtful since Hermann introduces only one constant of integration. He notes: 'Furthermore, it does not follow from your particular solution that it leads only to conic sections: after the first integration of your differentio-differential [i.e. second-order differential] equation you have forgotten to add in one or the other side a constant quantity; and for this reason, one might be left in doubt that there is

---

† Note that a factor 2 is missing before the integral sign.
‡ 'votre Solution paroît faite à dessein, accommodée à ce que vous cherchiez, & à ce que vous connoissiez déja'. Bernoulli, J. (1710): 521.

another kind of curve, other than the conic sections, that satisfies your problem.'†
Bernoulli's criticisms to Hermann seem unfair, since, as Jacopo Riccati – one of the
most talented eighteenth-century Italian mathematicians – showed, the variables
could be separated and the absence of a constant of integration was simply due to
the choice of coordinate reference system.‡

Bernoulli's *réponse* to Hermann reveals the values that directed his research.
According to Bernoulli:

- one should look for general analytical procedures without assuming the result sought;
- the general integral has to reveal that *all* the possible trajectories have been discovered.

The former requirement implies that the inverse problem of central forces has to be
tackled with a differential equation which should be integrated with standardized
techniques and applied to several laws of force. The latter implies that the
dependence of constants of integration upon initial conditions has to allow the
determination of a unique trajectory for any initial condition.

8.6.2.4 Bernoulli's criticism of Corollary 1

Bernoulli's attention to the generality of the mathematical procedures applied
to dynamics led him to criticize Newton's geometric solution of the inverse
problem given in Corollary 1 to Propositions 11–13. Bernoulli sharply states that
Newton's Corollary is just a supposition and not a demonstration. He remarks
that equiangular ('logarithmic') spiral trajectories ($\ln r = a\theta$) imply inverse cube
forces (as Newton had shown in Proposition 9, Book 1). An inverse cube force
is however compatible with other trajectories, such as hyperbolic (or 'inverse
Archimedean' spirals ($r\theta = a$)). According to Bernoulli this counter-example
reveals the weakness of Newton's Corollary: Newton's solution of the inverse
problem of inverse square forces is vitiated by a logical error. In fact it was not
safely applicable to other forces, such as inverse cube:

I said before, that M. Newton, after having proved that the central forces of a body, directed
towards one of the foci of any conic section described by that body, are always between
themselves in a ratio inverse to the squares of the distances of the body itself from the
given focus, supposes the inverse of this proposition without proving it [...] in order to
appreciate the necessity of the demonstration that I am going to give of this inverse one has
just to consider that a body, in order to move along a logarithmic spiral, requires central
forces in a reciprocal ratio of the cubes of the distances from the focus or centre of that
curve; it is not a consequence that with that force the body will always describe such a

---

† 'De plus il ne suit pas encore de votre Solution particuliere qu'elle ne convienne qu'aux seules Sections
Coniques: après la premiere intégration de votre équation differentio-differentielle vous avez oublié d'y
ajoûter de part ou d'autre une quantité constante; ce qui pourroit laisser quelqu'un en doute, si outre les
Sections Coniques, il n'y auroit point encore quelqu'autre genre de Courbes qui satisfist à votre question'.
ibid.: 522.

‡ Jacopo Riccati defended Hermann's solution in (1714). On this topic see Giuntini (1992).

curve; Since it is easy to be convinced, from the direct formulas of central forces, that a body will also have the forces in that ratio if it will describe an hyperbolic spiral.†

Bernoulli's very valid point is that Newton's argument, as it is tersely expressed in the first edition of the *Principia*, seems simply to state that since conic trajectories imply inverse square forces, then inverse square forces imply conic trajectories – clearly a *non sequitur*. We have now to turn to the efforts and the replies of the British.

8.6.2.5 *The analytical approach in the British school: De Moivre, Keill and Cotes*

At the turn of the century the Newtonians had achieved a complete understanding of how central forces could be expressed in terms of the analytical method of fluxions (§4.5). They employed the following representation for a central force:

$$F \propto \frac{r}{p^3 \cdot \rho}, \qquad (8.38)$$

where S is the centre of force, $r = SP$ is the radius vector, $p = SY$ is normal to the tangent at P and $\rho = PC$ is the radius of curvature at P (see fig. 8.5). This formula easily follows from Newton's formula (4.5), whose demonstration is given in §4.5.

The radius of curvature ρ, expressed in the coordinates r and p, is

$$\rho = \frac{r \cdot \dot{r}}{\dot{p}}. \qquad (8.39)$$

Equation (8.39) leads to the fluxional version of (8.38), i.e.

$$F \propto \frac{\dot{p}}{p^3 \cdot \dot{r}}. \qquad (8.40)$$

Formula (8.38) was discovered independently in 1705 by Abraham De Moivre. When he showed it to Newton, Newton replied that he had already obtained a similar formula. As De Moivre wrote in 1705 to Johann Bernoulli:

After having found this theorem, I showed it to M. Newton and I was proud to believe that it would have appeared new to him, but M. Newton had arrived at it before me; He showed this theorem to me among his papers that he is preparing for a second edition of his *Principia Mathematica*: all the difference is that, instead of expressing the law of centripetal force by means of the radius of concavity [rayon de la concavité], he expresses

---

† 'J'ai dit ci-devant que M. Newton, après avoir démontré que les forces centrales d'un corps, dirigées par un des foyers d'une Section Conique quelconque décrite par ce corps, sont toujours entr'elles en raison renversée des quarrées des distances de ce même corps à ce foyer, suppose l'inverse de cette proposition sans la démontrer [...] pour voir encore la necessité de la démonstration que je viens de donner de cette inverse, il n'y a qu' à considerer que de ce qu'un corps pour se mouvoir sur une spyrale logarithmique, requiert des forces centrales en raison réciproque des cubes des distances au foyer au centre de cette courbe; ce n'est pas une conséquence qu' avec de telles forces il decrivît toujours une telle courbe; Puisqu'il est aisé de se convaincre par les formules directes des forces centrales, que ce corps auroit aussi ces forces en cette raison s'il décrivoit une spyrale hyperbolique.' Bernoulli, J. (1710): 532–3.

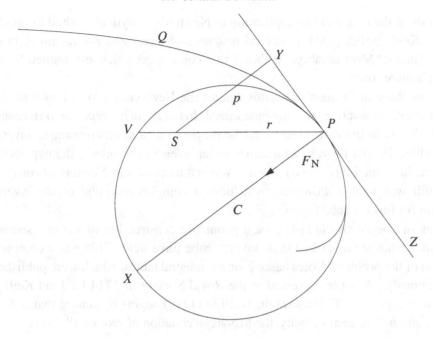

Fig. 8.5. Pedal coordinates of the trajectory. *S*, force centre. After Guicciardini (1995): 548

it by means of a chord inscribed in the circle of concavity: but he told me that it is better to express it by means of the radius, as I had done.†

In 1706 Johann Bernoulli in a reply to De Moivre presented his own demonstration of formulas (8.38) and (8.40), publishing them, without acknowledging De Moivre's priority, in the *Mémoires de l'Académie des Sciences* for the year 1710. Meanwhile, De Moivre had communicated his results to the Oxford group composed by Edmond Hallcy, David Gregory and John Keill.‡

It was John Keill who, in a paper published in the *Philosophical Transactions* for the year 1708, brought together the results on the analytical treatment of central forces achieved by the Newtonians. This paper is famous because in it Keill openly affirmed (it seems without Newton's knowledge) that Leibniz had plagiarized Newton's algorithm. This accusation was encapsulated within a work whose aim

† 'Après avoir trouvé ce théoreme je le montrai à M. Newton et je me flattois qu'il lui paroîtroit nouveau, mais M. Newton m'avoit prévenu; il me le fit voir dans les papiers qu'il prépare pour une seconde édition de ses *Principia Mathematica*: toute la différence qu'il y avoit, c'est qu'au lieu d'exprimer la loi de la force centripete par le moyen du rayon de la concavité, il l'exprimoit par le moyen d'une corde inscrite dans le cercle de la concavité: mais il me dit qu'il valoit mieux l'exprimer par le rayon, comme j'avois fait'. Wollenschläger (1902): 214. See also *Mathematical Papers*, **6**: 548, where Whiteside reconstructs the history of De Moivre's formula.

‡ Bernoulli, J. (1710): 521–33. David Gregory's own proof of De Moivre's result is in 'Codex E' in Christ Church, Oxford MS 346 (manuscript dated 1707). Varignon had already published an equivalent result (see formula (8.2)).

was to show the extent of the application of Newton's analytical method to central forces. Keill applies (8.40) to several instances of direct and inverse problem of central forces. Most notably, he shows that conic trajectories are implied by an inverse square force.

One of Bernoulli's main criticisms against the Newtonian procedures was that the generality of solutions was not guaranteed. In (1710) in his *réponse* to Hermann and in (1713) in the *Acta eruditorum* he proposed as a counter-example inverse cube orbits. He put forward an instance of an inverse cube orbit – the hyperbolic spiral, as he named it ($r\theta = a$) – which was not discussed in Newton's *Principia*. Bernoulli was unable, however, to achieve a complete solution of the inverse problem for inverse cube forces.

Cotes in *Logometria* (1714) gave a geometric construction of the five species of 'spirals' that are described in an inverse cube force field. This was a complete solution of the problem. Cotes based it on his integral tables, which were published posthumously. A paper presented at the Royal Society in 1714 by John Keill is devoted to a proof of Cotes's result. Keill in (1714) begins by stating that if $F = -k/r^3$, and $h/2$ is areal velocity, the Moivrean equation of motion (8.40) is

$$\frac{k}{r^3} = h^2 \frac{\dot{p}}{p^3 \cdot \dot{r}},$$
(8.41)

which by a first integration leads to

$$p^2 = \frac{h^2}{k} \frac{r^2}{1 + cr^2},$$
(8.42)

where c is constant of integration evaluable, since $p = h/v$, as $c = v_0^2/k - 1/r_0^2$.†
Five cases can be distinguished:

Case 1: $c = 0$, the equation $p^2 = (h^2/k)r^2$ is that of an equiangular spiral.‡ If c is different from 0, on integration Keill shows that four more cases are possible.

Case 2a: $c > 0$ and $h^2 > k$; case 2b: $c > 0$ and $h^2 = k$; case 2c: $c > 0$ and $h^2 < k$.

Case 3: $c < 0$ and $h^2 < k$.

Bernoulli's hyperbolic spiral $1/r = c\theta$ corresponds to case 2b. Cases 2a, 2b and 3 correspond to the Cotesian spirals: $r = \lambda \sec(\mu\theta)$, $r = \lambda \operatorname{cosech}(\mu\theta)$, $r = \lambda \operatorname{sech}(\mu\theta)$ (λ, μ constants).

Thus Cotes and Keill achieved, in analytical terms, the solution of the inverse problem of inverse cube central forces before the Continentals.

---

† Note that in terms of energy E – a concept not contemplated by Newtonians – cases 1, 2 and 3 below correspond to $E = 0$, $E > 0$, $E < 0$ respectively.

‡ This is the so-called 'pedal equation' of the spiral.

8.6.2.6 Keill and Bernoulli

In his 1714 paper on inverse cube orbits, Keill replied to Bernoulli's criticism of Newton expressed in the 'Réponse à M. Herman' of October 1710. He maintained that Bernoulli's solution of the inverse problem was equivalent to Newton's Propositions 39–41. He also affirmed that Newton had given a more concise and more elegant solution of the inverse problem for inverse square forces in Corollary 1 to Propositions 11–13. He was able to refer to the amplified version of Corollary 1 published in the second edition of the *Principia* (1713).†

One of the main points advanced by Keill against Bernoulli is that the Swiss mathematician had just translated Newton's Propositions 39–41 into a slightly different form. He wrote:

The solution of M. Bernoully differs from that of M. Newton only in the characters or symbols.‡

Keill also showed how, with a mere change of symbols, one could obtain one of the 'formulas' of the *Principia* (i.e. (8.30)) from Bernoulli's (8.35):

I saw that the Bernoullian formula completely coincides with the Newtonian; it differs only in the notation of quantities. In fact if one puts $ab - \int F\dot{x}$ for $ABGE$ [$\approx ABFD$], ac for Q, & x for A, a for CX, & $\dot{x}$ for IN [...] it will result that that formula does not differ from the Newtonian, more than the same words when written in Latin and in Greek.§

It is certainly true that Newton's Propositions 39–41 were ready for a translation into the language of the fluxional or the differential and integral calculi. Newton himself could perform such a translation easily. However, Newton gave the reader no hints on how to 'square' formula (8.30). It was Bernoulli who first published the necessary integration. Newton would have been able to perform such an integration, had he used his 'catalogues' of curves of 1671, but he did not publish it. Rather, he preferred to solve the inverse problem of inverse square forces in Corollary 1 to Propositions 11–13.

---

† Keill's arguments were restated in a 'Defénse du Chevalier Newton' published in 1716 in the *Journal litéraire*. Here Keill, referring to Bernoulli's claim of priority expressed in 'De motu corporum gravium' of February 1713, affirmed that his fluxional solution of 1708 was obtained before Bernoulli's 1710 'Réponse' to Hermann. Bernoulli replied anonymously to Keill's 1714 paper with an 'Epistola pro eminente mathematico' published in the *Acta Eruditorum* for July 1716. Bernoulli's pupil Johann Heinrich Krauss replied in the *Acta Eruditorum* for October 1718 to Keill's 1716 'Defénse'. A final 'Lettre de M. Jean Keill [...] a M. Jean Bernoulli' appeared in the *Journal litéraire* in 1719, and a Latin pamphlet *Epistola ad Virum Clarissimum Johannem Bernoulli* was printed in London in 1720.

‡ 'La Solution de M. Bernoully ne differant de celle de M. Newton, que dans les Caractères ou Symboles'. Keill (1716): 418.

§ 'vidi Bernoullianam formulam omnino cum Newtoniana coincidere; nec nisi in notatione quantitatum ab ea differre. Nam si pro $ab - \int F\dot{x}$ ponatur $ABGE$, pro ac ponatur Q, & x pro A, a pro CX, & $\dot{x}$ pro IN [...] constat formulam illam non magis a Newtoniana discrepare, quam verba latinis literis expressa differunt ab iisdem verbis scriptis in Graecis characteribus'. Keill (1714): 114.

In 1713 Keill wrote to Newton:

Since I left London I have considered Mr Bernoulli's solution of the Inverse Problem of Central Forces, and I am amused at his impudence [...] In his application of it to the particular case [of an inverse square force] he has with a great deal of labour showed that the curve described must be a Conick-Section when the thing may be demonstrated in a few lines.†

The concise demonstration to which Keill refers is, of course, Corollary 1 to Propositions 11–13:

But after all, M. Newton, at page 53 of the second edition of his *Principia*, has given us a demonstration of this particular case [i.e. $F \propto 1/r^2$] in three lines, where M. Bernoully employs 7 or 8 pages.‡

As we know, in 1710 Bernoulli had affirmed that Corollary 1 (1687) was not a demonstration. His point is that once you have shown that A implies B, you cannot state that the inverse is true. He therefore presented his counter-example: a logarithmic spiral implies inverse cube forces, but the inverse is untrue, since logarithmic spirals are not the only possible orbits for an inverse cube force. To this objection Keill replied as follows:

You can draw this conclusion, if you can always construct a logarithmic spiral such that if a body describes it in its motion, then that body must have at a given point, a given direction, a given speed and a given absolute centripetal force; and this is the case with conic sections; but since this is not possible with logarithmic spirals, we must draw a completely opposite conclusion.§

Keill's valid point here is that, in the amplified version of Corollary 1 (1713), Newton states that an inverse square force implies conic trajectories, since for *any* initial condition a conic can be constructed (through Proposition 17) which satisfies the equations of motion (as it is proved in Propositions 11–13) and which satisfies the given initial condition. It is untrue that for an inverse cube force and for *any* initial condition a logarithmic spiral can be so constructed.

8.6.2.7 *The acceptance of Corollary 1*

After several attempts to resist on this point, the Continentals eventually surrendered. Bernoulli, writing to Newton in 1719, accepted Corollary 1 as published in the second edition of *Principia* as a valid proof that conic sections are necessary orbits for an inverse square force. Of course, as he remarked, Newton's proof is

---

† *Correspondence*, **6**: 37.

‡ 'Mais après tout, M. Newton dans la Page 53 de la seconde Edition de ses Principes, nous a donné en trois lignes une Démonstration de ce cas particulier, au lieu que M. Bernoully y employe 7 à 8 Pages'. Keill (1716): 420.

§ 'Vous pouvez tirer cette consequence, si vous pouvez toûjours décrire une Spirale Logarithmique dans laquelle si un corps se meut, il doive avoir, à un point donnée, une direction donnée, une vitesse donnée & une force centripetale absoluë donnée; qui est le cas dans les Sectiones Coniques; mais parce qui cela ne se peut, dans la Spirale Logarithmique, nous en devons tirer une consequence toute contraire'. Keill (1719): 276–7.

an *a posteriori* proof. In Corollary 1 (and Proposition 17) Newton just checks that the orbits considered in Propositions 11–13 satisfy the equation of motion for any initial condition:

Gladly I believe what you say about the Addition to Corollary 1, Proposition 13, Book 1 of your incomparable work the *Principia*, that this was certainly done before these disputes began, nor have I any doubts that the demonstration of the inverse proposition, which you have merely stated in the first edition, is yours; I only said something against the form of that assertion, and wished that someone would give an analysis that led *a priori* to the truth of the inverse [proposition] and without supposing the direct [proposition] to be already known. This indeed, which I would not have said to your displeasure, I think was first put forward by me, at least so far as I know at present.†

During the eighteenth century Corollary 1 to Propositions 11–13 (plus Proposition 17) became a standard solution of the inverse problem of inverse square forces. Euler, in 1736, openly sided with Newton and against his master Bernoulli:

They [Johann Bernoulli and others] denied that it was proved satisfactorily by Newton that no curves other than conic sections satisfy the question, even though this is derived with sufficient clarity in Proposition 17 of the first Book of *Principia*.‡

At the end of the century, Pierre-Simon Laplace wrote:

This great geometer [Newton], generalizing his researches, showed that a projectile can move along any conic section, in virtue of a force directed towards a focus and reciprocal to the square of the distances: he derived the various properties of motion along such a kind of curves: he determined the necessary conditions which determine the curve to be a circle, an ellipse, a parabola or an hyperbola, conditions which depend only upon the primitive velocity and position of the body. Let this initial velocity, position, and direction have any chosen value, Newton assigned a conic section along which the body can move, and along which, therefore, it must move; and this answers the criticism of Jean Bernoulli, according to which Newton did not prove that conic sections are the only curves that can be described by a body acted upon by a force reciprocal to the square of the distances.§

† 'Lubens credo quod ais de aucto Corollario 1, Prop. XIII Lib. 1. Operis Tui incomparabilis Princip. Phil. hac nempe factum esse antequam hae lites coeperunt, neque dubitavi unquam, Tibi esse demonstrationem propositionis inversae quam nude asserueras in prima Opera Editione, aliquid dicebam tantum contra formam illius asserti, atque optabam, ut quis analysin daret, qua inversae veritatem inveniret a priori, ac non supposita directa jam cognita. Hoc vero, quod Te non invito dixerim, a me primo praestitum esse puto, quantum saltem hactenum mihi constat'. *Correspondence*, **7**: 76.

‡ 'Negabant enim a Neutono satis esse demonstratum praeter sectiones conicas nullam aliam curvam quaesito satisfacere, quamvis prop. XVII. lib. 1. *Princ. Philos. Naturalis* hoc satis clare evincere videatur'. Euler (1736): 221.

§ 'En généralisant ensuite ses recherches, ce grand géomètre fit voir qu'un projectile peut se mouvoir dans une section conique quelconque, en vertu d'une force dirigée vers son foyer, et réciproque au carré des distances: il développa les diverses propriétés du mouvement dans ce genre de courbes: il détermina les conditions nécessaires pour que la courbe soit un cercle, une ellipse, une parabole ou une hyperbole, conditions qui ne dépendent que de la vitesse et de la position primitive du corps. Quelles que soient, cette vitesse, cette position et la directon initiale du mouvement, Newton assigna une section conique que le corps peut décrire, et dans laquelle il doit conséquemment se mouvoir; ce qui répond au reproche que lui fit Jean Bernoulli, de n'avoir pas démontré que les sections coniques sont les seules courbes que puisse décrire un corps sollicité par une force réciproque au carré des distances'. Pierre-Simon Laplace, *Exposition du système du monde* (Paris, 1796). I quote from pages 521–2 of the 1835 edition, revised by the author and published by Fayard in 1984.

A modern reader might want to disagree with Newton, Euler and Laplace, and say that Corollary 1 (plus Proposition 17) lacks a uniqueness proof. When a proof is given that for any initial condition, a unique conic trajectory exists, this does not imply that conic sections are necessary orbits. However, neither Newton nor Bernoulli worried about uniqueness theorems.† The fact that, in the case of inverse square forces, only one orbit can exist for a given initial condition was assumed as intuitive by eighteenth-century mathematicians. Perhaps the clearest statement is to be found in a manuscript written by Newton in the late 1710s:

It has been objected that the Corollary of Proposition 13 of Book 1 in the first edition was not demonstrated by me. For they assert that a body P, going off a straight line given in position, at a given speed and from a given place under a centripetal force whose law is given [viz. reciprocally proportional to the square of the distance from the force-centre], can describe a great many curves. But they are deluded [Sed hallucinantur]. If either the position of the straight line or the speed of the body be changed, then the curve to be described can also be changed and from a circle become an ellipse, from an ellipse a parabola or hyperbola. But where the position of the straight line, the body's speed and the law of central force stay the same, differing curves cannot be described. And, in consequence, if from a given curve there be determined the centripetal force, there will conversely from the central force given be determined the curve. In the second edition [of the *Principia*] I touched on this in a few words merely; but in each [edition] I displayed a construction of this Corollary in Proposition 17 whereby its truth would adequately come to be apparent, and exhibited the problem generally solved in Proposition 41 of Book 1.‡

Newton shows that by gradually varying the initial conditions, the solutions vary smoothly, spanning all sorts of conic sections. This, for him, proves that he has found all the trajectories which answer the inverse problem.

The controversy on the inverse problem of central forces which began in 1710 ended ten years later. The two parties agreed on Newton's Corollary 1 (plus Proposition 17) as a valid proof. The British could also be credited with the solution of the inverse cube case. The agreement reached between Bernoulli and Newton could not, however, hide differences in research methods.

8.6.2.8 Moral

The Continentals belonging to the Basel and the Paris schools spent enormous efforts in the development of analytical methods in dynamics. Hermann, Varignon, the Bernoullis pursued a programme in dynamics which consisted of tackling more and more advanced problems with calculus techniques. Dynamics was reduced to a chapter of integral calculus. Great attention was paid to the role of constants of integration and to generality of solutions. The logical correctness of Corollary 1

---

† For a simple discussion of uniqueness in the solution of Newton's equation of motion see Dahr (1993). It should be noted that reduction to quadrature obtained in Prop. 41, Book 1, can be understood as the first step towards an existence and uniqueness proof. It seems to me, however, that the issue of uniqueness of solutions of the inverse central force problem was simply not approached by Newton and his contemporaries.

‡ *Mathematical Papers*, **8**: 457–8. Translation by Whiteside.

was eventually accepted, but it was understood that research had to be directed in other directions.   Bernoulli's 1710 reduction to quadratures of the general inverse problem was considered equivalent to Newton's Propositions 39–41 (e.g. by Hermann).†

Newton could have translated at least *part* of his *Principia* into the language of the analytical method of fluxions. He did so, but did not publish. Newton and his close adherents tended to underestimate the importance of such a translation. They were not so much interested in the generality of symbolical methods. They rather affirmed the superiority of geometrical methods. Keill's defence of the conciseness of Corollary 1 is representative of this Newtonian approach.

The difference between Newton's approach to central forces, defended by Keill, and Bernoulli's approach persisted well into the eighteenth century. We can see much of the divide between the British and the Continental eighteenth-century mathematical schools already characterized in the controversy on central forces that was fought in the decade following 1710.

8.6.3 Bernoulli's criticism of Proposition 10, Book 2

8.6.3.1 Proposition 10: the 1687 version

Another major criticism of Newton advanced by Johann and Niklaus Bernoulli concerns Proposition 10, Book 2. Let us now turn our attention to the version of this proposition as it appeared in the first edition of the *Principia*.

In Proposition 10 Newton considers a body which moves under the influence of a uniform constant gravity [g], and a resistance [$\mathcal{R}$] proportional to the product of the density of the medium [ς] and the square of the velocity [v^2]. Given the trajectory, it is required to find the density and the velocity at each point of the trajectory.‡

Let ACK be the given trajectory, drawn in a plane perpendicular to that of the horizon (see fig. 8.6). Newton assumes that a body describes this curve from A to K. He further imagines that the same body moves in the contrary direction, from K to A, as in a movie projected in the wrong temporal direction. So the speed of the body which progresses is the same as that of a regressing body at the same points. Furthermore, the resistance which retards the first body imparts a positive acceleration to the second body. After this introductory material, we are told that in a very small interval of time the first body describes the arc CG, while the second describes the arc Cg.§

† Hermann wrote: 'Hoc problema primum solutionem accepit a Cel. Newtono Prop. 41 Lib. *Princ. Phil. Nat. Math.* & postea a Perspicacissimo Geometra Joh. Bernoulli gemino modo.' Hermann (1716): 73.
‡ In subsection §8.6.3.1 I am basically following Galuzzi (1991).
§ 'Aequalibus autem temporibus describat corpus progrediens arcum quam minimum CG, & corpus regrediens arcum Cg'. *Variorum*: 377.

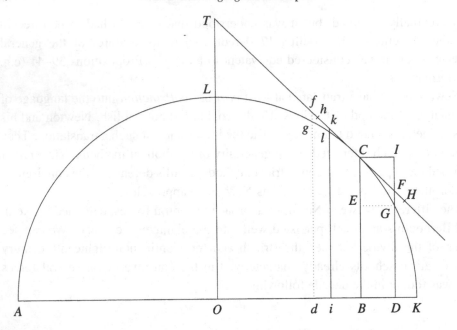

Fig. 8.6. Diagram for Proposition 10 (1687). After J. Bernoulli (1711), tab. XXI, fig. 1

In order to evaluate the effects of gravity and resistance, Newton divides motion into a tangential and a vertical component. The choice of these intrinsic coordinates allows Newton to state that during an infinitely little time, only resistance is acting along the tangent, and only gravity along the vertical. This choice of nonrectangular component directions is very confusing at a first reading. One in fact is tempted to object that resistance also acts along the vertical, and that there is a tangential component of gravity. However, once motion is decomposed into a tangential and a vertical component we have that resistance has no component in the vertical direction and gravity no component in the tangential direction.†

Newton states that without the action of the resistance and of gravity, the progressing body would describe the segment CH, while Ch $(= CH)$ is the segment that would be described by the regressing body. But considering the resistance, in the first case the body would arrive only at F, while the second

---

† Whiteside has remarked that 'here is born the confusion which blights Newton's succeeding argument: the 'linelet' FG is generated not merely by the force of vertically downward gravity, but also through the component (here negative) of the force of resistance to the motion along CFH'. *Mathematical Papers*, **8**: 374. However 'Newton knew that by the vector principle of Corollary 1 to his Laws of Motion this choice of nonrectangular component directions would permit him to think of the independent actions of the resistive force and [gravity] [...] The resistance to the motion is in the negative tangential direction. This resistance can be divided into components if neither component direction is in the tangential direction. In Newton's case the tangential direction *is* one of the chosen directions, so there is *no component of the resistive force in the [vertical] direction*'. Erlichson (1994a): 284.

would arrive at f. Newton states – invoking Lemma 10, Section 1, Book 1 (§3.3) – that FH is proportional to the resistance times the square of time. Furthermore, FG (a small Galilean fall generated by the constant gravity g) is proportional to the square of time. Therefore, resistance is proportional to FH/FG:

$$\mathcal{R} \propto \frac{FH}{FG}.\dagger \qquad (8.43)$$

By considering what happens in the opposite direction, Newton easily deduces that:

$$\mathcal{R} \propto \frac{Cf - CF}{FG}.\ddagger \qquad (8.44)$$

By hypothesis one can set $\mathcal{R} = \varsigma v^2$, while the velocity v can be expressed by $CF/\sqrt{FG}$.§ Summing up, the density of the medium is

$$\varsigma = \frac{Cf - CF}{CF^2}. \qquad (8.45)$$

This concludes Newton's demonstration. We have made the proportionality constants explicit in footnotes. Newton is obviously aware of these constants, but they do not appear in his published text since he talks in terms of proportions.

The problem arises when one wants to evaluate the above expressions (8.44) and (8.45) for a given curve. In the corollaries which follow, Newton deploys a power series representation for these curves (as we would say, he employs Taylor series). In order to do so he subdivides the abscissa into equal infinitesimal intervals $o = BD$. The curve's ordinate can thus be expressed by the series

$$P + Qo + Ro^2 + So^3 + \cdots. \qquad (8.46)$$

As Galuzzi observes : 'By this series it is obvious how to express CF and FG, but Cf creates a problem because it corresponds to a negative increment which is different in absolute value from o, as it corresponds to a length *described in the same time*'. In order to circumvent this difficulty, Newton takes on Cf the segment Ck equal to CF, and draws ki perpendicular to OK and intersecting the curve at l. After some manipulation he obtains

$$\varsigma = \frac{FG - kl}{(FG + kl)CF}. \qquad (8.47)$$

'Obviously, all quantities in this formula are easily calculated by Taylor series [since $iB = BD = o$], but it happens that this formula is wrong. In fact, in order

---

† Making proportionality constants explicit we have that for a body of unitary mass $m = 1$: $FH = (1/2)\mathcal{R}t^2$, $t^2 = (2/g)FG$, $\mathcal{R} = gFH/FG$.
‡ In fact $CH = Ch$ and $FH = hf$, therefore $2FH = Cf - CF$. Thus $\mathcal{R} = (g/2)(Cf - CF)/FG$.
§ Note that $v = CF/t = CF/\sqrt{2FG/g}$, so that $v^2 = (g/2)CF^2/FG$. This expression for velocity will be encountered again below.

to obtain it, Newton has substituted $\sqrt{fg}$ with $\sqrt{FG}$'. The difference between FG and fg is, however, a third-order infinitesimal, which cannot be ignored. As we will see below, Newton realized this mistake only during the process of preparing the second edition of the *Principia*.†

Newton observes that the first term of the above power series represents BC, the second represents IF, the third represents FG. He further notices that FG will be expressed by the third term, together with those that follow *in infinitum*. Newton's understanding of the geometric and kinematic meaning of the successive terms of a Taylor series is remarkable. It is now trivial to observe that $CF = \sqrt{o^2 + Q^2 o^2}$, $FG + kl = 2Ro^2$ and $FG - kl = 2So^3$.‡ In terms of Taylor coefficients we have

$$\varsigma = \frac{S}{R\sqrt{1 + Q^2}}. \tag{8.48}$$

Proposition 10 is followed by four 'Examples' handled by application of the above formula. The given trajectory is supposed to be a half circle (Example 1), a parabola (Example 2), an hyperbola (Example 3), and a generalized hyperbola (Example 4).

8.6.3.2 *An error is announced*

The dramatic succession of events related to Johann Bernoulli's discovery of a mistake in Proposition 10 has been told many times.§ In September 1712 Niklaus Bernoulli, Johann's nephew, arrived in London. He met Newton and informed him that Johann had detected a mistake in Proposition 10.¶ Actually Johann had noticed that from Newton's formula it was possible to derive a strange consequence. As was explained to Leibniz in a letter dated August 1710,‖ in the case of a semicircular path (i.e. Example 1) one had a resistance that was equal and opposite to the tangential component of gravity: the circumference would be described with constant speed! This is contrary to what Newton himself had demonstrated. The Bernoullis were working hard in order to find mistakes in the *Principia*. Two papers were soon presented to the Paris Academy: one by Johann, where the mistake was noticed and an alternative calculus solution was given, one by Niklaus where it was stated that the origin of this mistake was in Newton's misunderstanding of the

---

† Galuzzi (1991): 182.

‡ In fact, in Newton's notation, while $DG = P + Qo + Ro^2 + So^3 + \cdots$, one has that $il = P - Qo + Ro^2 - So^3 + \cdots$. Cf. *Variorum*: 382.

§ The best accounts are to be found in Hall (1980) and in Whiteside's commentary in *Mathematical Papers*, **8**: 48ff.

¶ De Moivre's letter to Johann Bernoulli dated 10 October 1710 in Wollenschläger (1902): 270–1.

‖ *LMS*, **3**: 854.

meaning of the higher-order infinitesimals appearing in the coefficients of a Taylor series.† In 1713 Johann also published a long paper on this matter in the *Acta*.‡

Newton recognized immediately that Niklaus was right. He was busy on the second edition of the *Principia*. Since Cotes had not noticed any error in Proposition 10, the pages with the unaltered 1687 version were already printed. Newton worked strenuously in order to reach an understanding of his mistake and produce a correct demonstration. What he achieved is presented in the next subsection, §8.6.3.3. Contrary to what the Bernoullis thought, his mistake was not related to a lack of understanding of the Taylor coefficients. As we noticed above Newton's mistake consisted in equating two linelets which differ by a third-order infinitesimal: it was a mistake with the geometry of higher-order infinitesimals, not with the Taylor coefficients.

However, the Bernoullis were insisting that Newton's error was proof of his lack of expertise in calculus. This is what Niklaus wrote in 1711 and what Johann wrote to Leibniz in June 1713.§ It is well known that Leibniz included Johann's criticism in the fly-sheet (called *Charta volans*) which was circulated in late 1713.¶ The *Charta volans* played an important role in the Leibnizian manoeuvres devised in order to counter the accusation of plagiarism just formally enunciated by the Royal Society's committee in the *Commercium epistolicum*. As we will see below, this was the beginning of a quarrel, conducted mainly by John Keill and Johann Bernoulli, which concerned the nature of Newton's error. Was it a proof of the inferiority of Newton's geometrical mathematical methods to Leibniz's analytical ones? Before considering the nature and content of this debate, let us devote some attention to Newton's corrected version of Proposition 10.

8.6.3.3 *The corrected 1713 version*

Newton at first attempted to salvage the original version of Proposition 10. He then realized his mistake with third-order infinitesimals and developed an alternative proof, which was included in the second edition of the *Principia*. The manuscripts of these hasty and nervous reworkings have been preserved. They have been edited and commented on masterfully.‖ There is little to be added to Whiteside's definitive analysis. Here I just sketch the essential steps of Newton's new version of Proposition 10.

Newton realized that his difficulties had originated in having taken into consideration quantities related to unequal increments of the abscissa (dB and BD), corresponding to equal increments of time. In the corrected argument Newton

---

† Bernoulli, J. (1711) and Bernoulli, N. (1711).
‡ Bernoulli, J. (1713).
§ *LMS*, **3**: 911.
¶ *Correspondence*, **6**: 15–17.
‖ *Mathematical Papers*, **8**: 312–424. See also Brinkley (1810).

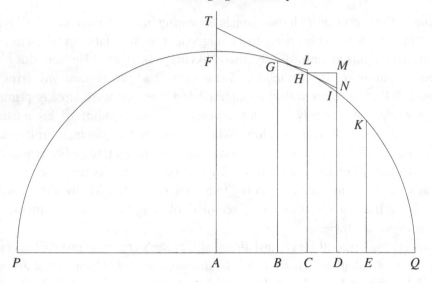

Fig. 8.7. Diagram for Proposition 10 (1713). After *Variorum*: 377

considers equal increments of the abscissa (BC, CD, DE) (see fig. 8.7). To these increments he relates the points G, H and I on the given trajectory and the tangents GL and HN.

The times T and t during which the body (whose mass is assumed unitary, $m = 1$) describes the arcs GH and HI are different. The decrement of speed occurring during time t is expressed by

$$\frac{GH}{T} - \frac{HI}{t}. \tag{8.49}$$

This is the decrement of speed attained in the infinitesimal time increment and is (by the second law of motion) equal to the tangential component of force.

Newton, in this revised version, adopts a considerably different model from his original. While the 1687 version is based on the usual Newtonian representation of force via the continuously accelerated deviation from inertial motion (Lemma 10, Section 1, Book 1, is invoked), here Newton aims at representing the infinitesimal variation of velocity. He thus considers two infinitesimal arcs GH and HI traversed by uniform motion, and uses (8.49) as an expression for the discontinuous infinitesimal change of speed. As we know, in the *Principia* Newton used both the continuous and the impulsive model, while Leibniz leaned towards the latter (§6.3.2).

In order to obtain the decrement of speed due to resistance only it is necessary to add to (8.49) the positive increment of speed due to the tangential component of gravity. Newton observes that 'In a body falling and describing in its fall the space NI, gravity generates a velocity by which twice that space could have been described in the same time, as Galileo proved, that is, the velocity $2NI/t$'.[†] The increment of speed due to the tangential component of gravity is $(2NI/t)(MI/HI)$.

Thus the resistance[‡] is

$$\frac{GH}{T} - \frac{HI}{t} + \frac{2NI}{t}\frac{MI}{HI}. \tag{8.50}$$

This is the basic relation which allows Newton to represent resistance, density and velocity.

Newton, in fact, continues as follows:

Since gravity generates the velocity $2NI/t$ in the same time in a falling body, the resistance will be to the gravity as [...]

$$GH\frac{t}{T} - HI + 2NI\frac{MI}{HI} \quad \text{to} \quad 2NI, \text{§}$$

a result that we can write as

$$\frac{\mathcal{R}}{g} = \left(GH\frac{t}{T} - HI + 2NI\frac{MI}{HI}\right)/2NI. \tag{8.51}$$

Note here that T and t are proportional to $\sqrt{LH}$ and $\sqrt{NI}$ respectively: 'the times in which the body describes arcs GH and HI will be as the square roots of the distances LH and NI which the body could describe in those times by falling from the tangents'.[¶] In fact, Newton conceives LH and NI as small Galilean falls due to gravity only.[‖] So the ratio t/T can be eliminated in terms of LH and NI.

Thus, if one denotes the constant increments of the abscissa by o, all the quantities occurring in (8.51) can be easily represented in terms of the coefficients of the power series:[††]

$$P + Qo + Ro^2 + So^3 + \cdots. \tag{8.52}$$

By substituting the appropriate Taylor coefficients it turns out that the ratio of resistance to gravity is

$$\frac{\mathcal{R}}{g} = \frac{3}{2}\frac{S\sqrt{1 + Q^2}}{2R^2}, \tag{8.53}$$

---

† *Variorum*: 378. Cf. *Principles*: 258.
‡ 'decrementum velocitatis ex resistentia sola oriundum'. *Variorum*: 379.
§ *Variorum*: 379. Cf. *Principles*: 258.
¶ *Variorum*: 377–8. Cf. *Principles*: 258.
‖ e.g. $LH = (1/2)gT^2$.
†† 'Et hinc, si curva linea $PFHQ$ definiatur per relationem inter basem seu abscissam AC & ordinatim applicatam CH, ut moris est; & valor ordinatim applicatae resolvatur in seriem convergentem: Problema per primos seriei terminos expedite solvetur ut in exemplis sequentibus.' *Variorum*: 380.

while the density is given by

$$\varsigma \propto \frac{3S}{2R\sqrt{1+Q^2}},$$ (8.54)

which differs by a factor $3/2$ from the 1687 result.

8.6.3.4 *The quarrel over Newton's understanding of higher-order infinitesimals*

Johann Bernoulli's solution Bernoulli's solution and his interpretation of Newton's mistake were sent in a letter to Varignon dated January 1711 and later published in the *Mémoires* of the Paris Academy.[†] In this letter he writes the differential equation of a body accelerated by a central force F and which moves in a medium exerting a resistance proportional to the density times some power of the velocity.

Bernoulli has recourse to formulas for force and curvature which were well established by 1711. He states that the tangential component of central force is given by

$$F_T = \frac{F\,\mathrm{d}r}{\mathrm{d}s} = \frac{v^2\mathrm{d}r}{\rho r\mathrm{d}\theta},$$ (8.55)

where r and θ are the polar coordinates, v is the speed (a scalar quantity), $\mathrm{d}s$ the infinitesimal arc and ρ the radius of curvature.[‡]

In order to write the differential equation of motion one has to 'add or subtract [to (8.55)] (according whether the body is moving away from the centre or whether it is approaching it) the medium's force of resistance, which is by hypothesis equal to ςv^n.'[§] On setting $\mathrm{d}y = r\mathrm{d}\theta$ one obtains

$$\frac{\mathrm{d}v}{\mathrm{d}t} = \frac{v^2\mathrm{d}r}{\rho\mathrm{d}y} \pm \varsigma v^n,$$ (8.56)

or

$$\frac{\mathrm{d}v}{v} + \frac{\mathrm{d}s\,\mathrm{d}r}{\rho\mathrm{d}y} + \varsigma v^{n-2}\mathrm{d}s = 0.$$ (8.57)

And 'since the relation between $\mathrm{d}r$, $\mathrm{d}y$, $\mathrm{d}s$, ρ and ς is given in r and y', setting $p = \mathrm{d}s/(\rho\mathrm{d}y)$ and $q = \varsigma\mathrm{d}s/\mathrm{d}r$ one has the following differential equation:

$$\frac{\mathrm{d}v}{v} + p\mathrm{d}r \pm v^{n-2}q\mathrm{d}r = 0.$$ (8.58)

This is the differential equation that Jacob Bernoulli had discussed in 1695. Johann

---

[†] Bernoulli, J. *Briefwechsel*, **3**: 341–53 and Bernoulli, J. (1711).
[‡] $F = F_N\mathrm{d}s/r\mathrm{d}\theta = (v^2/\rho)(\mathrm{d}s/r\mathrm{d}\theta)$. Notice that there is no normal component of the resistive force: it is thus legitimate to set $F_N = v^2/\rho$. As always the mass term m is 'absorbed' in these prototype equations/proportions so that, to all effects, $m = 1$.
[§] Bernoulli, J. (1711): 503.

had proposed a method of solution in 1697.† Applying this method Johann obtains v and, since $F = (v^2/\rho)(ds/rd\theta)$,

$$F = pc^{-2\int pdr} \cdot \sqrt{(1-\frac{1}{2}n)}(\mp 2 \pm n) \int c^{(2-n)\int pdr} q\,dr, \qquad (8.59)$$

where c stands for the base of the natural logarithm, which is (after Euler) usually denoted by the symbol e. Similar results were achieved by Verzaglia, Varignon and, in the most creative and independent way, by Hermann.‡

One cannot but appreciate the generality of Bernoulli's and Hermann's approach. In this application to motion in resisting media, stimulated by the problems faced in the second book of the *Principia*, they deploy the richness of the results achieved by the Basel school in the field of differential equations and the calculus of transcendental quantities. This approach differs from Newton's in several respects, which will be discussed below.

A mistake in Example 1 In Example 1 Newton considers the case in which the trajectory is a semicircumference (see fig. 8.6). He concludes that the ratio between resistance and gravity g is

$$\frac{\mathcal{R}}{g} = \frac{OB}{OK}. \qquad (8.60)$$

In fact

$$v = \frac{CF}{\sqrt{(2FG/g)}}, \qquad (8.61)$$

and thus (since $FG + kl \approx 2FG$, and invoking (8.47)):

$$\frac{\mathcal{R}}{g} = \frac{\varsigma v^2}{g} = \frac{FG - kl}{(FG + kl)CF}\frac{CF^2}{2FG} = \frac{(FG - kl)CF}{4FG^2} = \frac{S\sqrt{1 + Q^2}}{2R^2}. \qquad (8.62)$$

Development into Taylor series of the equation for the semicircumference and substitution of the appropriate coefficients leads to (8.60).

As Johann Bernoulli states in letters to Leibniz and Varignon this result 'leads to a contradiction'. In fact, according to (8.60) the magnitude of the tangential component of gravity should equal resistance:

$$\frac{\mathcal{R}}{g} = \frac{OB}{OK} = \frac{OB}{OC}. \qquad (8.63)$$

The semicircumference would be described with constant speed. But in Example 1 Newton shows that the speed is proportional to $\sqrt{BC}$. As Bernoulli triumphantly

† *Acta Eruditorum* (December 1695): 553 and (March 1697): 113–18.
‡ Mazzone & Roero (1997) for details.

observes, according to Newton the speed should be 'constant and variable; there-fore he is in contradiction'.[†]

From his formulas shown previously Bernoulli derives a result which differs from Newton by a factor of 3/2. He multiplies his differential equation (8.57) by v^2 and gets:

$$\frac{v^2 ds\,dr}{\rho\,dy} + \varsigma v^n ds = -v\,dv. \tag{8.64}$$

Thus, because of (8.55) and $dy = r\,d\theta$,

$$F\,dr + \varsigma v^n ds = -v\,dv. \tag{8.65}$$

Taking into consideration Example 1 (see fig. 8.6), Bernoulli sets $F = g$, $\rho = OC$, $\varsigma v^n = \mathcal{R}$, while r is the ordinate BC:

$$g\,dr + \mathcal{R}\,ds = -v\,dv. \tag{8.66}$$

But, since the component of gravity normal to the trajectory $g \cdot BC/OC$ is v^2/OC,

$$v^2 = g \cdot BC = gr, \tag{8.67}$$

and by differentiating:

$$v\,dv = \frac{1}{2}g\,dr. \tag{8.68}$$

It is thus possible to state

$$g\,dr + \mathcal{R}\,ds = -\frac{1}{2}g\,dr, \tag{8.69}$$

and thus

$$\frac{\mathcal{R}}{g} = -\frac{3}{2}\frac{dr}{ds} = -\frac{3}{2}\frac{CE}{CG} = -\frac{3}{2}\frac{OB}{OC} = -\frac{3}{2}\frac{OB}{OK}, \tag{8.70}$$

a result which differs from Newton's by a factor of 3/2.

Johann (and Niklaus) Bernoulli's explanation of Newton's error Niklaus Bernoulli's explanation of the origin of Newton's error appeared as an *Addition* to his uncle's paper.[‡] He stated the following. As we know, Newton had this formula for the ratio of resistance to gravity:

$$\frac{\mathcal{R}}{g} = \frac{S\sqrt{1 + Q^2}}{2R^2}. \tag{8.71}$$

According to Niklaus Bernoulli the geometric reasoning that led Newton to his formula (8.47) was perfectly correct. What Newton must have missed was a correct

---

† Bernoulli, J. *Briefwechsel*, **3**: 348–9.
‡ Bernoulli, N. (1711).

interpretation of the Taylor coefficients in terms of higher-order infinitesimals. If the trajectory is expressed as

$$P + Qo + Ro^2 + So^3 + \cdots, \tag{8.72}$$

the coefficients Q, R, S are not the first-, second- and third-order differentials, as Newton erroneously thought, according to the malicious interpretation of Niklaus Bernoulli. In order to get the correct expression for the higher-order differentials one has to multiply R by the factor 2 and S by the factor 6. Once this is done, (8.71) transforms into the correct result:

$$\frac{\mathcal{R}}{g} = \frac{6S\sqrt{1+Q^2}}{2(2R)^2} = \frac{3}{2}\frac{S\sqrt{1+Q^2}}{2R^2}. \tag{8.73}$$

The debate on Newton's error in Proposition 10 Niklaus's interpretation was endorsed by Johann in letters to Varignon, De Moivre and Leibniz. Johann also published it in his long paper in the *Acta* for 1713.† For instance, he wrote to Varignon:

My nephew has discovered that the origin of Mr. Newton's mistake does not consist in his method, which is perfectly correct, even though circonvoluted, but only in the fact that he has taken all the terms (other than the first two) of a power $(z + o)^n$ developed into an infinite series as equal to the differences of z^n.‡

His suggestion to Leibniz that this mistake was 'evidence by which we may allowably infer that the fluxional calculus was not born before the differential calculus' was inserted, without Johann's consent, in the *Charta volans*.§

The thesis that Newton did not have the means to calculate higher-order differentials occurred again and again during the quarrel over the priority in the invention of calculus. It was a quite extraordinary accusation: it amounted to saying that Newton did not know how to calculate the higher-order fluxions of $y = x^n$. It was easy for people such as Keill to reply convincingly to the attacks of the Bernoullis. Newton's mistake did not lie in his handling of the Taylor

---

† See his letter to Varignon of 3 August 1713 in *Briefwechsel*, **3**: 505–6, to De Moivre of 18 February 1713 in *Wollenschläger* (1902): 282 and to Leibniz of 7 June 1713 in *LMS*, **3**: 911.

‡ 'mon Neveu a decouvert que la source de l'erreur de Mr. Newton ne consist pas dans sa methode qui est fort bonne quoique embarassée, mais seulement en ce qu'il a pris tous les termes (hors les deux premiers) d'une puissance $(z + o)^n$ developpée, pour autant de differences de z^n.' Bernoulli, J. *Briefwechsel*, **3**: 505.

§ *LMS*, **3**: 911. Translation by Whiteside in *Mathematical Papers*, **8**: 57.

coefficient, but in his geometric reasoning, where he equated FG and fg.[†] Keill, who was writing under Newton's guidance, continued by attacking Leibniz for having made a mistake with higher-order differentials in the *Tentamen* (1689). It was a real comedy of errors. As Aiton has shown in his (1962), Newton's and Keill's accusation against Leibniz was as ungrounded as that of the Bernoullis against Newton.

Keill's battle against the Leibnizians culminated in a rather embarrassing defeat. In a letter to Montmort of January 1718 Keill proposed as a challenge the solution of the inverse problem of resisted motion, i.e. the problem of determining the trajectory of a body accelerated by constant gravity and by a resistance proportional to density and the square of velocity. Johann Bernoulli replied in papers published in 1719 and 1721 by solving an even more general problem: that of determining the trajectory of a body resisted by a force proportional to density and some power $2n$ of the velocity. This problem had indeed already been solved by Hermann in his *Phoronomia* (1716) thanks to the use of the differential equation (8.58) solved by Bernoulli in 1713.[‡] Nobody in Newton's camp could reply with a better result.

Even though the Leibnizian analysis of Newton's mistake in Proposition 10 was faulty, the confrontation with the Newtonians ended with Continental victory. This is in contrast to what we have seen with the criticisms of Book 1. I have demonstrated that in the case of topics such as motion in a central force field, the attraction of extended bodies or the perturbations of planetary motions, the two schools achieved comparable results in the period covered by my study. It is thus appropriate to enquire into the reasons which led to the Newtonian defeat in the field of motion in resisting media.[§]

8.6.3.5 Moral

The mathematical generality of the Leibnizian approach What Bernoulli did, together with other mathematicians influenced by the Basel circle such as Varignon and Hermann, was to address a very general problem by means of differential equations. The general problem was that of determining the central force, assuming as given the trajectory, the density and a resistance varying as any power of

---

[†] One of Keill's typical replies is the following: 'Dans l'année 1711 Messieurs Bernoully, Oncle & neveu, attaquerent Monsieur Newton su sujet d'une erreur par lui commise dans la 10 Proposition de son 2. Livre. La Cause de cette erreur provenoit de ce qu'il avoit prolongé une Tangente d'une certain arc de la Courbe au lieu de la prolonger de l'autre. Si l'on a égard au peu de tems qu'il avoit employé à écrire ce livre, à la difficulté & à la noveauté du sujet [...] il y aura lieu de s'étonner qu'il s'y soit trouvé un si petit nombre de fautes. Certainement il n'est pas d'un esprit généreux de triompher des méprises des grands Hommes.' Keill (1716): 428.

[‡] Hermann (1716): 345ff. Details on Bernoulli, J. (1719) and (1721) can be found in Blay (1992): 322–30.

[§] There are other mathematical demonstrations in Book 2 which were criticized, namely Propositions 15–16 (spiral motion due to inverse square central force plus resistance) and Proposition 26 (resisted cycloidal motion). For details on the criticisms to Proposition 26 advanced by Johann and Niklaus Bernoulli, Jacopo Riccati and James Stirling see *Mathematical Papers*, **6**: 441–2.

the velocity. Applying (8.55), Bernoulli could write the differential equation (8.57). By integration, v (and thus F) was obtained. The problem faced by Newton in Proposition 10 was dealt with as a very specific application of this general approach. Equation (8.57) could also be applied to other problems. For instance Newton, in Propositions 15 and 16, Book 2, had considered spiral motion generated by a central force plus a resistive force proportional to the density and the square of velocity. This problem was faced by Newton by means of a completely different geometrical strategy.† Bernoulli instead could deal with Proposition 10 and Propositions 15–16 in a unified way. His equations apply equally well to both problems: indeed he claimed to have found mistakes in Newton's analysis of spiral resisted motion. Even though the Bernoullis were wrong in stating that Newton's errors were due to lack of understanding of the Taylor coefficients, the feeling of superiority that they reveal in their publications and correspondence is, in this case, justified.

Newton's strategy in Proposition 10 was remarkably different. He too began with assuming the trajectory to be given. Unlike Bernoulli, he assumed that the force (in this case constant gravity g) was known. His problem was that of determining the resistance $\mathcal{R}$. He did so by considering small infinitesimal elements of the trajectory. Under certain assumptions about the mode of action of force and resistance during infinitesimal intervals of time (e.g. Galileo's law of fall can be applied to infinitesimal vertical fall), he could express resistance in terms of the infinitesimal elements of the trajectory. After this geometrical work, Taylor series were introduced: the power series representation of the trajectory was employed in order to achieve a representation of the infinitesimal elements of the trajectory, and thus of $\mathcal{R}$, in terms of the Taylor coefficients Q, R, S. While Bernoulli's approach consists in starting from the beginning with a differential equation (it is thus a problem in integration), Newton's approach consists in arriving at the end of his geometrical strategy at a symbolic expression in which the Taylor coefficients occur. In order to calculate the Taylor coefficients he has to find the higher-order fluxions: his problem belongs to the differential, not to the integral, calculus.

After this comparison between the approaches to projectile motion in resisting media followed in the two opposing camps, we now turn to some difficulties which are inherent not only in Proposition 10, but also in many other demonstrations of the *Principia*. We will use Proposition 10 as exemplar of many other parts Newton's *magnum opus*.

Representation of time One of the problems that Newton had to face was that of providing a geometrical representation of time. In the case of central force motion,

---

† Erlichson (1994a); Chandrasekhar (1995): 539–553.

this representation is provided by the Kepler area law: the area swept by the radius vector functions as a clock, and time can be represented by this area. In cases of perturbed Keplerian motion in void space, or motion in resisting media, such a simple representation of time is not available. This was to create problems for Newton, who was always seeking for a geometrical representation. Furthermore, Newton had problems in distinguishing what we might call a 'fluxional time' (the independent variable which flows uniformly) from 'physical time' (§3.2). In the 1687 version of Proposition 10 he began by considering the infinitesimal increments of the trajectory corresponding to *equal intervals of physical time.* This was a natural choice for Newton as he conceived the geometric quantities as generated by continuous flow. However, as we have discussed above, this choice of physical time as the independent variable led Newton into trouble. The geometric increments expressed in function of physical time are not easily representable in terms of the Taylor coefficients. In the corrected version Newton changed strategy and assumed *equal increments of the abscissa,* which becomes the independent variable. To these increments correspond unequal increments of physical time. In the fluxional jargon, now the abscissa, the 'fluxional time', is flowing uniformly (it increases by equal increments) while the flow of 'physical time' is *not* uniform. Physical time is thus a dependent variable which is eliminated from formula (8.51).

Representation of transcendental quantities As we know, Newton did not devote great efforts to improving the algorithm for transcendental quantities.† Rather, he represented them in geometric terms. We have seen the case of trigonometric quantities in the *Geometria curvilinea* (§2.3.2) and of logarithmic quantities in Book 2 of the *Principia* (§3.11, §3.13). Leibniz and his associates were instead developing a calculus of transcendental quantities expressing algorithmic rules for their differentials and integrals (§5.2 and §6.2.4).‡ Formula (8.59) is a typical result of the Leibnizian school in this field.

Representation of higher-order infinitesimals Another source of trouble, indeed the origin of Newton's mistake in Proposition 10, concerns the handling of higher-order geometrical infinitesimal linelets. As we have seen, the fatal step in the 1687 demonstration was the substitution of linelet FG by linelet fg. Newton's geometric procedure in Proposition 10 implies several approximations. The arcs are set equal to tangents, velocity is considered as constant along certain infinitesimal portions of the trajectory, linelets which differ by higher-order

---

† For Newton's symbolic approach to simple trigonometry see *Mathematical Papers,* **4**: 116, 134, 446, 460 and
  7: 455 *passim.*
‡ Breger (1986).

infinitesimals are equated. In these procedures it is difficult to control the degree of approximation. Is it legitimate to drop the higher-order elements in all the various approximations? How can we state the degree of approximation required by the problem, and how can we check the degree of approximation achieved? The use of symbolic calculus is, in this respect, extremely helpful. It is much easier to arrange that all infinitesimals up to a certain order are taken into consideration (see §3.6 and §3.15).

8.7 A remarkable improvement

The process of translation of the *Principia* into the language of calculus initiated by Varignon, Hermann and Johann Bernoulli was concluded only in the late 1730s by Leonhard Euler. In the *Mechanica* published in 1736 the talented former pupil of Johann Bernoulli was able to offer a general, uniform and well-regulated analytical method for approaching most of the problems faced by the Newton in the *Principia*. In the Preface of the *Mechanica* (1736) Euler wrote :

But what pertains to all the works composed without analysis, is particularly true with mechanics. In fact, the reader, even though he is persuaded about the truth of the things that are demonstrated, nonetheless cannot understand them clearly and distinctly. So he is hardly able to solve with his own strengths the same problems, when they are changed just a little, if he does not inspect them with the help of analysis and if he does not develop the same propositions into the analytical method. This is exactly what happened to me, when I began to study in detail Newton's *Principia* and Hermann's *Phoronomia*. In fact, even though I thought that I could understand the solutions to numerous problems well enough, I could not solve problems that were slightly different. Therefore I strove, as much as I could, to get at the analysis behind those synthetic methods in order, for my own purposes, to deal with those propositions in terms of analysis. Thanks to this procedure I perceived a remarkable improvement of my knowledge.†

In the two volumes of the *Mechanica* Euler deals with point mass dynamics. In the first volume he tackles rectilinear and curvilinear motion *in vacuo* and in resisting media. In the second volume one finds a treatment of (resisted and unresisted) motion of point masses constrained to move on lines or surfaces. This is only a part of Euler's programme of mathematization of mechanics. As he explains, his aim is to deal in further works with extended rigid, flexible and fluid bodies as well.‡ The full realization of this programme will occupy Euler all his life.

---

† 'Sed quod omnibus scriptis, quae sine analysi sunt composita, id potissimum Mechanicis obtingit, ut Lector, etiamsi de veritate eorum, quae proferuntur, convincatur, tamen non satis claram et distinctam eorum cognitionem assequatur, ita ut easdem quaestiones, si tantillum immutentur, proprio marte vix resolvere valeat, nisi ipse in analysin inquirat easdemque propositiones analytica methodo evolvat. Idem omnino mihi, cum Neutoni *Principia* et Hermanni *Phoronomiam* perlustrare coepissem, usu venit, ut, quamvis plurium problematum solutiones satis percepisse mihi viderer, tamen parum tantum discrepantia problemata resolvere non potuerim. Illo igitur iam tempore, quantum potui, conatus sum analysin ex synthetica illa methodo elicere easdemque propositiones ad meam utilitatem analytice pertractare, quo negotio insigne cognitionis meae augmentum percepi.' Euler (1736): 8.

‡ Euler (1736): 38–9.

The *Mechanica* can be defined as a work of systematization of results achieved mostly in the Bernoullian school. The whole work is a systematic application of a limited number of basic differential equations. After Euler's systematization of point mass dynamics in terms of differential and integral calculus, the Bernoullian style was to dominate the scene. The *Mechanica* showed clearly and extensively how 'remarkable' the application of differential calculus to the science of motion could be.

Among the first to notice the importance of Euler's *Mechanica* as an instrument for reading the *Principia* were Le Seur and Jacquier in the edition of the *Principia* published in Geneva with the supervision and interest of Jean-Louis Calandrini and with an extensive commentary. The so-called 'Jesuit edition' became the standard reference for having access to the *Principia*. It ran to several editions: in 1760 in Köln, in 1780–85 in Prague, in 1822 and 1833 in Glasgow. The commentary occupies as much space as the text and provides an explanation of every single page. In the *Monitum* to the reader prefaced to the first book (1739), Le Seur and Jacquier recognized their debt to David Gregory, Varignon, Hermann and Keill.†
In the *Monitum* prefaced to the second book (1740) we find the first reference to Euler's *Mechanica*. As a matter of fact, the *Mechanica* provided Le Seur and Jacquier with exactly the material they were looking for: a treatment in terms of differential equations of the Propositions of the *Principia*.‡

The edition of the *Principia* by Le Seur and Jacquier provides an encyclopedic *résumé* of the results on the applications of calculus to Newton's *Principia* achieved in the first four decades of the eighteenth century. Euler's *Mechanica* was singled out by the editors as the most important contribution in this field. After the

---

† 'Verum naturalis aequitas & mathematicus candor postulant, ut nos plurimum debere fateamur Doctissimis Viris, Davidi Gregorio, Varignonio, Jacobo Hermanno, Joanni Keillio, allisque multis, qui varias Newtonianae Philosophiae partes luculentis scriptis illustrarunt.' In fact we find many traces of Gregory's *Astronomiae*, of Varignon's papers on central forces and resisted motion, and of Hermann's *Phoronomia* and Keill's papers on central forces. The two editors also included long notes on the direct and inverse algorithm of calculus (as a commentary to Section 1, Book 1) and on infinite series. This allowed them to employ the calculus in the notes. For instance, Propositions 39–41, Book 1, are expressed in terms of calculus: the reference is to Varignon (1700), to Proposition 25 of the *Phoronomia* and to Keill (1714). In the annotations to Section 10, Book 1, we find references to the works on pendular motion by Huygens, Jacob and Johann Bernoulli, and Hermann, and to the work on double curvature by Clairaut. Sections 12 and 13, Book 1, on the attraction of extended bodies are commented on in terms of calculus: the reference is to Keill, 'sGravesande and, most notably, Maupertuis.

‡ The first three sections of Book 2 were thus annotated using Euler, together with Varignon's works, Johann Bernoulli (1713) and (1721), and the second book of the *Phoronomia*. In their commentary on Newton's treatment of the solid of least resistance, Le Seur and Jacquier refer to results by l'Hospital. They also refer to Fatio (1699), to the *Phoronomia* and to Johann Bernoulli's *Essay d'une nouvelle théorie de la manoeuvre des vaisseaux* (1714). In their commentary on Proposition 36, Book 2, on the flow of water from a hole, the editors take into consideration Daniel Bernoulli's *Hydrodynamica* (1738), as well as the controversial writings by Poleni, J. Riccati, Michelotti, Jurin and E. Manfredi. In their commentary on Section 9 Book 2, the two editors also refer to the attempts, most notably those by Johann Bernoulli, to defend the vortex theory of planetary motion from the criticisms raised in that section by Newton. The third book too is heavily annotated. Most notably, the essays on tides by Maclaurin, Euler and Daniel Bernoulli, which had just won the Paris Academy prize for 1740, are included. They occupy more than two hundred pages.

Mechanica the readers of the *Principia* turned their eyes to Le Seur and Jacquier's notes, rather than to Newton's text. By the middle of the eighteenth century, the style in which Newton had written his *magnum opus* belonged definitely to the past.†

8.8 Conclusion: confronting values

The Leibnizian programme of translation of the *Principia* into the language of the calculus is intertwined with the story of the attacks and counterattacks carried on during the priority dispute. The British were challenged to fight with the same weapons as their adversaries. The question to be answered was: were Newton and his acolytes able to perform such a translation? The British were able to answer positively as far as Book 1 was concerned. The Leibnizian criticisms to Book 2 proved to be much more difficult: the result was a British defeat. For the historian the interest of this confrontation lies in the fact that the Newtonians and the Leibnizians compared the advantages and disadvantages of their methods by testing them on specific issues. While doing so they rendered explicit the values that directed their researches.

---

† In broad outline the same considerations hold true for the French translation of the *Principia* by the Marquise du Chastelet which appeared in Paris in 1757–59. It ends with a long commentary, mainly due to Alexis-Claude Clairaut. It is of great interest because it shows how, in Paris at the middle of the eighteenth century, the *Principia* was read in terms of calculus. The discussion of Clairaut's commentary goes beyond the bounds of our research.

9

Conclusion: Newtonians, Leibnizians and Eulerians

9.1 A common heritage

Despite the chauvinistic divisions originated by the Newton–Leibniz controversy, the Newtonian and the Leibnizian schools shared a common mathematical method. They adopted two algorithms, the analytical method of fluxions and the differential and integral calculus, which were translatable one into the other. As far as foundations are concerned, the situation was much more fragmented than is usually thought. The Channel did not divide the friends of infinitesimals from the adherents to limits. One finds British who refer to the infinitesimalist side of the Newtonian method (i.e. to 'moments') and equate moments with differentials; one finds Continentals who support a theory of limits.

Rather than seeing the Newtonians and the Leibnizians as two groups practising different mathematical methods, it is more fruitful, and more adherent to historical evidence, to focus on the amount of shared knowledge between the two schools. The British and the Continentals shared a vast quantity of common algorithmic techniques, notational devices and foundational ideas. Newton and Leibniz, however, tried, with partial success, to orient their disciples along different research lines. The difference between the Newtonians and the Leibnizians consisted in the manner in which they oriented themselves in this common mathematical heritage: it consisted in the values that they adopted, in the purposes they had in mind, and in the ways in which they placed this heritage in historical perspective.

9.2 Equivalence

Passing from the fluxional to the differential notation was a triviality. For instance, David Gregory in his worksheets on fluxions easily adapted the proofs of the Bernoulli brothers published in the *Acta eruditorum* into Newtonian notation.† The

---

† Eagles (1977): 426ff.

correspondence between British and Continentals was generally carried on in the two notations: the British would use the dots and the Continental would reply with ds, translating all the formulas of the correspondent by simply changing $\dot{x}$ for dx. When Edmund Stone in 1730 translated l'Hospital's *Analyse* (1696) he followed an established practice. He wrote:

almost all the Foreigners represent the first Increment or Differential (as they call it) by the letter d, the second by dd, the third by ddd, etc.; the fluents, or Flowing Quantities, being called Integrals. But since this method in the Practice thereof, does not differ from that of Fluxions, and an Increment or Differential may be taken for a Fluxion; out of regard to Sir Isaac Newton, who invented the same before the year 1669, I have altered the Notation of our Author, and instead of d, dd, d^3, etc. put his Notation, viz. $\dot{x}, \ddot{x}, \dddot{x}$, etc. or some other of the last Letters of the Alphabet, printed thus, and called the infinitely small Increment, or Differential of a Magnitude the Fluxion of it.†

One would be wrong in thinking that the fluxional algorithm was practised by understanding $\dot{x}$ as 'the rate of flow of x'. This fluxional definition was operative only in some Prefaces of eighteenth-century British textbooks. Newton and his followers practised an algorithm in which $\dot{x}$ was handled as an infinitesimal increment of x. One difficulty that modern historians have had in reading the texts of the British fluxionists is caused by the fact they try to read fluxional formulas by translating $\dot{x}, \ddot{x}$, etc., as dx/dt, d^2x/dt^2, etc. The established practice was instead that indicated by Stone, a practice which guarantees the syntactic translatability between the two algorithms.

The main advantage of the Leibnizian calculus concerns the integral sign. With $\int y dx$ the integration-variable x is explicitly indicated. Newton's $\boxed{y}$, Qy and $\dot{y}$ need to be accompanied by verbal statements (such as 'supposing that x flows equably'). This has effects on integration techniques. In the Leibnizian calculus, integration by substitution and by parts can be performed in a more mechanical way. This advantage was recognized by the Newtonians, who often employed hybrid notations (which are a further proof of the syntactic equivalence between $\dot{x}$ and dx). For instance, in 1703 Cheyne employed $F : y\dot{x}$ for $\int y dx$, a notation employed also by Maclaurin in 1742.‡

One might maintain, however, that some deep foundational difference (e.g. in terms of Newtonian limits versus Leibnizian differentials) existed between the two schools. I have already discussed the equivalence between Newton's and Leibniz's approaches to foundation in §6.4. Here I would like to recall that in the case of their followers, too, no simple opposition is possible. We find Newtonians founding the calculus on infinitesimals as well as Leibnizians referring to limits as the true foundation of the calculus. As a matter of fact, at the turn of the century the British

---

† Stone (1730): xviii.

‡ Of course the capital 'F' stands for 'fluent of', while $\int$ means 'summa'. Cheyne (1703): 19ff. *Mathematical Papers*, **8**: 109. Maclaurin (1742): 665ff.

were learning the calculus from the *Acta eruditorum* and from l'Hospital's *Analyse* (1696), as well as from Newton's works. A few examples will suffice.

John Harris, a minor member of Newton's circle of mathematicians, who inserted an English translation of Newton's *De quadratura* in the second volume (1710) of his technical dictionary, did not feel the need to change the substance of the article on fluxions published in 1704 in the first volume of the work. The article reads as follows: 'By the *Doctrine of Fluxions*, we are to understand the Arithmetick of the *Infinitely small* Increments or Decrements of *Indeterminate* or *variable Quantities*, or as some call them the *Moments* or *Infinitely small Differences* of such variable Quantities. These Infinitely small Increments or Decrements, our Incomparable Mr. Isaac Newton, calls very properly by the Name of Fluxions'.† In the second volume Harris continued the article on fluxions with a statement concerning a 'general Method of finding the Fluxions of all Powers and Roots'. Harris wrote as follows: 'If a Quantity gradually increases or decreases, its immediate Increment or Decrement is called its *Fluxion*. Or the Fluxion of a Quantity is its Increase or Decrease indefinitely small'.‡

In the great majority of cases, the conception of continuity adopted by many early-eighteenth-century British was that of Newton's early works of 1669 and 1671; that is to say, it was based on Newton's analytical method in which infinitesimal/moments were employed. This rendered a transition towards Leibnizian treatment very easy. For instance, William Jones – who edited a collection of Newton's mathematical tracts in 1711 and served on the Royal Society committee set up during the priority dispute – published an introduction to fluxions in 1706 in which, paraphrasing l'Hospital (1696), he maintained that: 'all *Curved Lines* may be considered as composed of an Infinite Number of infinitely little right Lines: Any one of them Produced, only Touches the Curve, therefore is called the *Tangent* of that Point of the *Curve*.'§ A use of infinitesimals and a lack of any clear distinction between fluxions and differentials was, in fact, such a common occurrence that it would be easy to provide any number of such quotations.

On the Continent, on the other hand, we find Jacob Bernoulli, who in his polemic with Clüver backs an approach to calculus based on limits (see §6.4.2.3), or Varignon, who uses first and ultimate ratios in order to defend the calculus at the Académie.¶ Leibniz himself was extremely respectful towards the foundation of calculus in terms of first and ultimate ratios. In both schools it was agreed that limits provided the rigorous foundation, while infinitesimals were to be considered as useful abbreviations.

---

† Harris (1704): article 'Fluxions'.
‡ Harris (1710): article 'Fluxions'.
§ Jones (1706): 226. Compare l'Hospital (1696): 3.
¶ Peiffer (1988) and (1990).

When we consider the mathematization of the science of motion ('dynamics' in Leibnizian jargon) a major divide between Leibnizians and Newtonians seems to emerge. It is often maintained that while the British adhered to geometrical methods, the Continentals favoured symbolism. The *Principia* appears as the most evident proof of this divide: it is sufficient to open Newton's *magnum opus* and compare it to an early-eighteenth-century copy of the *Acta Eruditorum*. Diagrams and geometrical reasonings on one side, differential equations on the other. But this divide belongs to the published presentation of results. If we move from the public arena to the private side of correspondence and preparatory manuscripts, we discover a different story. Newtonians and Leibnizians differed mainly in their publication policies, rather than in their competences.†

9.3 Not equivalent in practice

As I argued in Chapters 6, 7 and 8, rather than looking for clear-cut distinctions between the two schools, one should refer to more subtle characteristics related to the mathematical practice. It is at this pragmatic level, rather than at the syntactic level of algorithmic manipulations, or at the semantic level of foundations, that relevant distinctions between Newtonians and Leibnizians can be found. It is at this level that we can state that the two schools pursued two methods which were 'not equivalent in practice'.

9.3.1 Inner validation criteria

Let us consider, to begin with, the notations of the two calculi. Despite the translatability between the two notations, and thus despite their formal equivalence, they were employed in different ways. While Leibnizians shared an interest in the improvement of notation and its independence from geometrical interpretation, Newtonians tended to underestimate it. Newton wasted no occasion of communicating to his disciples his indifference towards the algorithm. As we know, for him the true and 'more natural' method was the synthetic one. The algorithm had always to be understood as translatable into the geometric terms of the method of first and ultimate ratios. For instance in 1714, after having compared his notation with Leibniz's, he wrote to Keill:

but these are only ways of Notation & signify nothing to the method it self, wch may be without them.‡

---

† These optical illusions originated by a lack of distinction between published and unpublished sources are frequent. A remarkable case is offered by Galileo. Galilean scholars have learnt to attribute a greater importance to the role of experiment in Galileo's science after a more attentive analysis of Galileo's manuscript notes.

‡ Edleston (1850): 175.

While it is true that some British mathematicians, such as Brook Taylor and Roger Cotes, devoted efforts to improving the notation, it is on the Continent that most of the research on new notation was carried out. Undoubtedly, the Continentals' efforts ultimately led to relevant advantages. Schneider has noticed that in Leibniz's calculus the fundamental theorem is somehow 'built into' the notation itself. Indeed, the symbols d and ∫ suggest that differentiation and integration are operations and that they are the inverse of each other.†

One can also compare the two schools as far as integration techniques are concerned. As Scriba has observed in (1964), Newton emphasized the use of integration via infinite series. He expanded fluents into infinite series and 'integrated' termwise. This technique was, of course, also employed by Leibnizians. However, they preferred integration in 'closed' form: i.e. they looked for quadratures expressed not by infinite series, but by finite combination of 'functions'. Also Newtonians obtained results on 'closed' integrations (most notably Cotes), but it is certainly true that for them infinite series played a more prominent role than for Leibnizians. This contrast between the two schools is thus a matter of emphasis, i.e. it is a contrast related to the values which direct research along different lines.

The Continentals were investing more enthusiasm in the heuristic possibilities afforded by the mechanical manipulation of symbols. They admitted that calculation could be divorced from interpretability. The Newtonians insisted in considering rigorous those reasonings which could have an interpretation. As Newton said, his concepts of fluent and fluxion 'take place in the reality of physical nature' (see §2.3.2).‡ The algorithm of the analytical method was thus to be seen, according to the Newtonians, as embedded into the geometric context of the synthetic method. It is this embedding which guarantees continuity with ancient tradition as well as ontological content.§

We have seen in Chapters 7 and 8 that, while the Continentals were stressing the novelty and heuristic power of the Leibnizian calculus, the British insisted on the continuity between Newton's fluxional synthetic method and the geometry of the 'Ancients'. In Britain a deep interest in the 'geometry of the Ancients' was alive. Newton, David Gregory, Edmond Halley, Robert Simson, Colin Maclaurin and Matthew Stewart trod in the steps of those who, following the publication of Commandino's Latin edition of Pappus in 1588, received a new boost to their effort to 'restore' the Ancients' 'geometrical analysis'. The results of this interest were, for instance, the editions of Apollonius's *Conica* by Gregory and Halley, or the restoration of Euclid's *Porisms* by Simson and Stewart.¶ This

---

† Schneider (1988): 143.
‡ *Mathematical Papers*, **8**: 122–3.
§ On the importance of the question of ontology in the Newtonian response to Berkeley's *Analyst* (1734), in particular in Maclaurin's work, see Sageng (1989).
¶ Simson (1723), (1776).

philological activity was not an antiquarian enterprise for these mathematicians. They were convinced that the 'Ancients' had possessed a superior knowledge that had subsequently been lost. The geometrical style of the *Principia* was seen as in harmony with Archimedes' method of exhaustion. The relationship between the method of exhaustion and Newton's method of first and ultimate ratios was often underlined and formed, for instance, an important part of Maclaurin's defence of the Newtonian method in the Preface of *Treatise of fluxions* (1742). The Continentals reacted skeptically to the Newtonian mania for the *prisca*.

9.3.2 Policies

The comparison between Newton's and Leibniz's mathematizations of natural philosophy must also take into consideration the different policies pursued in the two schools. Newton established himself as the rediscoverer of a lost knowledge about the world. He wished to communicate this truth to an audience by and large unable to read equations and fluxions. It was easier to find some preparation in elementary geometry. We have seen how Halley's efforts were aimed at rendering the *Principia* understandable to the fellows of the Royal Society (§7.2). Also, the number of introductions to the *Principia* published in the first decades of the eighteenth century is proof of this need to propagate an understanding of universal gravitation to an audience not trained in mathematics (§7.4). The geometric language of Newton's work helped this programme of proselytism to be implemented. In particular, the first three sections of Book 1 were within the reach of any cultivated person willing to make a serious effort to go from Proposition 1 to Proposition 17. Furthermore, the classic façade of the *Principia* was useful to Newton's idea of presenting himself as a perpetuator of a lost tradition. Lectures given by early Newtonians in the Scottish and English universities included very little mathematics. Rather, they were concentrated on qualitative presentations of the System of the World and on subjects such as pneumatics, hydrostatics and optics. This is the case of Cotes's, Whiston's and Saunderson's lectures in Cambridge. The same holds true in broad outline for Maclaurin's lectures in Edinburgh and Gregory's in Oxford. Some effort was made to teach fluxions, but descriptive natural philosophy was predominant.

The complexity of Newton's publication policy deserves notice. After the methodological turn of the 1670s he continued to pursue mathematical researches in analytics. However, he always viewed analytics not as an independent algorithm, but rather as part of a broader geometrical method.† In his mature researches in the analytical method of fluxions he made it clear that a translation in terms of geometrically interpretable concepts was always at hand. Newton preferred to

---

† This thesis has been very convincingly defended in De Gandt (1986) and (1995): 217 *passim*.

publish his results in geometric form. He did so in the *Principia* as well as in lesser works such as in his paper on the brachistochrone problem. Newton's geometric style gives a flavour of solidity and unity to his mathematical work. We have learnt to appreciate how problematic and fractured, behind the appearances, is this style. At the same time, Newton wished to let something of his analytics be published. The oblique references in the *Principia* to a method whereby 'curvilinear figures can be squared', the correspondence favoured by Collins and Oldenburg – who acted as intermediaries between Newton and other leading mathematicians – and the edition of Wallis's *Opera* (1693–99) were the means that Newton chose, before the inception of the priority dispute, to let others know about the power of his analytical method of discovery. With his close disciples he proved to be more generous: mathematical manuscripts were shown to visitors, or even circulated.

Leibniz, on the other hand, conceived of himself as the promoter of new methods of reasoning. The calculus was just one successful example of the power of algorithmic thinking. The German diplomat was interested in promoting in Europe the formation of a group of intellectuals who could extend a universal knowledge achieved thanks to a new *characteristica*. He thus helped the formation of a school of mathematicians who distinguished themselves by their ability in handling the differentials and the integrals and by their innovative publication strategy. Often thanks to Leibniz's recommendation, they colonized chairs of mathematics all over Europe. The efficacy of this new Kabbala was affirmed to be independent from metaphysical or cosmological questions. The persons who practised it had to be professional mathematicians able to teach and propagate knowledge of calculus.

9.3.3 Research priorities

The field where the pragmatic difference between the two schools is more marked concerns the mathematization of natural philosophy, a field to which we have devoted this book. The reduction of dynamics to calculus constituted a constant stimulus for the Continentals. Mathematizing Nature meant for them writing and solving differential equations. Problems such as the isochrone, the brachistochrone, the catenary, the velaria, etc., were in fact interesting in so far they required improvements in integration techniques. The *Principia*, deemed meaningless from a cosmological point of view, was considered a source of interesting mathematical problems. To the Leibnizians it did not reveal the truth about the world; it was rather a gymnasium to test the muscles of the integral calculus.

The approach of the Newtonians could not be more different and, consequently, their mathematical methods for natural philosophy differed dramatically from those of the Leibnizians. For them, Newton had revealed the force which rules the heavenly motions. Mathematizing Nature meant dealing in mathematical terms with universal gravitation, mathematizing all the effects caused by the gravitational force, such as tides, or planetary shapes. The possibility of mathematically predicting these effects was crucial for the Newtonians. The acceptance of universal gravitation depended precisely on the success of such a mathematization. However, the analytical method of fluxions was not yet that powerful. The Newtonians found in their agenda a set of problems which could not be tackled with the calculi created by Newton and Leibniz. In the 1990s, research carried out by Chandrasekhar and Nauenberg has shown us how far could Newton go in dealing in analytical terms with lunar motion.† However, it seems to me that in these advanced topics the analytical method could be employed only sporadically. In several passages of their demonstrations the Newtonians were thus forced to turn to physical simplifications or to geometrical tools (see §3.15).

Leibniz and the Leibnizian mathematicians looked on the geometrical proofs of Newton's *Principia* with suspicion. One of their targets was to translate Newton's geometrical proofs into the language of the differential and integral calculus. Indeed dynamics proved to be a great source of inspiration for Leibnizians. It was by trying to develop new mathematical tools for the mechanics of extended bodies (rigid, elastic and fluid) that mathematicians such as Varignon, Johann and Daniel Bernoulli, Clairaut, Euler, d'Alembert and Lagrange enriched the calculus, developing new concepts and techniques. Such important results of the eighteenth-century calculus as trigonometric series, partial differential equations and the calculus of variations were to a great extent motivated by the analytical approach to dynamics that Leibniz had sought to promote. The eighteenth century was thus characterized by the analytical programme emphasized by the Leibnizian school, while the role attributed to geometry by Newton and his followers faded away.

9.4 Postscriptum: a decline of British mathematical science?

The extent and depth of the results achieved by eighteenth-century Continentals has shed a shadow on the British eighteenth-century fluxionists. Historians of mathematics devote little attention to losers. Since the Newtonian mathematical methods for natural philosophy have been superseded by those of Euler, the great champion of the Basel school, they have received little attention. The *Principia* is generally considered as cumbersome, difficult to read. These judgments are possible if one looks at the *Principia* through the eyes of Euler. We would not, however, accept a study of Locke based only on quotations from Leibniz! The aim of my research, since my book of 1989, has been that of stating the idea that at

---

† Chandrasekhar (1995), Nauenberg (1995), (1998a) and (1998b). See also Wilson (1995).

the beginning of the eighteenth century the choice between the Leibnizian and the Newtonian methods was open.

A depressing image of the eighteenth-century fluxional school is accepted in many histories of mathematics. According to this image, after Newton there was a decline caused by chauvinistic isolationism and by slavish adherence to a Newtonian method which was inherently inferior to the Leibnizian one. Only at the beginning of the nineteenth century were the British able to free themselves from the burden of the Newtonian heritage. Recent research has disproved the standard view on the decline of the British school. Now we are able to appreciate the contributions of mathematicians such as Cotes, De Moivre, Stirling, Maclaurin, Landen, Simpson and Waring.† These contributions were known, praised and used on the Continent: the fluxionists participated actively in the development of analytical techniques concerning, for example, series or ordinary differential equations. However, talking about a 'decline' of the British is still justified, provided that we pay attention to periodization and that we qualify the nature of the 'decline'.

The decline set in later than is usually thought. Until the 1740s the British were able to participate in new results and to communicate with the Continentals. They corresponded with no difficulty with the Leibnizians. Translations from and to British terminology and notation were not infrequent. We have discussed above, in fact, the equivalence between the two algorithms. It is from the 1740s that we find the British losing ground. The Continentals became less and less interested in what was happening across the Channel. At the end of the eighteenth century the two communities were almost completely separated. Typically, in 1796 J.J. Lalande observed that there was not a single first-rate analyst in England. Waring could only reply, rather petulantly, that he had given 'somewhere between three and four hundred new propositions of one kind or another'.‡ Other British mathematicians, often based in Scotland or in the English military schools – such as John Toplis, John Playfair, Thomas Leybourn, Robert Woodhouse – shared Lalande's evaluation. When the Analytical Society of Cambridge was founded in 1812 the time was ripe for a diagnosis which has been accepted by too many historians: the British disease was caused by a chauvinistic Newtonian myth and by the clumsy fluxional dot-notation. Thus the battle in 'favour of pure d-ism and against the Dot-Age of the University' – to cite once more Babbage's famous pun – was fought and won in the early Victorian era.

It is wrong, however, to locate the infection in Newton's method of fluxions. In the first decades of the eighteenth century what the British could do with fluxions and dots was comparable to what the Continentals did with differentials and ds.

---

† For an overview see Guicciardini (1989). The main references to secondary literature are on p. ix.
‡ Lalande (1796), Waring (1799).

What happened around the middle of the century? For a concise answer, the best thing is to look at the principal mathematicians of the Continental school in that period: Clairaut, d'Alembert and most notably Euler. They were the originators of a profound change in the history of the calculus. Their calculus became a theory concerning functions and derivatives, rather than Newtonian/Leibnizian quantities. It was a theory which progressively freed itself from geometrical interpretation. The new concept of function, a concept which was present neither in Leibniz's nor in Newton's calculus, was extended to include multivariate functions (e.g. by N. Bernoulli, Fontaine and Euler in the 1730s, 1740s) and partial derivatives (e.g. by d'Alembert, Euler and D. Bernoulli in the 1750s). Partial differential equations became one of the most important research areas. Partial differential equations and the calculus of variations proved to be powerful tools in the mathematization of mechanics. It was thanks to these new instruments that the Continentals, as Bos, Truesdell and Wilson have shown, were able to deal with the mechanics of rigid, flexible and fluid bodies or with advanced topics in celestial mechanics.† The mathematization of many problems posed prematurely by Newton in the *Principia* became possible (e.g. in the Moon theories of Euler, Clairaut and d'Alembert developed in the 1750s) thanks to a calculus which was neither Leibnizian nor Newtonian: it would be better to call it 'Eulerian calculus'.‡

The distance which separates the Eulerians from the British fluxionists in the 1740s is evident in Maclaurin's *Treatise of fluxions* (1742), one of the masterpieces of the British school. It is true that several of the results of the *Treatise* were referred to as important by the Continentals, even many years after Maclaurin's death in 1748.§ However, what was Maclaurin's contribution to the Eulerian calculus? He never uses functions, he has nothing to say on multivariate functions, his mechanics is squarely based on Newton's laws (he has no alternative principle such as, say, virtual velocity), he has no mechanics of rigid and elastic bodies, his variational techniques are rudimentary compared with Euler's. Of course, Maclaurin belongs to an earlier generation than Euler. The fact is that in the second half of the century the British remained within Maclaurin's horizon. The eighteenth century was a period rich in innovations, a period in which a new calculus and a new mechanics were created: the British missed this revolution.

The reasons why this failure occurred are too complex to be revealed by my study. The process of involution of the British eighteenth-century mathematical community is too vast a phenomenon to be encompassed by the interpretative results achieved in this book. However, my description of the values that directed the Newtonians in their mathematical researches in natural philosophy might

---

† Bos (1980b), Truesdell (1960), Wilson (1995). See also Todhunter (1873).
‡ See the excellent study Fraser (1997). See also Laugwitz (1997).
§ Grabiner (1997).

be useful in explaining the slow acceptance of the Eulerian calculus in Britain. These values (such as continuity with the past geometrical tradition, insistence on representability of mathematical symbols, distrust of algorithmic techniques) rendered many eighteenth-century British mathematicians little receptive towards the highly abstract, proudly innovative and progressively de-geometrized Eulerian calculus.†

For us it is difficult to understand the choice of the Newtonians. We are educated in a mathematical technique which renders the Eulerian calculus so simple to our eyes, while the Newtonian 'mathematical principles for natural philosophy' are nowadays difficult to read. Furthermore, we know that the winning move was Euler's. Johann Bernoulli's criticisms to Keill seem to us very well grounded. For us mechanics is written in differential equations. A modern reader considers the geometry of conic sections and proportion theory useless. When we formulate these evaluations we tend to forget that at the beginning of the eighteenth century the choice between mathematical methods for natural philosophy was not so obvious. The calculus appeared to many to be unfounded, uninterpretable, not powerful enough. In this book we have paid some attention to the reasons that led the Newtonians to move in a direction different from that taken by the Continentals.

Are there choices in the history of mathematics? Mathematics enjoys considerable stability through time: according to some, its results are eternal. It is thus tempting to describe its development in a cumulative and predictive way. This internal approach to history of mathematics has been widely followed, since mathematicians often share the idea that their discipline transcends the context. Because of this idea, an idea that contributes so much to the identity of the present-day mathematical community, any hint about the relationship of mathematics with its historical context is seen as heresy. It is probably this approach that induces many mathematicians to regard debates in the history of mathematics merely as sociological events; as events that have no relation with 'true' mathematics, a discipline which should leave no room for opinions, passions and choices. However, mathematics, as any other human enterprise, does develop in a context, in continuous relation with other scientific disciplines and with culture in general. It is applied to the real world for theoretical or technical purposes. It interacts with philosophy, religion, society. It is taught, defended, used. It is only when we consider mathematics in this broad cultural context that it becomes possible to write its history.

---

† Not all the eighteenth-century British mathematicians shared the values of the Newtonian school. Most notably Thomas Simpson, John Landen, William Spence, Thomas Leybourn and Edward Waring favoured the analytical method. See Guicciardini (1989).

Τὰ κοινὰ καινῶς, τὰ καινὰ κοινῶς.

Appendix

Frontispiece from Isaac Newton's *The method of fluxions and infinite series*, London 1736. This work is an English translation of a Latin tract written by Newton in 1670–71. Colson, who held the Lucasian Chair of Mathematics in Cambridge from 1739 to 1760, added a long commentary in which he deals *inter alia* with foundational problems. This image expresses values that Newton was able to communicate to some of his followers. The figure refers to a problem solved on pages 267–76. Actually the bucolic scene is superimposed on a geometric diagram. Two points 1 (top) and 2 (lower) 'flow' from left to right along two straight trajectories. The motion of 1 is retarded, the speed of 2 is constant. The motion of a point 3 along the trajectory LMN, to be found, is such that at each instant the two points 1 and 2 must lie on the tangent at 3. This is a typical 'inverse tangent problem': a curve, to be found, is defined by the properties of its tangent. These problems lead to fluxional (or 'differential' equations). One of Colson's preferred ideas was that Newton's fluents and fluxions exist in nature (in Newton's words ' Hae Geneses in rerum natura locum vere habent' (*Mathematical Papers*, **8**: 122–3)), while Leibniz's differentials are just fictions. According to Colson, the superiority of Newton's method over Leibniz's calculus – a point that Berkeley's *Analyst* (1734) would have missed – is that the fluxional symbols always refer to finite quantities which have an existence. In other words, as Newton wrote, 'fluxions are finite quantities and real, and consequently ought to have their own symbols; and each time it can conveniently so be done, it is preferable to express them by finite lines visible to the eye rather than by infinitely small ones' (*Mathematical Papers*, **8**: 113–5). Colson maintained that Newton's demonstrations are 'ocular demonstrations', a mathematical version of the *oculata demonstratio* defended by many natural philosophers and logicians. The fluxions (or instantaneous speeds) of the two flowing points/birds are visibly represented by finite segments. For instance the velocity at E is to the velocity at H as EL to HL, while the velocity at F is to the velocity at I as FM to IM. In symbols (see the papyrus at the bottom left): if $AG = y$ and $CK = x$, then $\dot{y}/\dot{x} = GN/KN$.

These ratios are the 'sensible measures of sensible velocities' referred to in the Latin motto at the top (Colson's demonstration and its methodological aspects are discussed in Sageng (1989): 281–98). The ancient mathematicians at the bottom left write the pertinent fluxional equation on papyri using fluxional dots. They express vividly the continuity with past tradition which was so important for Newton and his followers: their presence validates Newton's method and notation. The Greek motto at the bottom means: 'the common things in an unusual way, the unusual things in a common way'. The eye of a mathematician can see common things (the fowler tries to shoot two birds in one shot and moves closer and closer to the birds, 'not thinking to be sufficiently near', along a curve so that the rifle is properly aligned) in the unusual symbolic way of calculus. But these unusual symbols do not depart from the mathematical use inherited from past tradition.

References

Airy, George B. (1834). *Gravitation: an elementary explanation of the principal perturbations in the solar system*. London: C. Knight.

Aiton, Eric (1955). The contributions of Newton, Bernoulli and Euler to the theory of tides. *Annals of Science* **11**: 206–23.

Aiton, Eric (1960). The celestial mechanics of Leibniz. *Annals of Science* **16**: 65–81.

Aiton, Eric (1962). The celestial mechanics of Leibniz in the light of Newtonian criticism. *Annals of Science* **18**: 31–41.

Aiton, Eric (1964). The inverse problem of central forces. *Annals of Science* **20**: 81–99.

Aiton, Eric (1972a). *The vortex theory of planetary motions*. New York: Science History Publications.

Aiton, Eric (1972b). Leibniz on motion in a resisting medium. *Archive for History of Exact Sciences* **9**: 257–74.

Aiton, Eric (1984). The mathematical basis of Leibniz's theory of planetary motion. *Studia Leibnitiana* Sonderheft **13**: 209–25.

Aiton, Eric (1986). The application of the infinitesimal calculus to some physical problems by Leibniz and his friends. *Studia Leibnitiana* Sonderheft **14**: 133–43.

Aiton, Eric (1988). The solution of the inverse-problem of central forces in Newton's *Principia*. *Archives Internationales d'Histoire des Sciences* **38**: 271–6.

Aiton, Eric (1989a). Polygons and parabolas: some problems concerning the dynamics of planetary orbits. *Centaurus* **31**: 207–21.

Aiton, Eric (1989b). The contributions of Isaac Newton, Johann Bernoulli and Jakob Hermann to the inverse problem of central forces. *Studia Leibnitiana* Sonderheft **17**: 48–58.

Aoki, Shinko (1992). The Moon-Test in Newton's *Principia*: accuracy of inverse-square law of universal gravitaton. *Archive for History of Exact Sciences* **44**: 147–90.

Arnol'd, Vladimir I. (1990). *Huygens and Barrow, Newton and Hooke*. Basel, Boston, Berlin: Birkhäuser.

Arthur, Richard T. W. (1995). Newton's fluxions and equably flowing time. *Studies in History and Philosophy of Science* **26**: 323–51.

Axtell, James L. (1965). Locke's review of the *Principia*. *Notes and Records of the Royal Society of London* **20**: 152–61.

Axtell, James L. (1969). Locke, Newton and the two cultures. In *John Locke: problems and perspectives*, J. E. Yolton (ed.), pp. 169–82. Cambridge: Cambridge University Press.

Ball, W. W. Rouse (1893). *An Essay on Newton's Principia*. London, New York: Macmillan and Co.

Belaval, Yvon (1975). Note sur Leibniz et Platon. *Revue d'Histoire et Philosophie Religieuses* **55**: 49–54.

Bennett, Jim A. (1982). *The mathematical science of Christopher Wren*. Cambridge: Cambridge University Press.

Berkeley, George (1734). *The Analyst; or, a Discourse Addressed to an Infidel Mathematician. Wherein it is examined whether the Object, Principles, and Inferences of the modern Analysts are more distinctly conceived, or more evidently deduced, than Religious Mysteries and Points of Faith, by the Author of the Minute Philosopher*. London: for J. Tonson.

Bernoulli, Daniel (1738). *Hydrodynamica*. Argentorati: J. R. Dulseckeri.

Bernoulli, Johann *Briefwechsel = Der Briefwechsel von Johann I Bernoulli*. 3 vols. Basel, Boston, Berlin: Birkhäuser, 1955–92.

Bernoulli, Johann (1710). Extrait de la Réponse de M. Bernoulli à M. Herman, datée de Basle le 7 Octobre 1710. *Mémoires de l'Académie des Sciences*: 521–33.

Bernoulli, Johann (1711). Extrait d'une lettre de M. Bernoulli [...] touchant la manière de trouver les forces centrales dans les milieux resistans, en raisons composées de leur densités & des puissances quelconques des vitesses du mobile. *Mémoires de l'Académie des Sciences*: 47–53. In *Opera Omnia*, Lausanne & Genevae, 1742, **1**: 502–8.

Bernoulli, Johann (1713). De motu corporum gravium, pendulorum, & projectilium [...]. *Acta Eruditorum*: 77–95 + 115–32.

Bernoulli, Johann (1714). *Essay d'une nouvelle théorie de la manoeuvre des vaisseaux: avec quelques lettres sur le même sujet*. Basel: J. G. König.

[Bernoulli, Johann] (1716). Epistola pro eminente mathematico [...]. *Acta Eruditorum*: 296–315.

Bernoulli, Johann (1719). Responsio ad nonneminis provocationem, eiusque solutio quaestionis ipsi ab eodem propositae de invenienda linea curva quam describit projectile in medio resistente. *Acta Eruditorum*: 216–26. In *Opera Omnia*, Lausanne & Geneva, 1742, **2**: 393–402.

Bernoulli, Johann (1721). Operatio analytica per quam deducta est eiusdem solutio, quae extat in Actis Lips. 1719 M. Maii, Problematis de invenienda curva, quae describitur a projectili gravi in medio resistente. *Acta Eruditorum*: 228–30. In *Opera Omnia*, Lausanne & Geneva, 1742, **2**: 513–16.

Bernoulli, Niklaus (1711). Addition [to Bernoulli, J. (1711)]. *Mémoires de l'Académie des Sciences*: 54–6. In Johann Bernoulli *Opera Omnia*, Lausanne & Geneva, 1742, **1**: 509–10.

Bertoloni Meli, Domenico (1988). Leibniz's Excerpts from the *Principia Mathematica*. *Annals of Science* **45**: 477–505.

Bertoloni Meli, Domenico (1991). Public claims, private worries: Newton's *Principia* and Leibniz's theory of planetary motions. *Studies in History and Philosophy of Science* **22**: 415–49.

Bertoloni Meli, Domenico (1993a). *Equivalence and priority: Newton versus Leibniz*. Oxford: Clarendon Press.

Bertoloni Meli, Domenico (1993b). The emergence of reference frames and the transformation of mechanics in the Enlightenment. *Historical Studies in the Physical and Biological Sciences* **23**: 301–35.

Blay, Michel (1986). Deux moments de la critique du calcul infinitésimal: Michel Rolle et George Berkeley. *Revue d'Histoire des Sciences* **39**: 223–53.

Blay, Michel (1987). Le traitement newtonien du mouvement des projectiles dans les milieux résistants. *Revue d'Histoire des Sciences* **40**: 325–55.

Blay, Michel (1988). Varignon ou la théorie du mouvement des projectiles 'comprise en une Proposition générale'. *Annals of Science* **45**: 591–618.

Blay, Michel (1992). *La naissance de la mécanique analytique: la science du mouvement au tournant des XVIIe et XVIIIe siècles*. Paris: Presses Universitaires de France.

Bos, Henk J. M. (1972). Huygens, Christiaan. In *Dictionary of scientific biography*, 16 vols., C. Gillispie (ed.), 6, pp. 597–613. New York: Scribner, 1978–1980.

Bos, Henk J. M. (1974). Differentials, higher-order differentials and the derivative in the Leibnizian calculus. *Archive for History of Exact Sciences* **14**: 1–90.

Bos, Henk J. M. (1980a). Newton, Leibniz and the Leibnizian tradition. In *From the calculus to set theory, 1630–1910, an introductory history*, I. Grattan-Guinness (ed.), pp. 49–93. London: Duckworth.

Bos, Henk J. M. (1980b). Mathematics and rational mechanics. In *The ferment of knowledge: studies in the historiography of eighteenth century science*, G. S. Rousseau & R. Porter (eds.), pp. 327–55. Cambridge: Cambridge University Press.

Bos, Henk J. M. (1982). L'élaboration du calcul infinitésimal, Huygens entre Pascal et Leibniz. In *Huygens et la France*, R. Taton (ed.), pp. 115–122. Paris: Vrin.

Bos, Henk J. M. (1986). Introduction [to] Huygens, Christiaan. *The pendulum clock or geometrical demonstrations concerning the motion of pendula as applied to clocks*. Ames: Iowa State University Press: i–xxxii.

Bos, Henk J. M. (1996). Johann Bernoulli on exponential curves, ca. 1695: innovation and habituation in the transition from explicit constructions to implicit functions. *Nieuw Archief voor Wiskunde* Vierde serie **14**(1): 1–19.

Brackenridge, Bruce J. (1988). Newton's mature dynamics: revolutionary or reactionary? *Annals of Science* **45**: 451–76.

Brackenridge, Bruce J. (1990). Newton's unpublished dynamical principles: a study in simplicity. *Annals of Science* **47**: 3–31.

Brackenridge, Bruce J. (1992). The critical role of curvature in Newton's developing dynamics. In *The investigation of difficult things: essays on Newton and the history of the exact sciences in honour of D. T. Whiteside*, P. M. Harman & A. E. Shapiro (eds.), pp. 231–60. Cambridge: Cambridge University Press.

Brackenridge, Bruce J. (1995). *The Key to Newton's dynamics*. Berkeley, Los Angeles, London: California University Press.

Breger, Herbert (1986). Leibniz' Einführung des Transzendenten. *Studia Leibnitiana* Sonderheft **14**: 119–32.

Breger, H. (1990). Das Kontinuum bei Leibniz. In *L'infinito in Leibniz: problemi e terminologia*, A. Lamarra (ed.), pp. 53–67. Roma: Edizioni dell'Ateneo.

Brinkley, John (1810). On Sir Isaac Newton's first solution of the problem of finding the relation between resistance and gravity. *Transactions of the Royal Irish Academy* **11**: 45–59.

Brougham, Henry & Routh, Edward J. (1855). *Analytical view of Sir Isaac Newton's Principia*. London: Longman *et al.*

Caldo, Lorenzo (1929). La regola di Newton per la ricerca del moto degli apsidi nelle orbite prossime al cerchio. *Memorie della Società Astronomica Italiana* **5**: 9–17.

Carr, John (1821). *The first three sections of Newton's Principia*. London: for Baldwin *et al.*

Casini, Paolo (1981). Newton: gli scolii classici. *Giornale Critico della Filosofia Italiana* **60**: 7–53.

Casini, Paolo (1996). The Pythagorean myth: Copernicus to Newton. In *Copernico e la questione copernicana in Italia*, L. Pepe (ed.), pp. 183–99. Firenze: Olschki.

Chandler, Philip Prescott II (1975). *Newton and Clairaut on the motion of the Lunar Apse*.

Ph.D. thesis. San Diego: University of California.

Chandrasekhar, Subrahmanyan (1995). *Newton's Principia for the common reader.* Oxford: Clarendon Press.

Cheyne, George (1703). *Fluxionum Methodus Inversa: sive quantitatum fluentium Leges Generaliores.* London: by J. Matthew, sold by R. Smith.

Clarke, John (1730). *A demonstration of some of the principal sections of Sir Isaac Newton's Principles of Natural Philosophy: in which his peculiar method of treating that useful subject is explained, and applied to some of the chief phaenomena of the system of the world.* London: for James and John Knapton.

Cohen, I. Bernard (1971). *Introduction to Newton's Principia.* Cambridge: Cambridge University Press.

Cohen, I. Bernard (1980). *The Newtonian revolution.* Cambridge: Cambridge University Press.

Cook, Alan (1991). Edmond Halley's and Newton's *Principia*. *Notes and Records of the Royal Society of London* **45**: 129–38.

Cook, George L. (1850). *The first three sections and part of the seventh section of Newton's Principia.* Oxford: J. H. Parker.

Costabel, Pierre (1967). Newton's and Leibniz' dynamics. *The Texas Quarterly* **10**(3): 119–26.

Cotes, Roger (1714). Logometria. *Philosophical Transactions* **29**: 5–45.

Cotes, Roger (1722). *Harmonia Mensurarum.* Cambridge.

Couzin, Robert (1970). Leibniz, Freud and Kabbala. *Journal of the History of the Behavioural Sciences* **6**: 335–48.

Cushing, James T. (1982). Kepler's laws and universal gravitation in Newton's *Principia*. *American Journal of Physics* **50**: 617–28.

Dahr, Abhishek (1993). Nonuniqueness in the solution of Newton's equation of motion. *American Journal of Physics* **61**: 58–61.

Dawson, John (1769). *Four Propositions, &c. shewing, not only, that the Distance of the Sun, as attempted to be determined from the Theory of Gravity, By a late Author, is, upon his own Principles, Erroneous; but also, that is more than probable this Capital Question can never be satisfactorily answered by any Calculus of the kind.* Newcastle: by J. White & T. Saint, for W. Charnley.

De Gandt, François (1986). Le style mathématique des *Principia* de Newton. *Revue d'Histoire des Sciences* **39**: 195–222.

De Gandt, François (1987). Le problème inverse (prop. 39–41). *Revue d'Histoire des Sciences* **40**: 281–309.

De Gandt, François (1995). *Force and geometry in Newton's Principia.* Princeton: Princeton University Press.

Densmore, Dana (1995). *Newton's Principia, the central argument: translation, notes, and expanded proofs.* Translations and illustrations by William H. Donahue. Santa Fe, New Mexico: Green Lion Press.

Desaguliers, John T. (1734). *A course of experimental philosophy, vol. 1.* London: for J. Senex.

Descartes, René (1659–61). *Geometria, latine versa*, 2nd edn, 2 vols. Amsterdam.

Di Sieno, Simonetta & Galuzzi, Massimo (1989). La quinta sezione del primo libro dei *Principia*: Newton e il 'Problema di Pappo'. *Archives Internationales d'Histoire des Sciences* **39**: 51–68.

Ditton, Humphry (1705). *The General Laws of Nature and Motion; with their Application to Mechanicks. Also the Doctrine of Centripetal Forces, and velocities of Bodies, Describing any of the Conick Sections. Being a Part of the Great Mr. Newton's*

Principles. The whole illustrated with Variety of Useful Theorems and Problems, and Accomodated to the Use of the Younger Mathematicians. London: by T. Mead, for J. Seller, C. Price and J. Senex.

Dobbs, B. J. T. (1975). *The foundations of Newton's alchemy, or 'the hunting of the greene lyon'*. Cambridge: Cambridge University Press.

Domcke, George P. (1730). *Philosophiae mathematicae Newtonianae illustratae*. London: sold by T. Meighan and J. Batley.

Dupont, Pascal & Roero, Clara S. (1991). *Leibniz 84: il decollo enigmatico del calcolo differenziale*. Rende: Mediterranean Press.

Eagles, Christina M. (1977). *The Mathematical Work of David Gregory, 1659–1708*. Ph.D. thesis. Edinburgh University.

Edleston, Joseph (1850). *Correspondence of Sir Isaac Newton and Professor Cotes*. Cambridge, London: J. Deighton and J. W. Parker.

Edwards, C. H. (1979). *The historical development of the calculus*. New York, Berlin: Springer.

Emerson, William (1743). *The Doctrine of Fluxions*. London: by J. Bettenham, sold by W. Innys.

Emerson, William (1770). *A short comment on Sir Isaac Newton's Principia, containing notes upon some difficult places of that excellent book*. London: J. Nourse.

Erlichson, Herman (1990). Comment on 'Long-buried dismantling of a centuries-old myth: Newton's *Principia* and inverse-square orbits', by Robert Weinstock. *American Journal of Physics* **58**: 882–4.

Erlichson, Herman (1991a). How Newton went from a mathematical model to a physical model for the problem of a first power resistive force. *Centaurus* **34**: 272–83.

Erlichson, Herman (1991b). Motive force and centripetal force in Newton's mechanics. *American Journal of Physics* **59**: 842–9.

Erlichson, Herman (1992a). Newton and Hooke on centripetal force motion. *Centaurus* **35**: 46–63.

Erlichson, Herman (1992b). Newton's polygon model and the second order fallacy. *Centaurus* **35**: 243–58.

Erlichson, Hermann (1994a). Resisted inverse-square centripetal force motion along Newton's great 'look-alike', the equiangular spiral. *Centaurus* **37**: 279–303.

Erlichson, Herman (1994b). The visualization of quadratures in the mystery of Corollary 3 to Proposition 41 of Newton's *Principia*. *Historia Mathematica* **21**: 148–61.

Erlichson, Herman (1996). Evidence that Newton used the calculus to discover some of the propositions in his *Principia*. *Centaurus* **39**: 253–66.

Euclid *The Elements = The thirteen books of Euclid's Elements*. T. Heath (ed.). New York: Dover, 1956.

Euler, Leonhard (1736). *Mechanica, sive, Motus scientia analytice exposita*. St. Petersburg: Ex typographia Academiae Scientiarum = *Opera Omnia*, 2nd Series, vols. 1–2.

Evans, John H. (1837). *The first three sections of Newton's Principia, with an appendix, and the ninth and eleventh sections*. Cambridge: J. W. Parker, University Printer.

Farrell, Maureen (1981). *William Whiston*. New York: Arno Press.

Fatio de Duillier, Nicolas (1699). *Lineae Brevissimi Descensus Investigatio Geometrica Duplex. Cui addita est Investigatio Geometrica Solidi Rotundi, in quod Minima fiat Resistentia*. London: by R. Everingham, sold by J. Taylor.

Feigenbaum, Lenore (1992). The fragmentation of the European mathematical community. In *The investigation of difficult things: essays on Newton and the history of the exact sciences in honour of D. T. Whiteside*, P. M. Harman & A. E. Shapiro (eds.), pp. 383–97. Cambridge: Cambridge University Press.

Feingold, Mordechai (1993). Newton, Leibniz, and Barrow too: an attempt to a reinterpretation. *Isis* **84**: 310–38.

Fellmann, Emil A. (1988). The *Principia* and continental mathematicians. *Notes and Records of the Royal Society of London* **42**: 13–34.

Fichant, Michel (1994). *La réforme de la dynamique*. Paris: Vrin.

Figala, Karin (1977). Newton as alchemist. *History of Science* **15**: 102–37.

Fleckenstein, J. O. (1946). Johann I Bernoulli als Kritiker der *Principia* Newtons. *Elemente der Mathematik* **1**: 100–8.

Fleckenstein, J. O. (1948). Pierre Varignon und die mathematischen Wissenschaften im Zeitalter der Cartesianismus. *Archives Internationales d'Histoire des Sciences* **5**: 76–138.

Force, James E. (1985). *William Whiston, honest Newtonian*. Cambridge: Cambridge University Press.

Force, James E. & Popkin, Richard H. (1990). *Essays on the context, nature, and influence of Isaac Newton's theology*. Dordrecht, Boston, London: Kluwer.

Forsyth, A. R. (1927). Newton's problem of the solid of least resistance. In *Isaac Newton, 1642–1727, A memorial volume edited for the Mathematical Association*, W. J. Greenstreet (ed.), pp. 75–86. London: G. Bell and Sons.

Fraser, Craig G. (1985). D'Alembert's principle: the original formulation and application in Jean d'Alembert's *Traité de Dynamique*. *Centaurus* **28**: 31–61.

Fraser, Craig G. (1997). The background to and early emergence of Euler's analysis. In *Analysis and synthesis in mathematics: history and philosophy*, M. Otte & M. Panza (eds.), pp. 47–78. Dordrecht, Boston, London: Kluwer.

French, A. P. (1971). *Newtonian Mechanics*. New York: W. W. Norton & Co.

Frost, Percival (1854). *Newton's Principia, First Book, Sections I, II, III, with notes and illustrations and a collection of problems principally intended as examples of Newton's method*. Cambridge: Macmillan and Co.

Fueter, E. (1937). Isaak Newton und die schweizerischen Naturforscher seiner Zeit. *Beiblatt zur Vierteljahresschrift der Naturforschenden Gesellschaft in Zürich* No. 28, Jahrg. 82.

Galuzzi, Massimo (1991). Some considerations about motion in a resisting medium in Newton's *Principia*. In *Giornate di storia della matematica*, M. Galuzzi (ed.), pp. 171–89. Commenda di Rende: EditEl.

Gascoigne, John (1991). 'The wisdom of the Egyptians' and the secularisation of history in the age of Newton. In *The uses of antiquity: the scientific revolution and the classical tradition*, S. Gaukroger (ed.), pp. 171–212. Dordrecht, Boston, London: Kluwer.

Gaukroger, Stephen (1991). Introduction: the idea of antiquity. In *The uses of antiquity: the scientific revolution and the classical tradition*, S. Gaukroger (ed.), pp. ix–xvi. Dordrecht, Boston, London: Kluwer.

Giuntini, Sandra (1992). Jacopo Riccati e il problema inverso delle forze centrali. In *I Riccati e la cultura della Marca nel Settecento europeo*, G. Piaia & M. L. Soppelsa (eds.), pp. 127–49. Firenze: Olschki.

Giusti, Enrico (1993). *Euclides reformatus: la teoria delle proporzioni nella scuola galileiana*. Torino: Bollati Boringhieri.

Goldstine, Herman H. (1980). *A history of the calculus of variations from the 17th through the 19th Century*. New York, Heidelberg, Berlin: Springer.

Gouk, Penelope (1988). The harmonic roots of Newtonian science. In *Let Newton be! A new perspective on his life and works*, J. Fauvel, R. Flood, M. Shortland & R. Wilson (eds.), pp. 101–25. Oxford: Oxford University Press.

Gowing, Ronald (1983). *Roger Cotes: natural philosopher*. Cambridge: Cambridge University Press.

Gowing, Ronald (1992). A study of spirals: Cotes and Varignon. In *The investigation of difficult things: essays on Newton and the history of the exact sciences in honour of D. T. Whiteside*, P. M. Harman & A. E. Shapiro (eds.), pp. 371–81. Cambridge: Cambridge University Press.

Grabiner, Judith V. (1997). Was Newton's calculus a dead end? The continental influence of Maclaurin's *Treatise of fluxions*. *American Mathematical Monthly* **104**: 393–410.

Gravesande, Willem J. 's (1720–21). *Mathematical elements of Natural Philosophy confirmed by Experiments, or an Introduction to Sir Isaac Newton's Philosophy. Written in Latin by William James 'sGravesande, Doctor of Law and Philosophy, Professor of Mathematics and Astronomy at Leyden, and Fellow of the Royal Society of London. Translated into English by J. T. Desaguliers, L.L.D.* 2 vols. London: for J. Senex and W. Taylor.

Greenberg, John L. (1986). Mathematical physics in eighteenth-century France. *Isis* 77: 59–78.

Greenberg, John L. (1987). Isaac Newton et la théorie de la figure de la Terre. *Revue d'Histoire des Sciences* **40**: 357–66.

Greenberg, John L. (1995). *The problem of the Earth's shape from Newton to Clairaut: the rise of mathematical science in eighteenth-century Paris and the fall of 'normal science'*. New York: Cambridge University Press.

Grégoire de Saint-Vincent (1647). *Opus geometricum quadraturae circuli et sectionum coni*. Antwerp: Ioannem et Iacobum Meursios.

Gregory, David (1702). *Astronomiae Physicae & Geometricae Elementa*. Oxford: e Theatro Sheldoniano.

Gregory, David (1726). *The elements of physical and geometrical astronomy*. London: for D. Midwinter.

Grosholtz, Emily R. (1987). Some uses of proportion in Newton's *Principia*, Book 1: a case study in applied mathematics. *Studies in History and Philosophy of Science* **18**: 209–20.

Guicciardini, Niccolò (1989). *The development of Newtonian calculus in Britain, 1700–1800*. Cambridge: Cambridge University Press.

Guicciardini, Niccolò (1993). Newton and British Newtonians on the foundations of the calculus. In *Hegel and Newtonianism*, M. J. Petry (ed.), pp. 167–77. Dordrecht, Boston, London: Kluwer.

Guicciardini, Niccolò (1995). Johann Bernoulli, John Keill and the inverse problem of central forces. *Annals of Science* **52**: 537–75.

Guicciardini, Niccolò (1996). An Episode in the History of Dynamics: Jakob Hermann's Proof (1716–1717) of Proposition 1, Book 1, of Newton's *Principia*. *Historia Mathematica* **23**: 167–81.

Guicciardini, Niccolò (1998a). Did Newton use his calculus in the *Principia*? *Centaurus* **40**: 303–44.

Guicciardini, Niccolò (1998b). *Newton: un filosofo della natura e il sistema del mondo*. Milano: Le Scienze (German transl. by M. Spang, *Spektrum der Wissenschaft*, 1999).

Guicciardini, Niccolò (1999). Newtons Methode und Leibniz' Kalkül. In *Geschichte der Analysis*, N. Jahnke (ed.). Heidelberg: Spektrum Akademischer Verlag.

Hall, Alfred R. (1952). *Ballistics in the seventeenth century*. Cambridge: Cambridge University Press.

Hall, Alfred R. (1958). Correcting the Principia. *Osiris* 13: 291–326.

Hall, Alfred R. (1980). *Philosophers at war: the quarrel between Newton and Leibniz.* Cambridge: Cambridge University Press.

Hall, Alfred R. (1992). *Isaac Newton: adventurer in thought.* Oxford: Blackwell.

Hall, Alfred R. & Boas Hall, Marie (1962). *Unpublished scientific papers of Isaac Newton.* Cambridge: Cambridge University Press.

Halley, Edmond (1686). A Discourse concerning Gravity, and its Properties, wherein the Descent of Heavy Bodies, and the Motion of Projects is briefly, but fully handled: Together with the Solution of a Problem of great Use in Gunnery. *Philosophical Transactions* **16**: 3–21.

Halley, Edmond (1687). [Letter to King James II]. London. (Printed copy of letter by Halley accompanying the copy of the *Principia* presented to the King. Reprinted as Halley (1697).

Halley, Edmond (1695). A Proposition of General Use in the Art of Gunnery, shewing the Rule for Laying a Mortar to pass, in order to strike any Object above or below the Horizon. *Philosophical Transactions* **19**: 68–72.

Halley, Edmond (1697). The true Theory of the Tides, extracted from that admired Treatise of Mr. Isaac Newton, Intituled, *Philosophiae Naturalis Principia Mathematica*; being a Discourse presented with that Book to the late King James. *Philosophical Transactions* **19**: 445–57.

Harris, John (1704–1710). *Lexicon Technicum: or, an Universal English Dictionary of Arts and Sciences.* 2 vols. London: for D. Brown *et al.*

Harrison, John (1978). *The library of Isaac Newton.* Cambridge: Cambridge University Press.

Hayes, Charles (1704). *A Treatise of Fluxions: or, an Introduction to Mathematical Philosophy.* London: by E. Midwinter, for D. Midwinter and T. Leigh.

Heinekamp, Albert (1982). Christiaan Huygens vu par Leibniz. In *Huygens et la France*, R. Taton (ed.), pp. 99–114. Paris: Vrin.

Herivel, John W. (1965). *The background to Newton's Principia: a study of Newton's dynamical researches in the years 1664–1684.* Oxford: Clarendon Press.

Hermann, Jacob (1700). *Responsio ad clarissimi viri Bernh. Nieuwentijt considerationes secundas circa calculi differentialis principia editas.* Basel.

Hermann, Jacob (1710). Extrait d'une lettre de M. Herman à M. Bernoulli. *Mémoires de l'Académie des Sciences*: 519–521.

Hermann, Jacob (1716). *Phoronomia, sive de viribus et motibus corporum solidorum et fluidorum libri duo.* Amsterdam: Rod. & Gerh. Wetstenios.

Hermann, Jacob (1717). Lettre de M. Herman. *Journal Litéraire* **9**: 406–15.

Hiscock, W. G. (1937). *David Gregory, Isaac Newton and their circle: extracts from David Gregory's Memoranda, 1677–1708.* Oxford: for the editor.

Hofmann, Joseph E. (1974). *Leibniz in Paris 1672–1676: his growth to mathematical maturity.* Cambridge: Cambridge University Press.

Hoskin, Michael (1977). Newton, providence and the universe of stars. *Journal for the History of Astronomy* **8**: 77–101.

Hospital, Guillaume F. A. de l' (1696). *Analyse des infiniment petits pour l'intelligence des lignes courbes.* Paris: Imprimerie Royale.

Huygens, Christiaan *Oeuvres = Oeuvres Complètes*, 22 vols. Den Haag: M. Nijhoff, 1888–1950.

Iliffe, Robert (1995). 'Is he like other men?': the meaning of the *Principia Mathematica* and the author as idol. In *Culture and society in Stuart Restoration: literature, drama, history*, G. Maclean (ed.), pp. 159–76. Cambridge: Cambridge University Press.

Jesseph, Douglas M. (1989). Philosophical theory and mathematical practice in the Seventeenth Century. *Studies in History and Philosophy of Science* **20**: 215–44.

Jones, William (1706). *Synopsis palmariorum matheseos*. London: by J. Matthews, for J. Wale.

Jourdain, Philip E. B. (1920). The analytical treatment of Newton's problems. *The Monist* **30**: 19–36.

Keill, John (1702). *Introductio ad veram physicam*. Oxford: e Theatro Sheldoniano.

Keill, John (1708). Epistola ad Clarissimum Virum Edmundum Halleium Geometriae Professorem Savilianum, de Legibus Virium Centripetarum. *Philosophical Transactions* **26**: 174–8.

Keill, John (1714). Observationes [...] de inverso problemate virium centripetarum. *Philosophical Transactions* **29**: 91–111.

Keill, John (1716). Défense du Chevalier Newton. *Journal Litéraire* **8**: 418–33.

Keill, John (1718). *Introductio ad veram Astronomiam*. Oxford: e Theatro Sheldoniano.

Keill, John (1719). Lettre de Monsieur Jean Keyll [...] à Monsieur Jean Bernoulli. *Journal Litéraire* **10**: 261–87.

Kirsanov, Vladimir S. (1992). The earliest copy in Russia of Newton's *Principia*: is it David Gregory's annotated copy? *Notes and Records of the Royal Society of London* **46**: 203–18.

Kitcher, Philip (1973). Fluxions, limits and infinite littlenesse: a study of Newton's presentation of the calculus. *Isis* **64**: 33–49.

Knobloch, Eberhard (1989). Leibniz et son manuscrit inédite sur la quadrature des sections coniques. In *The Leibniz Renaissance*, pp. 127–51. Firenze: Olschki.

Krauss, Johann Heinrich (1718). Responsio ad Cl. Viri Johannis Keil. *Acta Eruditorum*: 454–66.

Kuhn, Thomas (1957). *The Copernican revolution: planetary astronomy in the development of western thought*. Cambridge (Ma.): Harvard University Press.

Lai, Tyrone (1975). Did Newton renounce infinitesimals? *Historia Mathematica* **2**: 127–36.

Lalande, Jerome J. (1796). Notice sur la vie de Condorcet. *Mercure de France* (20 January): 143.

Landen, John (1771). *Animadversions on Dr. Stewart's computation of the Sun's distance from the Earth*. London: by G. Bigg, for the Author, sold by J. Nourse.

Laplace, Pierre-Simon de (1796). *Exposition du système du monde*. Paris: De l'Imprimerie du Cercle-Social.

Laugwitz, Detlef (1997). On the historical development of infinitesimal mathematics I: the algorithmic thinking of Leibniz and Euler. *American Mathematical Monthly* **104**: 447–55.

Leibniz, Gottfried W. *LMS = Leibnizens mathematische Schriften*. 7 vols., C. I. Gerhardt (ed.). Berlin, Halle: Asher & Co., Schmidt, 1849–1863. [Reprinted Hildesheim: Olms, 1971]

Leibniz, Gottfried W. *LPS = Die philosophischen Schriften von G. W. Leibniz*. 7 vols., C. I. Gerhardt (ed.). Berlin: Weidmannische Buchhandlung, 1875–1890. [Reprinted Hildesheim: Olms, 1978]

Leibniz, Gottfried W. *Mathematikern = Der Briefwechsel von G. W. Leibniz mit Mathematikern*. C. I. Gerhardt (ed.). Berlin: Weidmannische Buchhandlung, 1889.

Leibniz, Gottfried W. (1973). *Marginalia in Newtoni Principia Mathematica*. E. A. Fellmann (ed.). Paris: Vrin.

Leibniz, Gottfried W. (1976). *Philosophical papers and letters*. L.E. Loemker (trans. and ed.), 2nd edn. Dordrecht: Reidel.

Leibniz, Gottfried W. (1993). *De quadratura arithmetica circuli ellipseos et hyperbolae cujus corollarium est trigonometria sine tabulis*. E. Knobloch (ed.). Göttingen: Vandenhoeck & Ruprecht.

MacDonald Ross, George (1978). Leibniz and alchemy. *Studia Leibnitiana* Sonderheft **7**: 166–77.

MacDonald Ross, George (1983). Leibniz and Renaissance Neoplatonism. *Studia Leibnitiana* Supplementa **23**: 125–34.

Maclaurin, Colin (1742). *A treatise of fluxions*. Edinburgh: Ruddimans.

MacPike, Eugene F. (1932). *Correspondence and papers of Edmond Halley, preceded by an unpublished memoir of his life by one of his contemporaries and the 'Éloge' by D'Ortous de Mairan*. Oxford: Clarendon Press (edition used: new issue by Taylor & Francis, London, 1937).

Maffioli, Cesare S. (1994). *Out of Galileo: the science of waters 1628–1718*. Rotterdam: Erasmus Publishing.

Mahoney, Michael S. (1993). Algebraic vs. geometric techniques in Newton's determination of planetary orbits. In *Action and reaction: proceedings of a symposium to commemorate the trecentenary of Newton's Principia*, P. Theerman & A.F. Seeff (eds.), pp. 183–205. Newark, London and Toronto: University of Delaware Press/Associated University Presses.

Malebranche, Nicolas (1967–1978). *Oeuvres complètes*. 2nd edn, 20 vols., A. Robinet (ed.). Paris: Vrin.

Mancosu, Paolo (1989). The metaphysics of the calculus: a foundational debate in the Paris Academy of Sciences, 1700–1706. *Historia Mathematica* **16**: 224–48.

Mancosu, Paolo & Vailati, Enrico (1990). Detleff Clüver: an early opponent of the infinitesimal calculus. *Centaurus* **33**: 325–44.

Mandelbrote, Scott (1993). 'A duty of the greatest moment': Isaac Newton and the writing of biblical criticism. *British Journal for the History of Science* **26**: 281–302.

Manuel, Frank E. (1963). *Isaac Newton, historian*. Cambridge: Cambridge University Press.

Manuel, Frank E. (1974). *The religion of Isaac Newton*. Oxford: Clarendon Press.

Marie, Maximilien (1883–88). *Histoire des sciences mathématiques et physiques*. 12 vols. Paris: Gauthier-Villars.

Martins, Roberto de A. (1993). Huygens' reaction to Newton's gravitational theory. In *Renaissance and Revolution: humanists, scholars, craftsmen and natural philosophers in early modern Europe*. J. V. Field & F. A. J. L. James (eds.), pp. 203–13. Cambridge: Cambridge University Press.

Mazzone, Silvia & Roero, Clara S. (eds.) (1992). *Guido Grandi – Jacob Hermann. Carteggio (1708–1714)*. Firenze: Olschki.

Mazzone, Silvia & Roero, Clara S. (1997). *Jacob Hermann and the diffusion of the Leibnizian calculus in Italy*. Firenze: Olschki.

McGuire, J. E. & Rattansi, M. (1966). Newton and the 'Pipes of Pan'. *Notes and Records of the Royal Society of London* **21**: 108–43.

Mignard, François (1988). The theory of the figure of the Earth according to Newton and Huygens. *Vistas in Astronomy* **30**: 291–311.

Milnes, James (1702). *Sectionum conicarum elementa nova methodo demonstrata*. Oxford: e Theatro Sheldoniano, Impensis Anth. Peisley.

Nagel, Fritz (1991). Johann Bernoulli und Giuseppe Verzaglia, Monstrum Italicum aut Basiliense. *Basler Zeitschrift für Geschichte und Altertumskunde* **91**: 83–105.

Nauenberg, Michael (1994). Newton's early computational method for dynamics. *Archive for History of Exact Sciences* **46**: 221–52.

Nauenberg, Michael (1995). Newton's perturbation methods for the 3-body problem and its application to lunar motion. (Lecture given at the Dibner Institute. To be published.)

Nauenberg, Michael (1998a). Newton's unpublished perturbation method for the lunar motion. *International Journal of Engineering Science*.

Nauenberg, Michael (1998b). Newton's Portsmouth perturbation method for the three-body problem and its application to lunar motion. In *The Foundations of Newtonian Scholarship*, R. Dalitz & M. Nauenberg (eds.). Singapore: World Scientific.

Nauenberg, Michael (1998c). The mathematical principles underlying the *Principia* revisited. *Journal for the History of Astronomy* **29**: 286–300.

Needham, Tristan (1993). Newton and the transmutation of force. *American Mathematical Monthly* **100**: 119–37.

Newton, Isaac *Arithmetick = Universal arithmetick, or, a treatise of arithmetical composition and resolution*. J. Raphson (transl.) & S. Cunn (ed.). London: for J. Senex, W. Taylor, T. Warner and J. Osborn, 1720.

Newton, Isaac *Correspondence = The correspondence of Isaac Newton*. 7 vols., H. W. Turnbull *et al.* (eds.). Cambridge: Cambridge University Press, 1959–1977.

Newton, Isaac *Mathematical Papers = The mathematical papers of Isaac Newton*. 8 vols., D. T. Whiteside *et al.* (eds.). Cambridge: Cambridge University Press, 1967–1981.

Newton, Isaac *Principles = Sir Isaac Newton's Mathematical Principles of Natural Philosophy and his System of the World*. Translation by Andrew Motte (1729), revision by Florian Cajori (1734). 5th printing. Berkeley and Los Angeles: University of California Press, 1962.

Newton, Isaac *Variorum = Philosophiae naturalis principia mathematica. The third edition (1726) with variant readings assembled and edited by Alexandre Koyré and I. Bernard Cohen, with the assistance of Anne Whitman*. Cambridge: Cambridge University Press, 1972.

Newton, Isaac (1962). *Unpublished scientific papers of Isaac Newton: a selection from the Portsmouth Collection in the University Library, Cambridge*. A. R. Hall & M. Boas Hall (eds.). Cambridge: Cambridge University Press.

Newton, Isaac (1975). *Isaac Newton's Theory of the Moon's Motion (1702)*. I. B. Cohen (ed.). Folkestone: Dawson.

Newton, Isaac (1989). *The preliminary manuscripts for Isaac Newton's 1687 Principia 1684–1686 (with an introduction by D. T. Whiteside)*. Cambridge: Cambridge University Press.

Newton, Isaac (1995). *Trattato sull'Apocalisse*. M. Mamiani (ed.). Torino: Bollati Boringhieri.

Newton, Isaac (1999). *Philosophiae Naturalis Principia Mathematica, Mathematical Principles of Natural Philosophy, Translated by I. Bernard Cohen and Anne Whitman with the assistance of Julia Budenz, preceded by a Guide to Newton's Principia by I. Bernard Cohen*. Berkeley, Los Angeles, London: University of California Press.

Oughtred, William (1631). *Clavis mathematicae*. Oxford.

Pascal, Blaise (1659). *Lettres de A. Dettonville contenant quelques unes de ses inventions de géométrie*. Paris.

Pasini, Enrico (1988). Die private Kontroverse des G. W. Leibniz mit sich selbst. Handschriften über die Infinitesimalrechnung im Jahre 1702. In *Leibniz: Tradition und Aktualität. V. internationaler Leibniz-Kongreß*, pp. 695–709. Hannover: Gottfried-Wilhelm-Leibniz-Gesellschaft.

Pasini, Enrico (1993). *Il reale e l'immaginario: la fondazione del calcolo infinitesimale nel pensiero di Leibniz*. Torino: Sonda.

Pemberton, Henry (1728). *A view of Sir Isaac Newton's philosophy*. London: by S. Palmer.

Peiffer, Jeanne (1988). La conception de l'infiniment petit chez Pierre Varignon, lecteur de Leibniz et de Newton. In *Leibniz: Tradition und Aktualität. V. internationaler Leibniz-Kongreß*, pp. 710–17. Hannover: Gottfried-Wilhelm-Leibniz-Gesellschaft.

Peiffer, Jeanne (1990). Pierre Varignon, lecteur de Leibniz et de Newton. *Studia Leibnitiana* Supplementa **27**: 244–66.

Politella, Joseph (1938). *Platonism, Aristotelianism and Cabalism in the Philosophy of Leibniz*. Ph.D. thesis. University of Philadelphia.

Pourciau, Bruce (1991). On Newton's proof that inverse-square orbits must be conics. *Annals of Science* **48**: 159–72.

Pourciau, Bruce (1992a). Newton's solution of the one-body problem. *Archive for History of Exact Sciences* **44**: 125–46.

Pourciau, Bruce (1992b). Radical Principia. *Archive for History of Exact Sciences* **44**: 331–63.

Pourciau, Bruce (1998). The preliminary mathematical lemmas of Newton's *Principia*. *Archive for History of Exact Sciences* **52**: 279–95.

Proudman, J. (1927). Newton's work on the theory of the tides. In *Isaac Newton, 1642–1727, A memorial volume edited for the Mathematical Association*, W.J. Greenstreet (ed.), pp. 87–95. London: G. Bell and Sons.

Pycior, Helena M. (1997). *Symbols, impossible numbers, and geometric entanglements: British algebra through the commentaries on Newton's Universal arithmetick*. Cambridge: Cambridge University Press.

Riccati, Jacopo (1714). Risposta ad alcune opposizioni fatte dal Sig. Giovanni Bernoulli. *Giornale dei Letterati d'Italia* **19**: 185–210.

Rigaud, Stephen P. (1838). *Historical essay on the first publication of Sir Isaac Newton's Principia*. Oxford: Oxford University Press (reprinted with an introduction by I. B. Cohen, Johnson Reprint, 1972).

Rigaud, Stephen P. (1841). *Correspondence of scientific men of the seventeenth century*. 2 vols. Oxford: Oxford University Press.

Robinet, André (1960). Le groupe malebranchiste introducteur du calcul infinitésimal en France. *Revue d'Histoire des Sciences* **13**: 287–308.

Robinet, André (1988). *G. W. Leibniz. Iter Italicum (Mars 1689–Mars 1690): la dynamique de la République des Lettres*. Firenze: Olschki.

Robinet, André (1991). *L'empire leibnizien: la conquête de la chaire de mathématiques de l'Université de Padoue. Jacob Hermann et Nicolas Bernoulli*. Trieste: Lint.

Ronan, Colin A. (1970). *Edmond Halley, Genius in Eclipse*. London: Macdonald.

Sageng, Erik (1989). *Colin MacLaurin and the foundations of the method of fluxions*. Ph.D. thesis. Princeton University.

Schmitt, Charles B. (1966). Perennial Philosophy: from Agostino Steuco to Leibniz. *Journal for the History of Ideas* **27**: 505–32.

Schneider, Ivo (1968). Der Mathematiker Abraham De Moivre (1667–1754). *Archive for History of Exact Sciences* **5**: 177–317.

Schneider, Ivo (1988). *Isaac Newton*. München: C. H. Beck.

Scriba, Christoph J. (1964). The inverse method of tangents: a dialogue between Leibniz and Newton (1675–1677). *Archive for History of Exact Sciences* **2**: 112–37.

Sigurdsson, Skuli (1992). Equivalence, pragmatic platonism, and discovery of the calculus. In *The invention of physical science*, M. J. Nye *et al.* (eds.), pp. 97–116.

Dordrecht, Boston, London: Kluwer.

Simpson, Thomas (1743). *Mathematical Dissertations on a variety of Physical and Analytical Subjects*. London: for T. Woodward.

Simpson, Thomas (1750). *The Doctrine and Application of Fluxions*. London: for J. Nourse.

Simpson, Thomas (1757). *Miscellaneous Tracts on Some Curious, and very interesting Subjects in Mechanics, Physical-Astronomy, and Speculative Mathematics*. London: for J. Nourse.

Simson, Robert (1723). Pappi Alexandrini Propositiones duae Generales. *Philosophical Transactions* **32**: 330–40.

Simson, Robert (1776). *Opéra Quaedam Reliqua*. Glasgow: R. & A. Foulis.

Snelders, H. A. M. (1989). Christiaan Huygens and Newton's theory of gravitation. *Notes and Records of the Royal Society of London* **43**: 209–22.

Speiser, David (1988). Le 'Horologium Oscillatorium' de Huygens et les *Principia*. *Revue Philosophique de Louvain* **86**: 485–504.

Stewart, Matthew (1761). *Tracts, Physical and Mathematical. Containing, An Explication of several important points in Physical Astronomy; and, A new method for ascertaining the Sun's distance from the Earth, by the Theory of Gravity*. Edinburgh: for A. Millar, J. Nourse, W. Sands, A. Kincaid and J. Bell.

Stewart, Matthew (1763). *The distance of the Sun from the Earth determined by the theory of gravity*. Edinburgh: for A. Millar, J. Nourse, D. Wilson, W. Sands, A. Kincaid and J. Bell.

Stone, Edmund (1730). *The Method of Fluxions both Direct and Inverse. The Former being a Translation from the Celebrated Marquis De L'Hospital's Analyse des Infinemens Petits: and the Latter Supply'd by the Translator*. London: for W. Innys.

Sylla, Edith (1984). Compounding ratios, Bradwardine, Oresme, and the first edition of Newton's *Principia*. In *Transformation and tradition in the sciences: essays in honor of I. Bernard Cohen*, E. Mendelsohn (ed.), pp. 11–43. Cambridge: Cambridge University Press.

Thorp, Robert (1777). [Commentaries on] Newton, Isaac, *Mathematical Principles of Natural Philosophy [. . .] Translated into English, and illustrated with a commentary, by Robert Thorp, M.A. volume the first*. London: for W. Strahan and T. Cadell.

Todhunter, Isaac (1873). *A history of the mathematical theories of attraction and the figure of the Earth: from the time of Newton to that of Laplace*. London: Macmillan and Co.

Trompf, Garry W. (1991). On Newtonian history. In *The uses of antiquity: the scientific revolution and the classical tradition*, S. Gaukroger (ed.), pp. 213–49. Dordrecht, Boston, London: Kluwer.

Truesdell, Clifford (1958). The new Bernoulli edition. *Isis* **49**: 54–62.

Truesdell, Clifford (1960). A program toward rediscovering the rational mechanics of the Age of Reason. *Archive for History of Exact Sciences* **1**: 3–36.

Varignon, Pierre (1700). Des forces centrales, ou des pesanteurs necessaires aux planetes pour leur faire décrire les orbes qu'on leur a supposés jusqu'icy. *Mémoires de l'Académie des Sciences*: 218–37.

Varignon, Pierre (1706). Comparaison des forces centrales avec les pésanteurs absoluës des corps mûs de vitesses variées à discrétion le long de telles courbes qu'on voudra. *Mémoires de l'Académie des Sciences*: 178–235.

Varignon, Pierre (1707). Des mouvemens faits dans des milieux qui leur résistent en raison quelconque. *Mémoires de l'Académie des Sciences*: 382–476.

Varignon, Pierre (1710). Des forces centrales inverses. *Mémoires de l'Académie des*

Sciences: 533–44.

Varignon, Pierre (1725). *Éclaircissemens sur l'analyse des infiniment petits*. Paris: Rollin.

Vermij, Rienk H. & van Maanen, Jan A. (1992). An unpublished autograph by Christiaan Huygens: his letter to David Gregory of 19 January 1694. *Annals of Science* **49**: 507–23.

Viète, François (1646). *Opera mathematica: in unum volumen congesta, ac recognita*. F. van Schooten (ed.). Leiden: Elzeviriorum.

Waff, Craig B. (1975). *Universal gravitation and the motion of the Moon's apogee: the establishment and reception of Newton's inverse-square law*. Ph.D. thesis. Baltimore, Maryland: The Johns Hopkins University.

Waff, Craig B. (1976). Isaac Newton, the motion of the lunar apogee, and the establishment of the inverse square law. *Vistas in Astronomy* **20**: 99–103.

Wallis, John (1656). *Arithmetica infinitorum*. Oxford: L. Lichfield and T. Robinson.

Wallis, John (1685). *A treatise of algebra, both historical and practical*. London: by J. Playford, for R. Davis.

Wallis, John (1693–99). *Opera mathematica*. 3 vols. Oxford: e Theatro Sheldoniano.

Waring, Edward (1799). Original letter of the late Dr. Waring. *Monthly Magazine* May: 306–10.

Westfall, Richard S. (1971). *Force in Newton's physics: the science of dynamics in the seventeenth century*. New York: American Elsevier.

Westfall, Richard S. (1980). *Never at rest: a biography of Isaac Newton*. Cambridge: Cambridge University Press.

Whewell, William (1837). *History of the inductive sciences*. Cambridge, London: J. W. Parker, J. and J. J. Deighton.

Whiston, William (1707). *Praelectiones Astronomicae Cantabrigiae in Scholis Publicis Habitae*. Cambridge: Typis Academicis.

Whiston, William (1710). *Praelectiones Physico-Mathematicae Cantabrigiae in Scholis Publicis Habitae*. Cambridge: Typis Academicis.

Whiston, William (1716). *Sir Isaac Newton's Mathematick Philosophy more easily Demonstrated*. London: for J. Senex.

Whiston, William (1749). *Memoirs of the life and writings of Mr. William Whiston*. London: for the author.

Whiteside, Derek T. (1970). The mathematical principles underlying Newton's *Principia Mathematica*. *Journal for the History of Astronomy* **1**: 116–38.

Whiteside, Derek T. (1975). Newton's lunar theory: from high hope to disenchantment. *Vistas in Astronomy* **19**: 317–28.

Whiteside, Derek T. (1991a). How forceful has a force proof to be? Newton's *Principia*, Book 1: Corollary 1 to Propositions 11–13. *Physis* **28**: 727–49.

Whiteside, Derek T. (1991b). The prehistory of the *Principia* from 1664 to 1686. *Notes and Records of the Royal Society of London* **45**: 11–61.

Wightman, William P. D. (1953). Gregory's 'Notae in Isaaci Newtoni Principia Philosophiae'. *Nature* **172**: 690.

Wightman, William P. D. (1957). David Gregory's commentary on Newton's *Principia*. *Nature* **179**: 393–4.

Wilson, Curtis A. (1989). The Newtonian achievement in astronomy. In *The general history of astronomy*, vol. 2A = *Planetary Astronomy from the Renaissance to the rise of astrophysics, Part A: Tycho Brahe to Newton*, R. Taton & C. Wilson (eds.), pp. 233–74. Cambridge: Cambridge University Press.

Wilson, Curtis A. (1994). Newton on the equiangular spiral: an addendum to Erlichson's account. *Historia Mathematica* **21**: 196–203.

Wilson, Curtis A. (1995). Newton on the Moon's variation and absidal motion: the need for a newer 'new analysis'. (Lecture given at the Dibner Institute. To be published.)

Wolff, Christian (1732–41). *Elementa Matheseos Universae.* 5 vols. Généve: apud Marcum-Michaelem Bousquet & Socios.

Wollenschläger, Kurt (1902). Der mathematische Briefwechsel zwischen Johann I Bérnoulli und Abraham De Moivre. *Verhandlungen der Naturforschenden Gesellschaft in Basel* **43**: 151–317.

Wright, John [?] M. F. (1830). *The Principia of Newton, with notes, examples, and deductions: containing all that is read at the University of Cambridge.* Cambridge: W. P. Grant.

Yates, Frances (1977). Did Newton connect his maths and alchemy? *Times Higher Education Supplement* (18 March) **282**: 13.

Yoder, Joella G. (1988). *Unrolling time: Christiaan Huygens and the mathematization of nature.* Cambridge: Cambridge University Press.

Zehe, Horst (1983). Die Gravitationstheorie des Nicolas Fatio de Duillier. *Archive for History of Exact Sciences* **28**: 1–23.

Index

281